Topics in Applied Physics
Volume 119

Topics in Applied Physics

Topics in Applied Physics is a well-established series of review books, each of which presents a comprehensive survey of a selected topic within the broad area of applied physics. Edited and written by leading research scientists in the field concerned, each volume contains review contributions covering the various aspects of the topic. Together these provide an overview of the state of the art in the respective field, extending from an introduction to the subject right up to the frontiers of contemporary research.

Topics in Applied Physics is addressed to all scientists at universities and in industry who wish to obtain an overview and to keep abreast of advances in applied physics. The series also provides easy but comprehensive access to the fields for newcomers starting research.

Contributions are specially commissioned. The Managing Editors are open to any suggestions for topics coming from the community of applied physicists no matter what the field and encourage prospective editors to approach them with ideas.

Managing Editor

Dr. Claus E. Ascheron

Springer-Verlag GmbH
Tiergartenstr. 17
69121 Heidelberg
Germany
Email:claus.ascheron@springer.com

Assistant Editor

Adelheid H. Duhm

Springer-Verlag GmbH
Tiergartenstr. 17
69121 Heidelberg
Germany
Email:adelheid.duhm@springer.com

David J. Lockwood · Lorenzo Pavesi (Eds.)

Silicon Photonics II

Components and Integration

With 196 Figures

Springer

Editors

Dr. David J. Lockwood
National Research Council of Canada
Institute for Microstructural Sciences
1200 Montreal Road
Bldg. M-50
Ottawa Ontario K1A 0R6
Canada
david.lockwood@nrc-cnrc.gc.ca

Prof. Lorenzo Pavesi
Università di Trento
Dipartimento di Fisica
Laboratorio Nanoscienze
Via Sommarive 14
38050 Povo
Italy
lorenzo.pavesi@unitn.it

ISSN 0303-4216 e-ISSN 1437-0859
ISBN 978-3-642-10505-0 e-ISBN 978-3-642-10506-7
DOI 10.1007/978-3-642-10506-7
Springer Heidelberg Dordrecht London New York

Library of Congress Control Number: 2004046611

Cover design: Integra Software Services Pvt. Ltd., Pondicherry

Printed on acid-free paper

Springer is part of Springer Science+Business Media (www.springer.com)

In memory of Ulrich Gösele (1949–2009), a pioneer and leading light of silicon nanotechnology.

Preface

Optoelectronics and photonics are playing an essential role in many aspects of daily life, including information and communication technologies, environmental and green technologies, mechanical and chemical sensing, consumer electronics, and biomedicine. So far, the use of optical components in communication systems has been mainly limited to direct replacement of electrical cables by optical cables. With the continual increase in link bit rates, optical cables are now replacing electrical cables for shorter and shorter interconnect lengths. Optoelectronic and photonic technologies are becoming less costly and more integrated, and there is currently an opportunity for optics to move "inside the box" and change the interconnect topology at all levels.

Currently, most optoelectronic devices are fabricated as discrete components. This approach is based on serial (e.g., step-by-step) fabrication and packaging, and it makes optoelectronic technology drastically different compared to microelectronics, where the domination of parallel fabrication made possible ultra-large-scale integration with the price of individual devices below $\$10^{-8}$. Also, discrete assembly reduces the optoelectronic system reliability and decreases the manufacturing yield. Additional complications arise due to materials issues: in microelectronics the major material is elemental Si, while traditional semiconductor materials for optoelectronics are III–V alloys with much more complex technological requirements. Finally, optical waveguides and waveguide based devices are very bulky compared to electron devices; thus, the densities of electron devices in integrated circuits are many orders of magnitude greater compared to that in integrated optoelectronic systems.

Silicon photonics, where photonics devices are fabricated by using silicon or silicon compatible materials and where the manufacturing is based on the available microelectronics infrastructure, is emerging as the technology that can face all these challenges. The first volume of this series of books on silicon photonics appeared in 2004. We stated then that "This book is aimed at presenting the fascinating picture of the state of-the-art in silicon photonics and providing perspectives on what can be expected in the near future." Many of the visionary concepts reported there have been surpassed by reality. Silicon photonics is booming and growing at an incredible pace with many breakthroughs appearing day by day. Speed, integration density, active components, logic, nonlinear optics, etc., are all surpassed frontiers, which silicon photonics has continuously moved apart. At the beginning of the second decade of the new millennium, several devices enabled by silicon photonics are already on the market and new ones are emerging day by day.

Due to the broad nature and importance of the topic, as well as the rapid progress in this field, it is quite appropriate to publish this sequel to our first book on *Silicon Photonics* in the Springer Topics in Applied Physics Series. We envisaged this book as the second of a series of books on silicon photonics, i.e., we are willing to produce further books on this topic as the field develops.

The present book is focused on components and integration and opens with a Chapter by Yamada, which reviews the fundamental characteristics and basic applications of such waveguides. Some passive devices, such as branches and wavelength filters, and dynamic devices based on the thermo-optic effect or carrier plasma effect have been developed by using silicon photonic wire waveguides. These waveguides also offer an efficient media for nonlinear optical functions, such as wavelength conversion. Their optical polarization characteristics can be a serious obstacle to some practical applications, but such difficulties can be eliminated by using a monolithically integrated polarization diversity system. Chapter 2 by Xu explores further the problems associated with mode polarization in waveguides. This chapter reviews the characteristics of silicon-on-insulator ridge waveguide birefringence, as governed by the waveguide cross-section geometry, the cladding stress level, and cladding thickness. Typical stress levels in dielectric cladding films such as silicon dioxide and silicon nitride are such that the stress-induced birefringence is of comparable magnitude to the waveguide geometrical birefringence. Hence the total waveguide birefringence can be precisely controlled by counter balancing these two factors. The application of this technique for achieving polarization independence and polarization splitting in photonic components is described using passive and active tuning of the stress-induced birefringence. In Chapter 3 Roelkens and Van Thourhout elaborate on the use of diffraction gratings to achieve an efficient, compact, alignment-tolerant, polarization-independent, and broadband optical coupling between an optical fiber and optoelectronic components. An optical probe based on a diffraction grating integrated on the facet of a single-mode fiber is described that enables testing individual components in a silicon-on-insulator nanophotonic integrated circuit. Chapter 4 by Chang et al. tackles the difficult technical problem of providing on-chip light sources, which are a critical component for integrated silicon photonics but lag other photonic components in their level of development. Erbium is an optical dopant that can be employed as a viable means for on-chip light generation, which also has the advantage of being compatible with long-distance telecommunications transmission wavelengths. In this chapter, Er-doped silicon-rich silicon nitride and Er silicates are introduced as promising host materials for compact on-chip light sources. Germanium on silicon is an enabler of silicon photonics as well as high-speed CMOS electronics and recently germanium has played a significant role in integrating materials such as III–Vs on silicon. In Chapter 5 Ichikawa et al. describe an ultra-thin germanium buffer layer technology that has created entirely new fields for applications such as high-efficiency cost-effective tandem solar cells using silicon as the cell as well as the mechanical substrate. Such solar cells have successfully reproduced their ideal external quantum efficiency and prove that it is possible to successfully integrate silicon and GaAs.

The remaining four chapters focus on system integration. First, Scandurra in Chapter 6 describes possible applications of silicon photonics to the system-on-chip (SoC) domain. The higher and higher integration density is becoming such that many issues arise when a SoC has to be integrated, and electrical limits of interconnect wires are a limiting factor for performance that could be overcome through the use of the optical interconnect. Today, many semiconductor industries are investigating such a novel field and a number of projects are currently in progress to demonstrate the feasibility of such a revolutionary on-chip communication system relying on both CMOS technology and photonics. Chapter 7 develops this concept further. Liao et al. highlight a recent demonstration of a silicon photonic integrated chip that is capable of transmitting data at an aggregate rate of 200 Gb/s. It is based on wavelength division multiplexing where an array of eight high-speed silicon optical modulators is monolithically integrated with a demultiplexer and a multiplexer. This demonstration represents a key milestone on the way to fabricating terabit per second transceiver chips to meet future demands. In Chapter 8 Pinguet et al. describe the intimate relationship between process, devices, and system design by examining the development of Luxtera's CMOS Photonics technology. They address the challenges of integrating optoeletronic elements, including germanium photodiodes, in a commercial CMOS process without significantly affecting the electronics performance and the manufacturability of the process. A complete monolithically integrated wavelength division multiplexed 40-Gbps transceiver chip is fabricated as an example of the application of their complete technology platform and to demonstrate its capabilities for optoelectronic integration. Lastly, Fedeli et al. in Chapter 9 address different ways to merge photonics devices on an electronic circuit with microelectronics tools on large size wafers. The preferred route is above-integrated-circuit fabrication, with two options ruled by thermal constraints. The high-temperature option is based on wafer bonding on an optical silicon-on-insulator module, and the low-temperature option relies on the heterogeneous integration of III–V devices.

Finally, we dedicate this book to the memory of Ulrich Gösele (1949–2009), our colleague who recently and prematurely passed away. He was among the first who initiated the field by discovering the quantum nature of porous silicon and the possibility of its use in optoelectronics. Moreover, he pioneered the technique of wafer bonding, which is nowadays a common practice to realize hybrid devices.

Ottawa, Canada *David J. Lockwood*
Trento, Italy *Lorenzo Pavesi*
June 2010

Contents

Chapter 1
Silicon Photonic Wire Waveguides: Fundamentals and Applications

Koji Yamada

Abstract This chapter reviews the fundamental characteristics and basic applications of the silicon photonic wire waveguide. Thanks to its ultra-small geometrical structures and compatibility with the silicon electronics, the silicon photonic wire waveguide provides us with a highly integrated platform for electronic–photonic convergence. For the practical achievement of this platform, however, we must search for ways to reduce the propagation loss and coupling loss to external fibers and overcome the polarization dependence. Progress has been made by applying state-of-the-art technologies specially tuned to the fabrication of nanometer structures, and the fundamental propagation performance has already become a practical standard. Some passive devices, such as branches and wavelength filters, and dynamic devices based on the thermo-optic effect or carrier plasma effect have been developed by using silicon photonic wire waveguides. These waveguides also offer an efficient media for nonlinear optical functions, such as wavelength conversion. Although polarization dependence remains a serious obstacle to the practical applications of these waveguides, waveguide-based polarization manipulation devices provide us with effective solutions, such as a polarization diversity system.

1.1 Introduction

In recent years, silicon photonics has attracted attention as an emerging technology for optical telecommunications and for optical interconnects in microelectronics. Based on highly sophisticated silicon semiconductor technology, silicon photonics would provide us with an inexpensive highly integrated electronic–photonic platform, in which ultra-compact photonic devices and electronic circuits are converged.

Similar to the existing silica-based or III–V semiconductor-based photonic systems, silicon photonics also requires an optical waveguide system. The waveguide

K. Yamada (✉)
NTT Microsystem Integration Laboratories, NTT Corporation,
Atsugi, 243-0198 Japan
e-mail: kyamada@aecl.ntt.co.jp

D.J. Lockwood, L. Pavesi (Eds.): *Silicon Photonics II*.
Topics in Applied Physics **119**, 1–29 (2011)
DOI 10.1007/978-3-642-10506-7_1

must have features that allow us to accommodate passive and dynamic photonic devices such as wavelength filters and modulators. The waveguide system must also be flexible enough to allow active functions, such as light emission and detection to be implemented. Of course, it should guarantee a sufficiently low propagation loss for constructing and integrating these photonic functions. Furthermore, for monolithic electronic–photonic convergence, the most important advantage of silicon photonics, the following requirements must be met:

(1) Waveguides should be constructed on silicon substrates or be constructed together with silicon electronic devices.
(2) Waveguide fabrication processes should not damage electronic devices.
(3) Waveguides should not be damaged by the processes for fabricating electronic devices.
(4) Waveguide materials must not be hazardous to silicon electronics.
(5) Geometrical criteria, such as the layout of photonic devices, should not interfere with the electronic circuit layout.

These requirements are very difficult to meet with conventional waveguide systems. For example, the fabrication process for conventional silica-based waveguide requires high temperatures exceeding 1,000 °C, which would seriously damage electronic devices. Moreover, silica-based waveguides have a large bending radius on the order of millimeters or even centimeters, making it impossible to integrate photonic circuits on an electronic chip, whose typical size is a few centimeters square. The III–V compound semiconductor-based waveguides and photonic devices have geometries smaller than those in the silica-based system; however, on a silicon substrate it is very difficult to epitaxially grow the high-quality III–V materials needed for the construction of practical photonic devices. Etching and other fabrication procedures are completely different from silicon processes. Moreover, III–V material contamination must be eliminated from silicon electronics. At present, it is therefore very difficult to introduce III–V compounds into silicon electronics. Polymer waveguides made of organic materials cause less damage to electronic devices. However, their use is limited to the uppermost layers formed after the electronic circuits are completed or to other regions separated from the electronic devices because they cannot withstand the temperatures used in electronic device fabrication.

Recently, silicon oxynitride (SiON) and silicon nitride (Si_3N_4) waveguides have also been proposed as a compact waveguide system [1, 2]; however, these silicon nitride waveguides might be not suitable for electronic–photonic convergence for infrared light at wavelengths around 1,500 nm. Silicon nitride materials formed by low-temperature deposition methods generally contain N–H bond residues, which readily absorb infrared light at around 1,500 nm. A high-temperature (1,000 °C or more) annealing can reduce the absorption, but the high-temperature process is not compatible with electronic devices.

From the viewpoint of material, a silicon-based waveguide is obviously preferable for electronic photonic convergence. So far, several types of silicon waveguides have been proposed such as rib-type waveguides with core dimensions of a few micrometers [3, 4] and photonic wire waveguides with core dimensions of several hundreds of nanometers [5–10]. The latter are especially promising for electronic–photonic convergence because their ultra-small core dimensions and micrometer bending sections match the dimensions of electronic circuits. The waveguides are constructed on silicon-on-insulator (SOI) substrates, where the uppermost SOI layer is used as the waveguide core so that there is no need to form the core material. The cladding material is a silica-based material formed by a low-temperature process such as plasma-enhanced chemical vapor deposition (PE–CVD) [11].

Besides the advantages of silicon photonic wire waveguides in electronic–photonic convergence, highly integrated ultra-compact photonic circuits based on photonic wire waveguides guarantee low power consumption and low packaging cost. Furthermore, the waveguides also offer advanced functionality through the use of its semiconductor characteristics of silicon.

1.2 Fundamental Design of Silicon Photonic Wire Waveguides

1.2.1 Guided Modes

A schematic of s silicon photonic wire waveguide is shown in Fig. 1.1a. The waveguide consist of a silicon core and silica-based cladding. The core dimension should be determined so that a single-mode condition is fulfilled. The single-mode condition is very important in constructing practical functional devices because whether or not we can implement a desired function depends on the fundamental guided mode. The core dimension of single-mode silicon photonic wire waveguides is significantly smaller than that of conventional single-mode silica waveguides. In the waveguide system consisting of a silicon core and silica claddings, the refractive

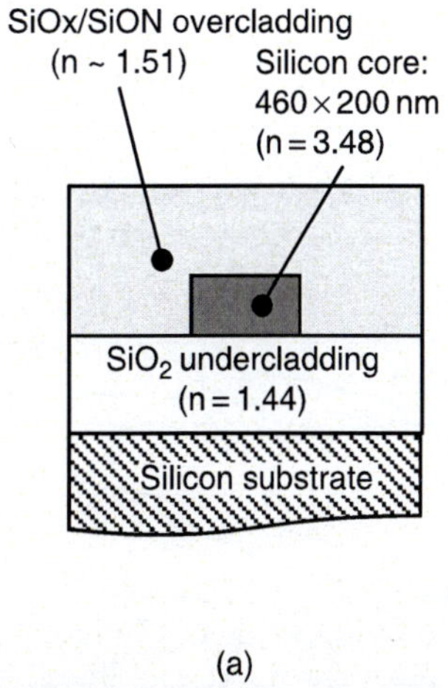

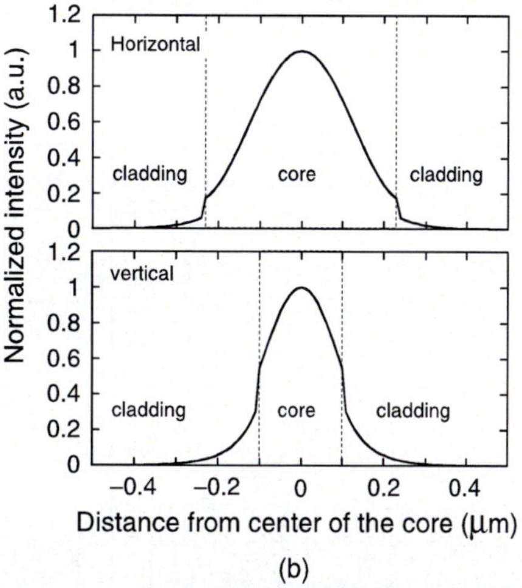

Fig. 1.1 (**a**) Cross-sectional structure and (**b**) optical intensity distribution of a typical silicon photonic wire waveguide

index contrast Δ between the core and cladding is as large as 40%, which allows a total internal reflection with a very large incident angle of 60°. This situation is similar to that in metallic rectangular waveguides, whose waveguide dimensions are smaller than or comparable to a half-wavelength of the guided electromagnetic waves. In silicon photonic wire waveguides, therefore, the core dimension that fulfills a single-mode condition should also be smaller than or comparable to a half-wavelength of a guided wave in silicon. Since the refractive index of silicon is about 3.5 for photon energies below the band-gap energy, the core dimension of a silicon photonic wire waveguide should be less than or comparable to 400 nm for 1,310∼1,550-nm telecommunications-band infrared light. Generally, the core shape is made flat along the substrate to reduce the etching depth in practical fabrications. In many cases, the height of the core is typically half of the width. Thus, a typical core geometry is a $400 \times 200\text{-nm}^2$ rectangle.

A detailed analysis of the guided modes can be performed by various numerical methods such as the finite difference method (FDM) [12], finite element method (FEM) [13], and film mode matching method (FMM) [14]. Figures 1.2a and b show calculated effective indices n_{eff} of guided modes for 1,550-nm infrared light in various core geometries. Calculations were performed by the FMM and the indices of silicon and silica were set at 3.477 and 1.444, respectively. The mode notations are taken from [15], in which E^x and E^y modes represent the transverse electric (TE)-like and transverse magnetic (TM)-like modes, respectively. As shown Fig. 1.2a for waveguides of 200-nm silicon thickness, single-mode conditions are fulfilled when the core width is less than 460 nm for TE-like guided modes in which the dominant electric field is parallel to the substrate. The mode field profile of the TE-like fundamental mode is shown in Fig. 1.1b for a $460 \times 200\text{-nm}^2$ core. For a TM-like mode in which the dominant electric field is perpendicular to the substrate,

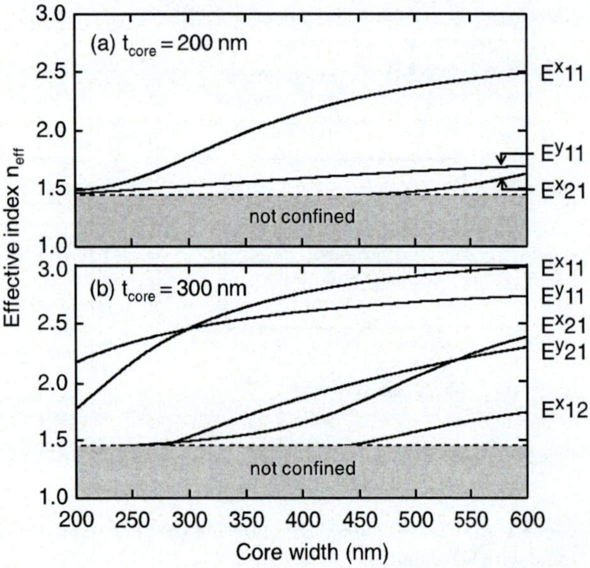

Fig. 1.2 Core width dependence of the effective indices of silicon photonic wire waveguides

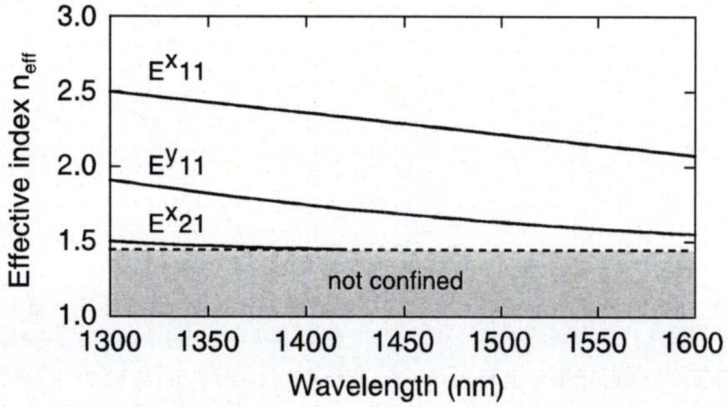

Fig. 1.3 Wavelength dependence of the effective indices of silicon photonic wire waveguides

the single-mode condition is fulfilled in a core larger than that for the TE-like mode. The effective indices of TE and TM fundamental modes show a large difference. In other words, the 200-nm-thick flat core produces a large polarization dependence. For waveguides with 300-nm-thick silicon as shown in Fig. 1.2b, core widths satisfying single-mode conditions are smaller than those for 200-nm-thick silicon. In a 300-nm^2 core, the refractive indices are identical for the TE and TM fundamental modes: that is, the polarization dependence can be eliminated.

Figure 1.3 shows the calculated wavelength dependence of the effective refractive indices for the waveguides with a 400 × 200-nm^2 core. Calculations were performed by the FMM and the material dispersions of refractive indices were considered. As shown in Fig. 1.3, the single-mode condition is violated in the wavelength region below 1,420 nm for the TE-like mode. For the 1,310-nm telecommunications wavelength band, therefore, a smaller core should be used for satisfying the single-mode condition.

1.2.2 Effect of Geometrical Errors and Birefringence

As shown in Fig. 1.2, the effective indices of silicon photonic wire waveguide are extremely sensitive to the core geometries. The group index n_g, which is an essential parameter in designing delay-based devices such as optical filters, is also affected significantly by the core geometry. Figures 1.4a and b show calculated group indices and their sensitivities to the core width $dn_g/n_g dw$. For TE-like modes, for which most of the photonic functions are designed, the sensitivities to the core width $dn_g/n_g dw$ are around $2\times10^{-4}\,\text{nm}^{-1}$ for a 400 × 200-nm^2 core and much higher for a 300-nm^2 core. For wavelength filters for dense wavelength division multiplexing (DWDM), the group index should be controlled to on the order of 1×10^{-4} or less. The index restriction corresponds to a core width accuracy of 0.5 nm or less, which is essentially unattainable with current micro-fabrication technology. Fortunately, there are optimum geometries giving very low sensitivities to the core width. For example, 385 × 200-nm^2 and 325 × 300-nm^2 cores are robust against the errors in core width. Waveguides with a very wide core are also robust against the errors in core width. In an arrayed waveguide grating (AWG) filter, waveguides with 750 × 200-nm^2 cores are used to reduce phase errors due to the variation of

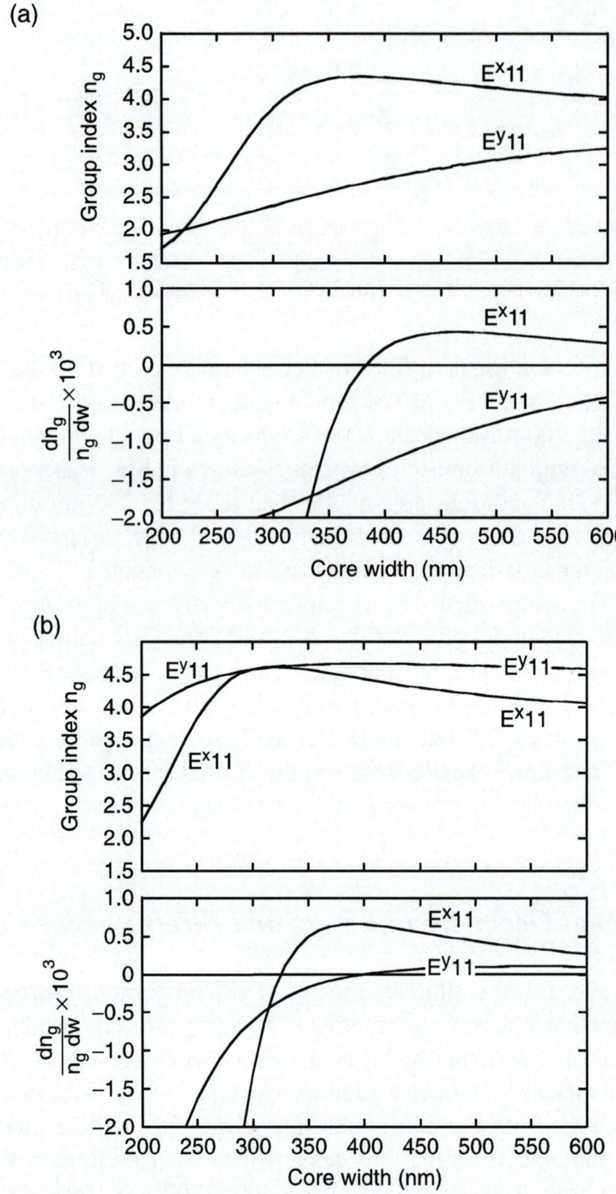

Fig. 1.4 (**a**) Core width dependence of the group indices and their derivatives for waveguides with 200-nm-thick cores (**b**) Core width dependence of the group indices and their derivatives for waveguides with 300-nm-thick cores

core width [16]. When we use such a wide-core waveguide, however, higher order modes stimulated in bending and other asymmetric structures become a concern.

Figure 1.4 also that the structural birefringence is incredibly large and that the problem of the polarization dependence in a silicon photonic wire waveguide is practically unsolvable. In a waveguide with a 400 × 200-nm^2 core, the group indices are 4.33 for the TE-like fundamental mode and 2.78 for the TM-like fundamental mode. The difference in the group indices gives a polarization mode dispersion (PMD) of 51.7 ps/cm, which seriously limits the applicable bandwidth in high-speed data transmission. The polarization-dependent wavelength (PDλ) in delay-based filter devices, such as AWGs and ring resonators, would be incredibly large. A square core would not be a solution to this problem because the group index is very sensitive to the core geometry, as mentioned before. In other words, polarization diversity is necessary for eliminating the polarization dependence in photonic devices based on silicon photonic wire waveguides.

1.2.3 Propagation Loss and Radiation Loss in Bending

The intrinsic loss in undoped silicon is very low for the photon energies below the band gap (~1.1 eV); therefore, the propagation loss of photonic wire waveguides is mainly determined by scattering due to surface roughness of the core. The effect of the surface roughness on the scattering loss in dielectric waveguides has been theoretically studied and formulated by Payne and Lacey [17]. The formula is represented by a complicated function characterized by a root-mean square roughness σ and the correlation length of the surface roughness structure l_c; the upper bound of the scattering loss α_{max}, as given in [18], is shown below.

$$\alpha_{max} = \frac{\sigma^2 \kappa}{k_0 d^4 n_1}, \tag{1.1}$$

where, k_0, d, and n_1 are wavevector of the light in vacuum, the half-width of the core, and effective index of a silicon slab with the same thickness as the core, respectively. The factor κ depends on the waveguide geometry and the statistical distribution (Gaussian, exponential, etc.) of the roughness, in which the correlation length l_c is included. According to [17], κ is on the order of unity for most practical waveguide geometries. Thus, the scattering loss is inversely proportional to the fourth power of d. In other words, it will seriously increase in photonic wire waveguides with an ultra-small core. A roughness of only 5 nm, for instance, would cause a 60-dB/cm scattering loss in a 400-nm-wide core made of a 200-nm-thick silicon slab whose effective index is 2.7. To achieve a practical scattering loss of a few decibels per centimeter, the surface roughness should be about 1 nm or less.

Radiation losses in the bending section can be calculated by applying cylindrical coordinates in numerical mode solvers. Figure 1.5 shows calculated radiation losses for a 90° bend in various waveguides. The calculations were performed by using a commercially available FMM mode solver [19]. As shown in this figure, the bending performance varies with polarization. For the TE-like mode in the waveguide with a 400×200-nm^2 core, the radiation loss is negligible even if the bending radius is as small as 2.5 μm. For the TM-like mode, however, a bending radius of over a few tens of micrometers is needed in order to achieve negligible radiation loss. The bending loss also varies with core dimensions. As shown in Fig. 1.5, for TE-like modes, a waveguide with a 300-nm^2 core requires a larger bending radius than a flat waveguide.

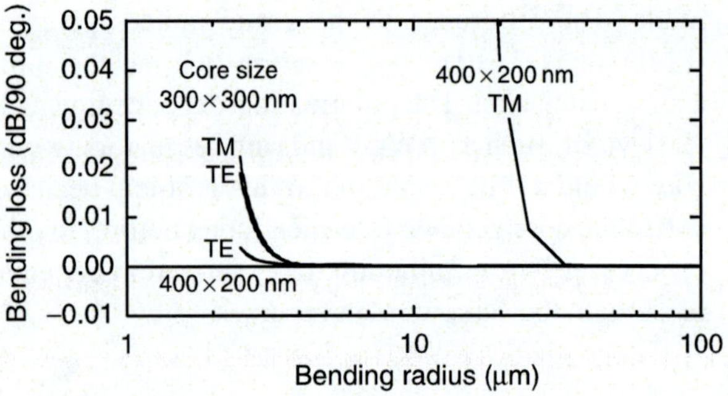

Fig. 1.5 Calculated radiation losses for a 90° bend in various waveguides

1.2.4 Coupling to External Fiber

Since a silicon photonic wire waveguide has a very small mode profile, spot-size conversion is essential for connecting it to external circuits such as single-mode optical fibers. A highly efficient spot size converter (SSC) with a silicon reverse adiabatic taper has already been proposed [20]. As shown in Fig. 1.6, it has a double-core structure consisting of a thin silicon taper and silica-based waveguides. In a typical design for 1,550-nm-wavelength infrared light, the tip of the taper should be ultimately reduced to less than 100 nm and the silica-based waveguide has a 3-μm^2 core with a 2.5% index contrast to the cladding. In such a double-core structure, light leaking from the silicon taper is captured by a silica-based waveguide, which guarantees efficient optical coupling to external optical fibers.

Figure 1.7 shows the calculated conversion efficiencies between a silicon photonic wire waveguide with a 400 × 200-nm^2 core and a silica-based waveguide with a 3-μm^2 core. As shown in this figure, a 200-μm-long taper with a 80-nm tip gives a conversion loss of around 0.1 dB for both polarizations. A shorter and thicker taper would give lower coupling efficiencies. Since the coupling loss between a silica-based waveguide with a 3-μm^2 core and a high-numerical-aperture (NA) fiber with a 4.3-μm mode field diameter is about 0.1 dB and the conversion loss between

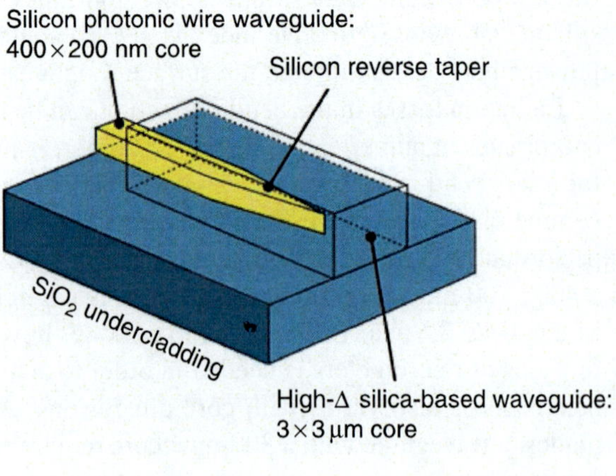

Fig. 1.6 Schematic of the spot size converter

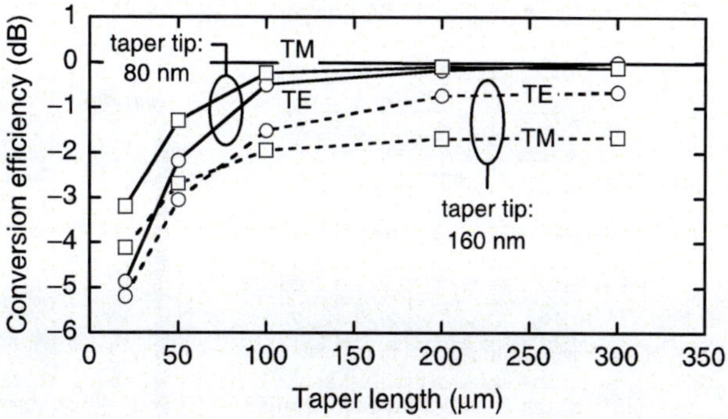

Fig. 1.7 Calculated conversion efficiencies in spot size converters with various geometries

the high-NA fiber and the ordinary single mode can be reduced to about 0.1 dB by applying a thermally expanded core (TEC) technology [21], a total coupling loss of less than 0.5 dB can be achieved between a photonic wire waveguide and a single-mode fiber.

A grating coupler input/output structure has also been proposed [22], for which the coupling efficiency between a photonic wire waveguide and an ordinary single-mode fiber is calculated to be −5 to −3 dB.

1.3 Fundamental Propagation Performance

1.3.1 Fabrication

Figure 1.8 shows a typical fabrication process for a Si-wire waveguide. First, a hard mask layer and resist mask layer are formed on a SOI substrate. The hard mask is used to improve the selectivity of Si etching and is often made of SiO_2. Next, waveguide patterns are defined by using electron beam (EB) lithography or excimer laser deep ultraviolet (DUV) lithography [9], which are capable of forming 100-nm patterns. Ordinarily, EB and DUV lithography technologies are used in the fabrication of electronic circuits where they are optimized for patterning of straight and intersecting line patterns. Therefore, no consideration has been given to curves and roughness in the pattern edges, which are important factors in fabricating low-loss optical waveguides. To reduce propagation losses of the waveguides, it is necessary to reduce the edge roughness to around 1 nm or less. This means that particular care must be taken in the data preparation for EB shots or DUV masks. Figure 1.9 shows an ultra-small ring resonator with and without special treatment for EB data preparation, where a drastic reduction in side-wall roughness is observed as a result of the special treatment [23]. The writing speed of the EB lithography must also be considered in practical fabrication. For practical purposes, it is probably necessary to use EB lithography with a variable-shaped beam.

After resist development and SiO_2 etching for a hard mask, the silicon core is formed by low-pressure plasma etching with an electron-cyclotron resonance (ECR) plasma or inductive coupled plasma. To ensure the edge roughness of the side walls

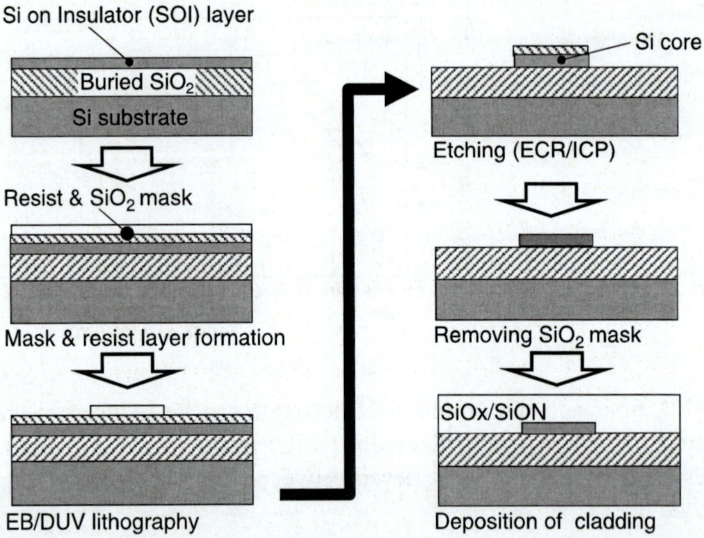

Fig. 1.8 Typical fabrication process of a silicon photonic wire waveguide

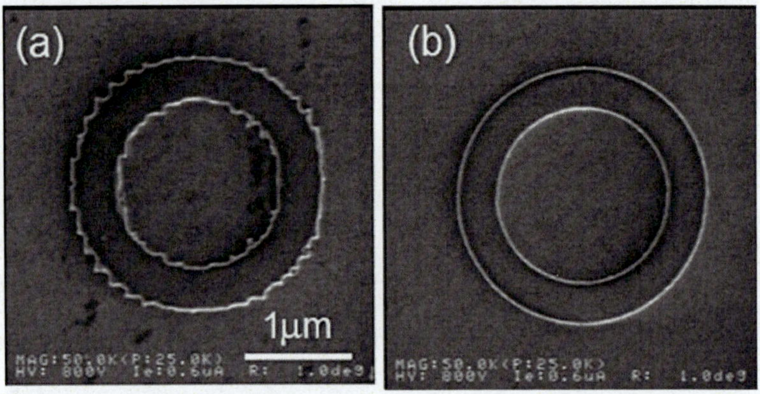

Fig. 1.9 SEM images of ring resonators (**a**) without and (**b**) with EB data optimizations

is at sub-nanometer levels, the plasma conditions and the selection of etching gases must be tuned for individual plasma equipment.

Finally, an overcladding layer is formed with a SiO_2-based material or polymer resin material. To avoid damaging the silicon layer, the cladding layers must be deposited by a low-temperature process, such as the plasma-enhanced chemical vapor deposition (PE–CVD) method [11]. In particular, for waveguides associated with electronic structure, it is essential to use a low-temperature process so as not to damage the electronic devices.

Figure 1.10a shows a scanning electron microscope (SEM) image of the core of a silicon photonic wire waveguide with a cross-section of $400 \times 200\,\text{nm}^2$ [8]. The geometrical shape closely matches the design values, and the perpendicularity of the sidewalls is also very good. Figure 1.10b shows a photograph of the 80-nm-wide

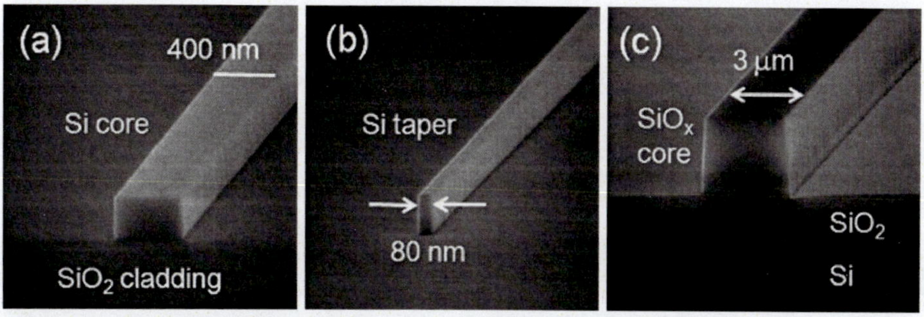

Fig. 1.10 SEM images of a silicon photonic wire waveguide system. (**a**) Core of silicon photonic wire waveguide, (**b**) silicon taper, and (**c**) core of SiOx waveguide for spot size converter

taper tip in the SSC. The taper and waveguide core were constructed by using a common fabrication process. Figure 1.10c shows an SEM photograph of the core of the silica waveguide for the SSC. The silica waveguide core was fabricated by depositing SiO_2-based material by the PE–CVD method and etched by reactive ion etching (RIE). By adjusting the deposition conditions, the refractive index of the material is tuned to be 2.5% higher than that of an ordinary thermal oxide. In the final product, a 7-μm SiO_2 overcladding layer covers the whole structure.

1.3.2 Propagation Performance

Figure 1.11 shows a typical transmission loss of silicon photonic wire waveguides with SSC, fabricated in the manner mentioned above. High-NA optical fibers with 4.3-μm mode field diameter (MFD) are used for external coupling. As shown in this figure, the propagation loss for the TE-like mode has been improved to be around 1 dB/cm. In the waveguide with flat cores, the propagation losses for TM-like modes are generally better than those for TE-like modes. Oxidation of the core sidewalls may further reduce the propagation losses [24]. The propagation loss of around 1 dB/cm is already at a practical level, since photonic devices based on silicon photonic wire waveguides are typically smaller than 1 mm. Besides the sidewall

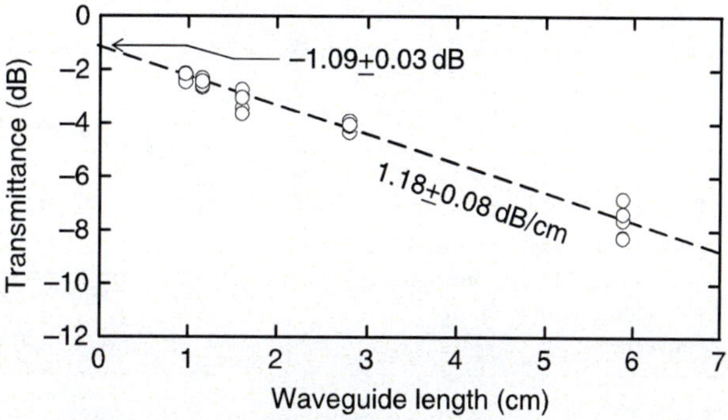

Fig. 1.11 Measured propagation loss of a silicon photonic wire waveguide with spot size converters

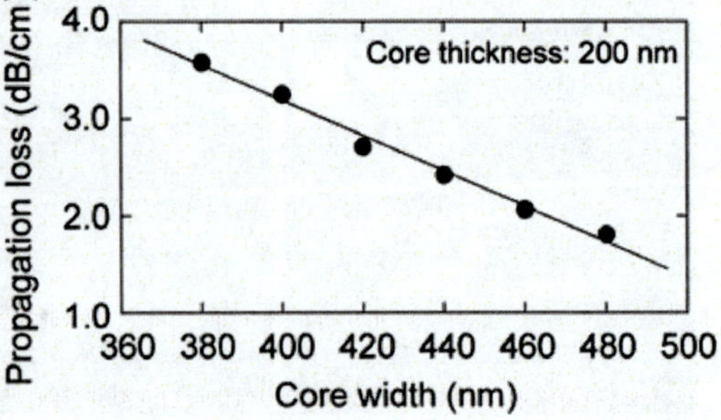

Fig. 1.12 Relation between measured propagation losses and core width

roughness of the core, the core width affects propagation losses as well. Figure 1.12 shows the relation between measured propagation losses and core widths. As shown, the propagation loss is reduced by increasing the core width because the effect of sidewall roughness is reduced in a wide core. When the core width exceeds 460 nm, the waveguide can also guide a higher order mode which may degrade the performances of some photonic devices.

The coupling loss between optical fiber and silicon photonic wire waveguide is represented by the intercept of the vertical axis in Fig. 1.11. The loss value at the intercept includes two waveguide/fiber interfaces; therefore in this case, one interface has a 0.5 dB coupling loss at a wavelength of 1,550 nm.

Figure 1.13 shows the transmission spectrum of a silicon photonic wire waveguide with SSCs. The spectrum remains flat over a 200-nm wide bandwidth, and no absorption dip is observed. The flat spectrum means that the SiO_2-based material used in the SSC does not contain impurities with N–H bonds. Although absorption by residual O–H bonds exists at wavelengths of around 1,400 nm, the resulting losses are not large. It is also possible to eliminate O–H bonds by heat treatment.

Figure 1.14 shows bending losses of single mode waveguides for TE-like modes. For the bending radius of over 5 μm, bending losses are negligible. Even for an ultra-small bending radius of around 2 μm, a waveguide with a flat core maintains

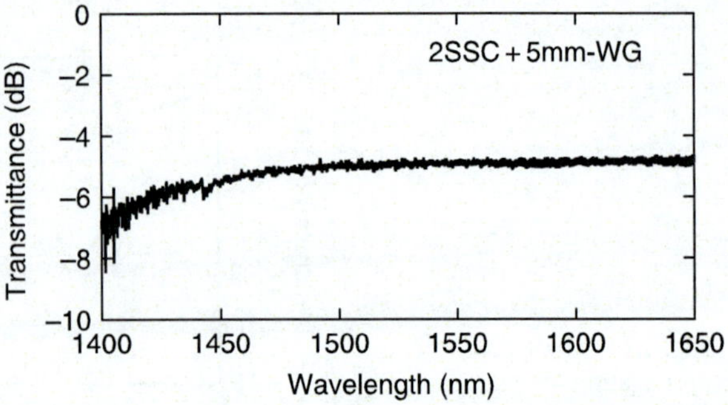

Fig. 1.13 Measured transmission spectrum of a typical silicon photonic wire waveguide with spot size converters

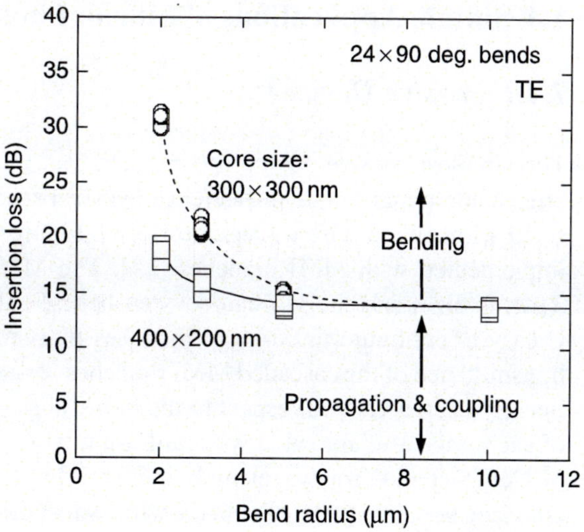

Fig. 1.14 Relation between measured insertion losses of the waveguides and bending radius

a low loss of below 0.1 dB per 90° bend. A waveguide with a square core shows a larger bending loss for a bending radius below 5 μm. For TM-like modes, especially in waveguides with flat cores, bending losses are generally larger than those for TE-like modes. Bending losses measured in various research groups are summarized in [10].

The birefringence of the waveguide can be evaluated by the free spectral ranges (FSRs) of ring resonators. Figure 1.15 shows measured transmission spectra at around 197 THz ($\lambda = 1.514\,\mu$m) for a ring resonator of 10-μm radius. In this figure, the FSR in TM-like modes (1.67 THz) is significantly larger than that in TE-like modes (1.11 THz). Using the FSR, we can roughly express the group index of the waveguide by $n_g = c/2\pi R\Delta f$, where c, R, and Δf are the speed of light in a vacuum, the radius of the ring resonator, and the FSR in hertz, respectively. Thus, the group indices are estimated to be 4.30 and 2.86 for TE and TM modes, respectively, which agree well with the design values in the previous section.

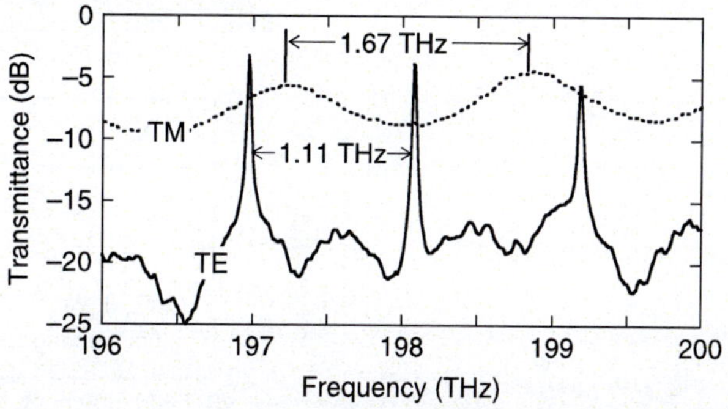

Fig. 1.15 Measured drop port spectra of a ring resonator with 10-mm radius

1.4 Simple Applications of Silicon Photonic Wire Waveguides

1.4.1 Passive Devices

The simplest passive device is a branch. Various branching devices, such as a Y branch and a multimode interference (MMI) branch, have been proposed and fabricated with silicon-wire waveguides [6, 7]. Figure 1.16a shows a multi-stage branch implemented with MMI branches [25]. The MMI branch units are no more than 3 μm in size, and the waveguides can bend light with a micrometer radius, so it is easy to configure multi-stage branches in a small area. Figure 1.16b shows the transmittance of the cascaded MMI branches for each output port. The transmittance linearly decreases with respect to the branching order. In other words, the branches of each stage are fabricated with uniform quality.

Compact add/drop wavelength filters are also being developed by using the ultra-compact bending parts of silicon-wire waveguides. Various ring-resonator-based filters have been developed. A single resonator is very compact and suitable for high-density integration; however, its Lorentzian resonance, as shown in Fig. 1.15, is not suitable for filters for telecommunications applications, which require flat pass bands. For a flat pass band, cascaded ring resonators have recently been developed [26].

Complex wavelength filters, such AWG filters [16, 27] and the lattice filters [28, 29], have also been developed. A photograph of a lattice filter is shown in Fig. 1.17a. In such filters, the pass-band spectrum can be fine-tuned by applying various optimization techniques, such as apodization. As shown in Fig. 1.17b, for

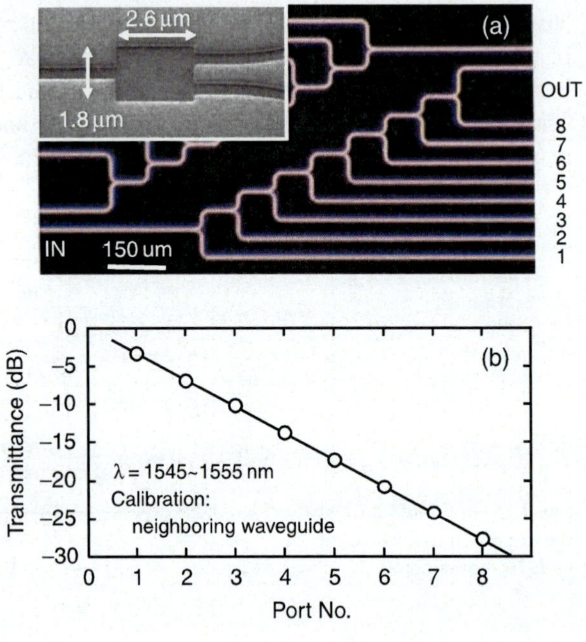

Fig. 1.16 (**a**) Cascaded MMI branches and (**b**) their transmission characteristics

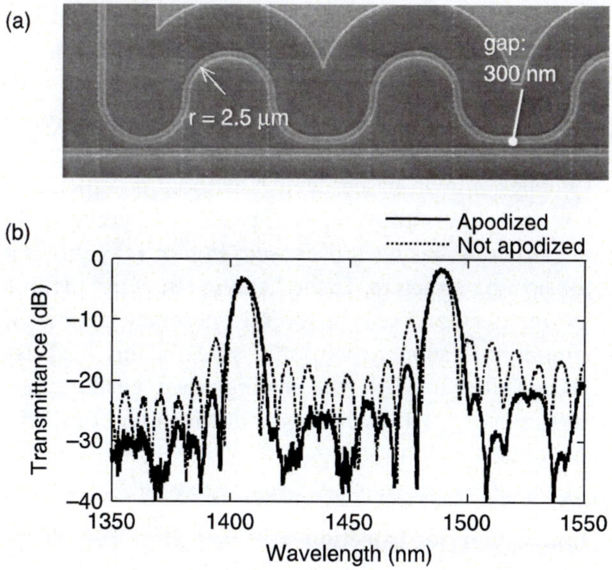

Fig. 1.17 (**a**) A SEM photograph of a part of a lattice filter and (**b**) measured drop port spectra

example, adjacent channel crosstalk is reduced to be about −20 dB in the lattice filters for which apodization of the coupling coefficients in the directional couplers has been performed [29]. The shape of the pass band could also be controlled by apodization in these sophisticated filters.

Wavelength tuning and polarization dependence are serious obstacles to the practical applications of these wavelength filters. As shown in the previous sections, effective indices of silicon photonic wire waveguides are extremely sensitive to geometrical errors in the core shape. The indices are also sensitive to the environmental temperature. Therefore, index tuning by temperature control or carrier injection is necessary for fixing the pass band to a certain wavelength. As described in the previous sections, the structural birefringence is also very sensitive to the core geometry, and it is virtually impossible to eliminate the polarization dependence of passive devices based on silicon photonic wire waveguides. A practical solution for the polarization dependence is the polarization diversity, whose details are described in Sect. 1.5.

1.4.2 Dynamic Devices

1.4.2.1 Thermo-optic Effect

The thermo-optic (TO) effect in silicon is ten times or more stronger than that in silica [30] and the volume of the core in silicon photonic wire waveguides is smaller than that of the core in silica-based waveguides. Therefore, power consumption in silicon-based TO devices can be one-tenth of that in silica-based devices. To achieve such a low power consumption, it is very important to concentrate the heating power in the waveguide core. However, heat expansion in the thick (∼3 μm) and laterally wide cladding layer is an obstacle to the heat convergence. In the early applications of the TO effect in optical switches based on Mach-Zehndar interferometers (MZIs), heaters were constructed on a thick overcladding without lateral thermal insulation. In such TO switches, a few tens of milliwatts of heater power is necessary for a π phase shift in one arm of a MZI [11]. Compared to typical silica-based TO switches, silicon-based devices without an efficient heating structure do not show a remarkable improvement in power consumption. Recently, a very efficient heating structure has been proposed. Figure 1.18 shows a cross-sectional schematic of the heating structure. In the heating structure, the undercladding and side-cladding are removed and only a several-micrometer-wide silica cladding containing a silicon core is supported by thin silica beams [31]. By applying this structure as a phase shifter in MZIs, the heating power has been reduced to be less than 1 mW, as shown in Fig. 1.19, which is about 1/100th of that of conventional silica-based TO switches.

1.4.2.2 Carrier Injection

In the infrared and visible regions, the optical properties of silicon are affected by free carriers, especially for photon energies below the band gap (~1.1 eV). By applying the Drude approximation and assuming the carrier plasma frequency is

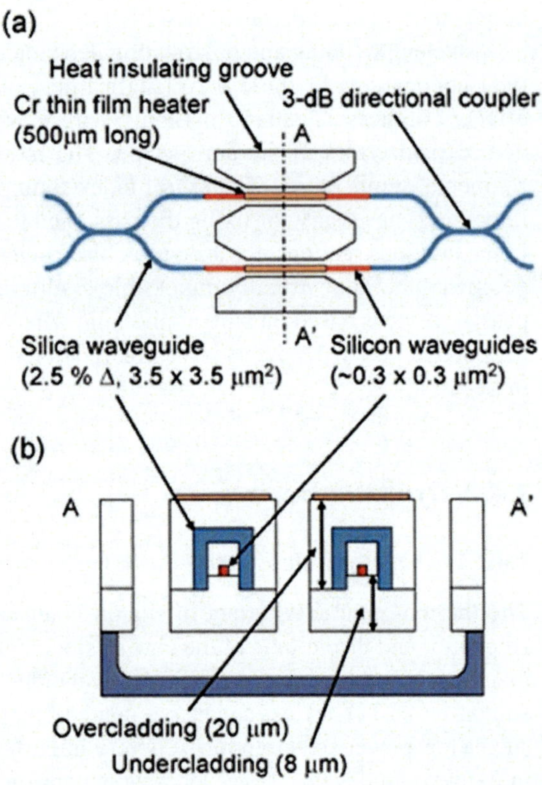

Fig. 1.18 Schematic structure of the low power consumption switch: (**a**) Top view and (**b**) cross-sectional view along line A–A′ in (**a**) [31]

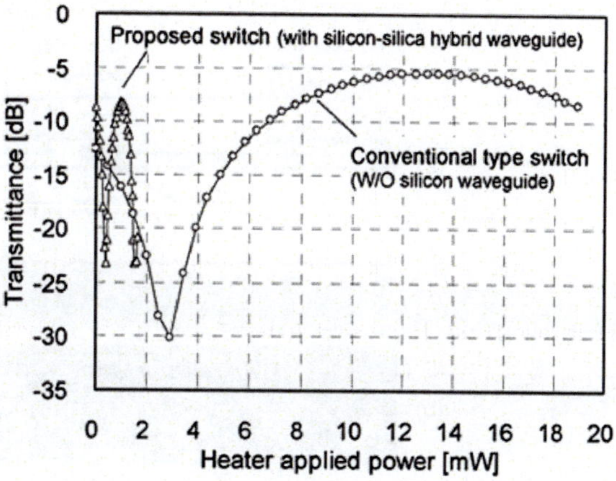

Fig. 1.19 Switching characteristics of low-power and conventional switches for TE polarization [31]

significantly lower than the optical frequency, the changes of refractive index Δn and absorption coefficient $\Delta\alpha$ are expressed by the following formulas [32]:

$$\Delta n = \frac{-e^2\lambda^2 N}{8\pi^2 c^2 \varepsilon_0 n m^*} \tag{1.2}$$

$$\Delta a = \frac{e^3\lambda^2 N}{4\pi^2 c^3 \varepsilon_0 n \mu m^{*2}} \tag{1.3}$$

where e, c, ε_0, and n are the elementary charge, the speed of light in a vacuum, the vacuum permittivity, and the refractive index of silicon, respectively. N, m^*, and μ are the injected carrier density, the carrier effective mass, and carrier mobility, respectively.

To control the concentration of free carriers in the waveguide core, silicon photonic wire waveguides with a p-i-n carrier injection structure have been developed [33–35]. A Typical cross-section of a waveguide with a p-i-n carrier injection structure is shown in Fig. 1.20a. The typical core size of waveguides with a p-i-n structure is 400–600 × 200 nm^2. At both sides of the core, the waveguide has slab regions for carrier injection. Although the core shape resembles that of conventional silicon rib waveguides [3], its geometrical size is far smaller than in the conventional ones. Moreover, the bending radius is less than a hundredth of that in conventional rib waveguides. These reductions in geometries are greatly advantageous for applications of silicon photonic wire waveguides with the carrier injection structures. For example, the device length would be significantly reduced because the carrier concentration can be increased in an ultra-small core. In such a short device with an ultra-small core, the electric power consumption can be significantly reduced. Moreover, the operation speed can also be increased because a small core guarantees a fast carrier transit time.

By applying these advantages of silicon photonic wire waveguides with p-i-n structures, various dynamic devices, such as high-speed modulators [33, 35] and variable optical attenuators (VOA) [34], have been developed. Among them, in this section, the application of the p-i-n carrier injection structure to a VOA is described.

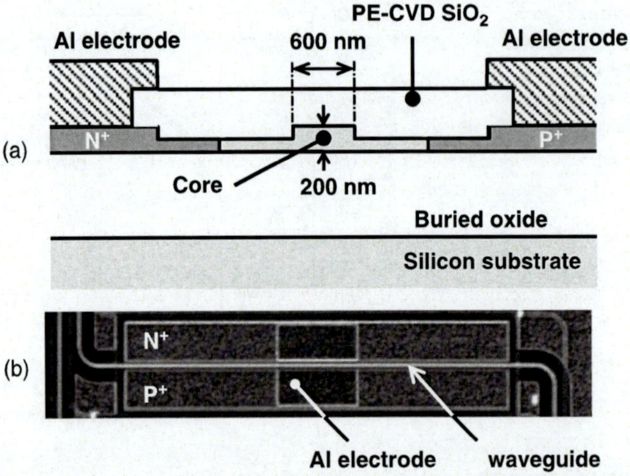

Fig. 1.20 (**a**) Cross-sectional view and (**b**) top view of a silicon photonic wire-based carrier injection structure

As show in Fig. 1.20a, the waveguide is a rib-type one with a 600×200-nm^2 core and 100-nm-thick slab. This structure satisfies the single-mode condition for 1,550-nm infrared light. The slab is thicker than those reported in [33] and [35]. The thick slab is important for reducing propagation loss and device impedance.

The core was fabricated in same manner as for passive waveguides. After thermal oxidation for surface passivation, the n^+ and p^+ regions were defined in the slab section by lithography and implanted with phosphorus and boron as dopants. After annealing for the activation and diffusion of dopants, a silicon dioxide layer was deposited. The dopant density was about 10^{20} cm^{-3} in each doped region. Aluminum electrodes were then constructed on the doped regions.

Figure 1.20b shows a fabricated waveguide with the carrier injection structure in which the pair of contact electrodes can be seen. The propagation loss is typically less than 2 dB/cm. Figure 1.21 shows typical electrical characteristics of a 1-mm-long carrier injection structure. In spite of the device length, the reverse current is very low, around pico-amperes. This means the leakage of the device is small, so that injected carriers will effectively interact with guided light.

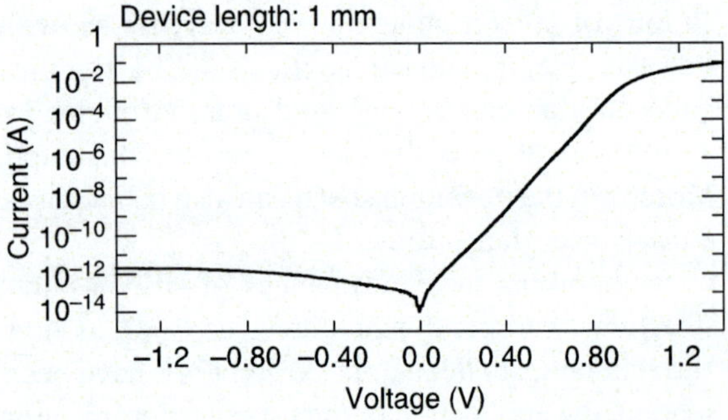

Fig. 1.21 Measured current-voltage relation of the carrier injection structure

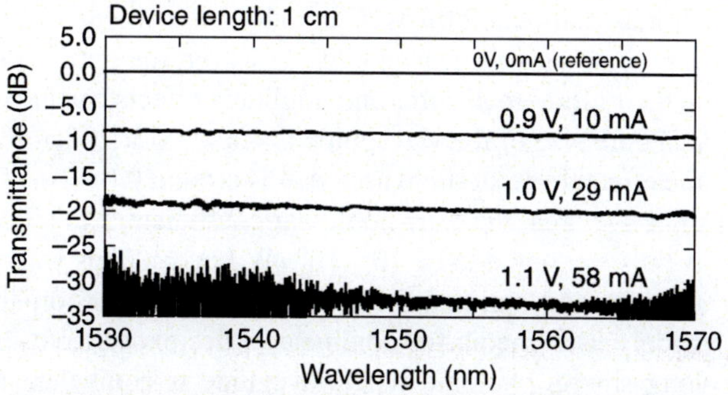

Fig. 1.22 Measured transmission characteristics of the variable optical attenuator

By injecting carriers into the core, optical attenuation is performed through the free carrier plasma effect mentioned above. Figure 1.22 shows transmission spectra of a 10-mm-long waveguide for various injected currents. As the injected current increases, guided light is absorbed by the injected carriers. The wavelength dependence of the attenuation is very flat in the measured 40-nm bandwidth. Operation with a 30-dB attenuation requires a power of about 55 mW, which is about a half of the power consumption of conventional rib-type devices [34]. The power consumption can be reduced by a series connection of the devices, which is a similar to the technique used in conventional devices. Figure 1.23 shows the temporal response of optical attenuation in a 1-mm-long device with a 10-ns injection pulse. In a 2-V reverse bias operation, the response time is about 2 ns, which is about 100 times faster than that in conventional devices [34].

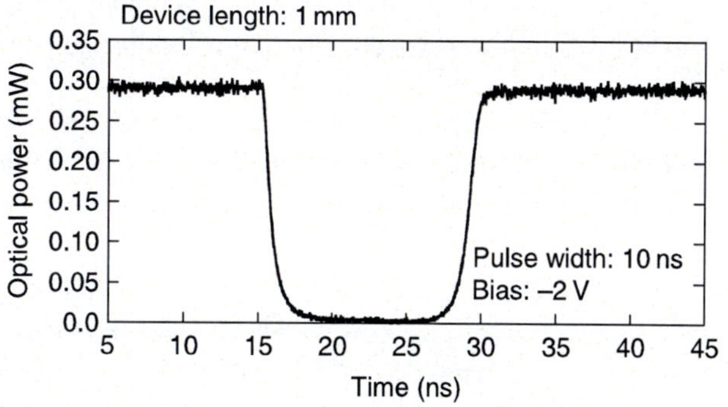

Fig. 1.23 Measured temporal response of the variable optical attenuator

Considering the linear propagation loss of the waveguide with the carrier injection structure, the on-chip insertion loss is estimated to be 2 dB in a 10-mm device and 0.2 dB in a 1-mm one. These values can satisfy required practical standards.

1.4.3 Nonlinear Functions

In addition to its use as a platform for passive and dynamic functions, a silicon photonic wire waveguide is also a promising platform for active functions based on various nonlinear effects. Generally, bulk silicon shows weak nonlinearities. In silicon photonic wire waveguides, however, the optical power density can be increased in their ultra-small cores and nonlinear effects can be remarkably enhanced. The typical core size of the waveguide is 400–500 × 200 nm^2 at 1,550-nm wavelength, and the effective mode-field area A_{eff} is around 0.05 μm^2 [36]. Thanks to the ultra-small effective area, the Kerr nonlinear coefficient γ in silicon photonic wire waveguides is thus estimated to be 10^5–10^6 (W km)$^{-1}$. This value is about a hundred-thousand times larger than that for a single-mode dispersion-shifted fiber (DSF), which is a widely used media for nonlinear optics experiments. In other words, a centimeter-long silicon photonic wire waveguide is equivalent to a 100-m long DSF, giving practical efficiencies in various nonlinear functions. Although the interaction length and efficient power injection have been obstacles to obtaining practical efficiencies, the propagation loss of the waveguides has recently been improved to be 1–2 dB/cm and the coupling loss to external fibers has been reduced to be 0.5 dB by using a special SSC [20]. Thus, a power density of around 300 MW/cm^2 could be attained using commercially available 20-dBm CW light sources. The power density is high enough to obtain efficient nonlinear functions.

By applying the enhancement of nonlinear effects in a silicon photonic wire waveguide, various nonlinear functions, such as all-optical modulation based on the two photon absorption (TPA) effect [37], wavelength conversion and parametric amplification based on four wave mixing (FWM) effects [36, 38, 39] and stimulated Raman scattering (SRS) [4, 40], have been tested. Very recently, entangled-photon-pair creation has also been tested by using spontaneous FWM [41–43].

Among these functions, FWM-based wavelength conversion is here described as a typical nonlinear function giving an efficiency comparable to that in conventional devices [38]. The experimental setup for the wavelength conversion is shown in Fig. 1.24. A signal light with 10-Gbps non-return-to-zero (NRZ) modulation and a CW-operated pump light were injected collinearly into a 2.8-cm-long waveguide with SSCs. The polarization was adjusted to a TE-like mode, and the propagation loss of the waveguide was about 2.3 dB/cm. At the waveguide's entrance, the pump

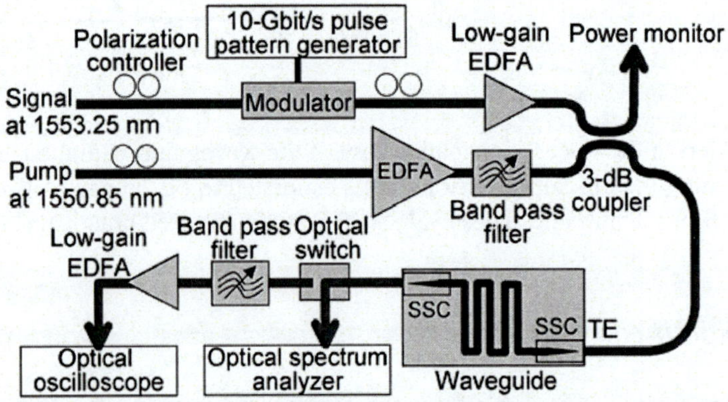

Fig. 1.24 Experimental setup for wavelength conversion using the four-wave mixing effect

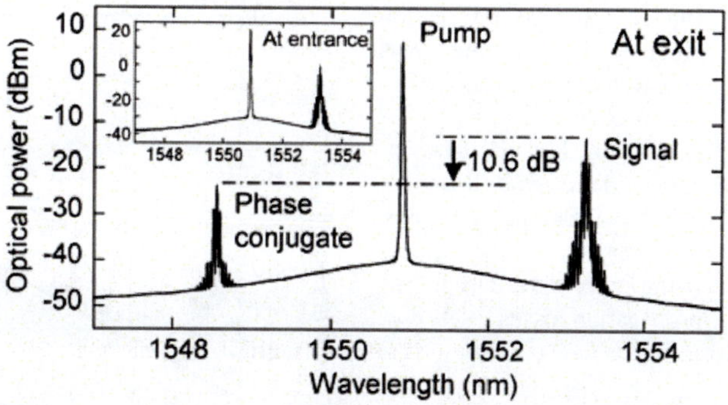

Fig. 1.25 Experimental result of wavelength conversion using the four-wave mixing effect

power was 160 mW CW, which gives a power density of around 430 MW/cm^2 in the waveguide.

The output spectrum at the exit of the waveguide is shown in Fig. 1.25. Besides the peaks of the injected pump and signal light, a peak of the phase conjugated light, or converted light, can be seen in the spectrum. The internal conversion efficiency is around -10 dB [38], which is comparable to that observed in periodically poled $LiNbO_3$ devices a few years ago [44], and large enough for practical data processing. As shown in Fig. 1.26, clear eye patterns of a 10-Gbps pseudorandom binary sequence (PRBS) data stream are observed in the converted light. The bandwidth giving a -3-dB efficiency degradation was measured to be over 20 nm [36]. The bandwidth was limited by the phase mismatching among waves. In silicon photonic wire waveguides, waveguide dispersion can be easily controlled by adjusting the core shape; therefore, the bandwidth could be further improved [39].

The conversion efficiency with respect to the pump power is shown in Fig. 1.27. Efficiency saturation due to the free carrier absorption effect can obviously be seen. The free carriers are generated through the TPA effect whose efficiency is proportional to the second power of the optical power density. Therefore, the simplest way to eliminate FCA is to reduce the power density. A theoretical estimation shows that a waveguide with 0.5-dB/cm propagation loss (dashed line) would provide an efficiency of -4 dB even if the pump power were 80 mW, where FCA is well suppressed [38]. Carrier sweep-out using a p-i-n structure, such as described in the previous section, is also a promising way to prevent FCA.

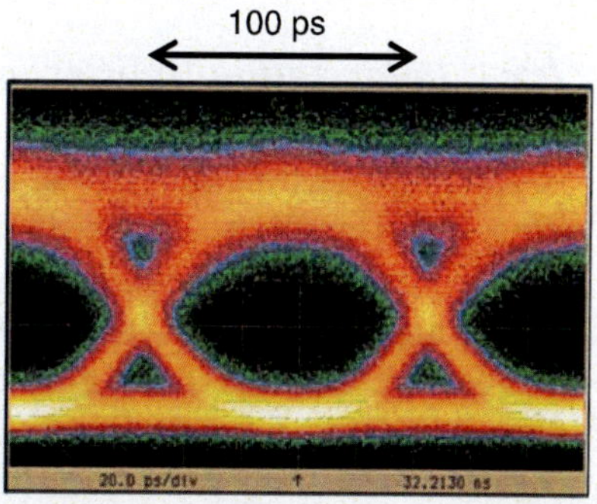

Fig. 1.26 Eye pattern of the signal of phase conjugated light

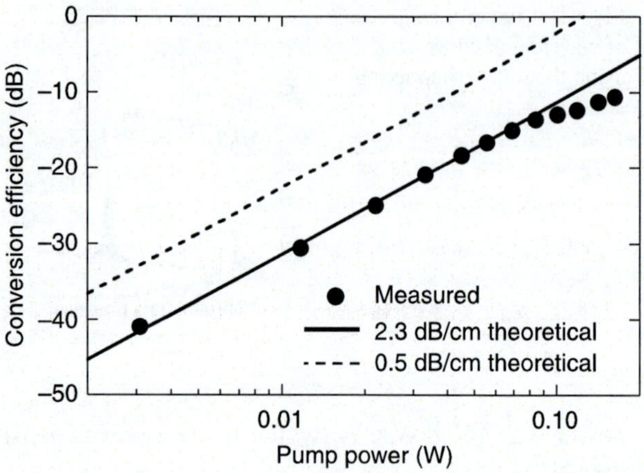

Fig. 1.27 Internal conversion efficiency versus pump power in wavelength conversion. The *dots* represent experimental results. The *solid* and *dashed lines* are theoretical estimations under the same conditions as those in the experiment except for propagation losses

1.5 Polarization Manipulation

As described in Sect. 1.2, the polarization dependence is one of the most serious obstacles to the practical application of silicon photonic wire waveguides, and polarization diversity is a unique solution to this problem. Generally, polarization diversity is achieved by using polarization splitters and rotators. If TE-like and TM-like modes can be separated by polarization splitters, and the TM-like modes can be converted into TE-like ones by rotators, we need only consider optical functions for the TE-like modes, not for both components. Thus, the polarization dependence could be eliminated by using a single-polarization (TE) platform. Such a polarization diversity system might be constructed by using free-space optical devices and optical fibers; however, this would require too much area and would be costly. For the polarization diversity circuit to be practical, it should be monolithically constructed on a chip. Recently, two types of on-chip polarization diversity circuits have been proposed [45, 46]. One is based on a two-dimensional grating coupler [45]. The other is based on a polarization splitter and rotator with silicon photonic wire waveguides [46]. Details of the latter are described in this section.

1.5.1 Polarization Splitter and Rotator

A scanning-electron micrograph of the fabricated polarization splitter is shown in Fig. 1.28. This device is based on a simple directional coupler with silicon photonic wire waveguides [47]. The coupling length for the TM-like modes is much shorter than for the TE-like modes. Thus, the TM-like modes can traverse the coupler, whereas the TE-like modes propagate along the initial waveguide. In a design for 1,550-nm infrared light, a directional coupler consisting of 200-nm-high and 400-nm-wide cores with a 480-nm gap can separate two orthogonal polariza-

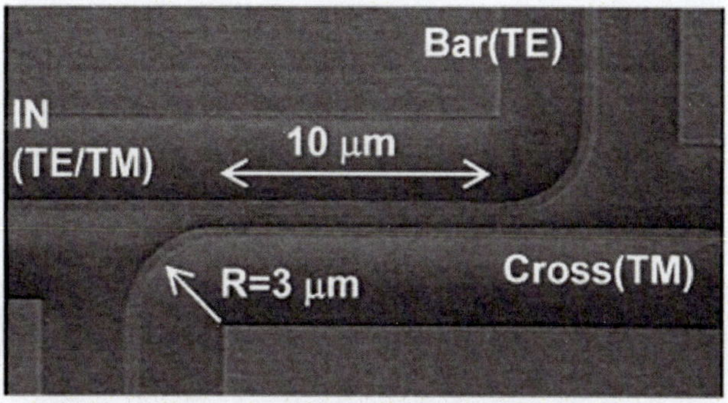

Fig. 1.28 A SEM image of the polarization splitter

tions through only a 10-μm-long propagation. For the suppression of additional coupling at the bending sections and for high-density integration, the ends of the coupler are tightly bent with a 3-μm radius. In the 200×400-nm^2 rectangular core, TE-like modes can propagate through the tight bend, but TM-like modes can not. Therefore, the exit waveguide for the TM-like modes is straight along the coupler.

The measured transmission of the splitter for various incident polarizations is shown in Fig. 1.29. The polarization was controlled by rotating a half-wave plate installed in front of the single-mode fiber for input. In the cross port, TM-like modes are dominant and TE modes are suppressed; in the bar port, vice versa. The maximum polarization extinction ratios (PERs) are 26 and 14 dB for bar and cross ports, respectively. Cascading two splitters improves the PER to about 23 dB for the cross port. A detailed study has shows that the applicable bandwidth is over 100 nm for a −10 dB PER [47].

A schematic of the fabricated rotator is shown in Fig. 1.30a [48]. The device has an off-axis double-core structure consisting of a 200-nm^2 silicon core, a 840-nm^2 silicon oxynitride (SiON) core, and a silica overcladding. The right and bottom edges of the thin silicon core overlap the corresponding edges of the second core. The SiON layer, with a refractive index of 1.60, and the silica layer are deposited by the PE–CVD method.

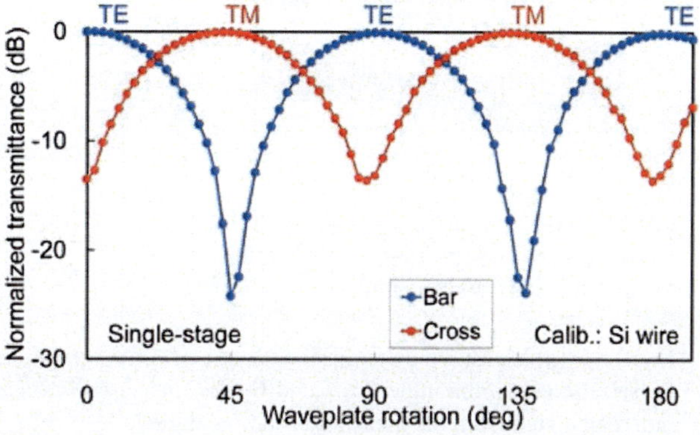

Fig. 1.29 Measured transmission of a single-stage polarization splitter for various incident polarizations

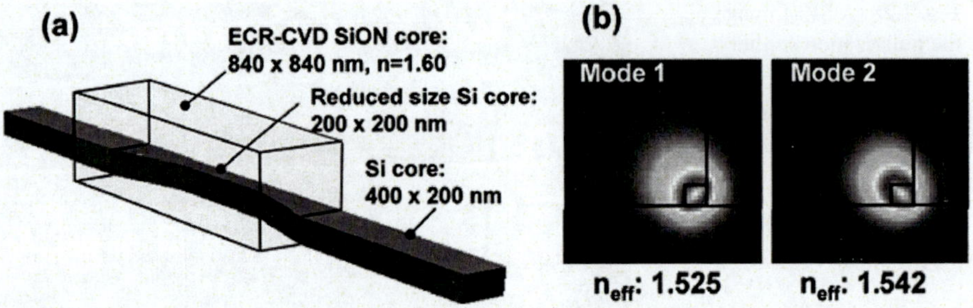

Fig. 1.30 (**a**) Schematic of the polarization rotation structure and (**b**) its eigenmodes

As shown in Fig. 1.30b, the double-core structure has two orthogonal eigenmodes with the eigenaxes tilted toward the substrate. Thus, propagation through the waveguide produces a rotation of the polarization plane when the polarization of incident light is just TE-like or TM-like modes. The polarization rotation angle is described by $\phi = \pi L \Delta n/\lambda$, where L is the propagation length, Δn the difference in the effective indices between two orthogonal eigenmodes, and λ the operation wavelength. The rotator length for 90° rotation $L_{\pi/2}$ is given by $\lambda/2\Delta n$.

The measured output spectra for both polarization components of a 35-μm-long device are shown in Fig. 1.31a. Although the incident light has a TM-like component, the TM-like component after passing through the device is suppressed and the TE-like component becomes dominant. In other words, the incident TM component is converted into TE. The excess loss through the device is about 1 dB. The spectral ripples are caused by the polarization rotation in the normal silicon waveguides used for input and output, and they complicate the estimation of the actual rotation angle. A Poincare-mapping measurement gives a more accurate rotation angle and also the extinction ratio. As shown in Fig. 1.31b, the maximum rotation angle is 72°, and the maximum extinction ratio is about 11 dB, which is obtained for a 35-μm-long device [48].

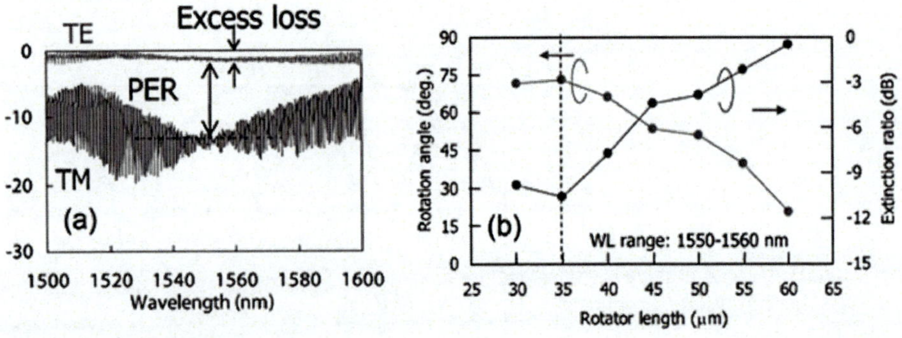

Fig. 1.31 (**a**) Measured transmission spectra of the polarization rotator with a length of 35 μm; (**b**) Polarization rotation angle estimated by the measured Poincare map and polarization extinction ratio calculated from the measured rotation angle

1.5.2 Polarization Diversity

A microscope image of the fabricated polarization diversity circuit is shown in Fig. 1.32 [46]. The circuit consists of polarization splitters, rotators, and a ring resonator wavelength filter with a 10-μm radius. The ring resonator has an extraordinary polarization dependence, which is shown in Fig. 1.15. In this diversity circuit, a splitter for the input separates the TE and TM components and feeds them into two waveguides. For the waveguide for the TM-like mode, the polarization is rotated into the TE-like mode. Thus, the polarization-dependent ring resonator can be used for both polarization inputs. There are two output waveguides for the ring resonator. In one of the waveguides, the TE-like mode is rotated into the TM-like mode and finally combined with another TE-like mode. In such a geometry, there is no PDλ, because this system uses only one functional device working for the TE mode. Moreover, the symmetry of the circuit eliminates the PDL and PMD originating from the normal waveguiding section [46]

Transmission spectra for wavelength filters with and without the polarization diversity are shown in Fig. 1.33. Without the diversity, the ring resonator works

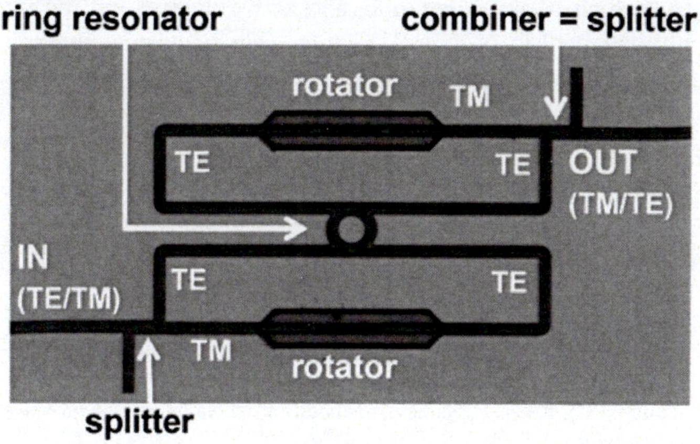

Fig. 1.32 Photograph of the polarization diversity circuit for a ring resonator filter

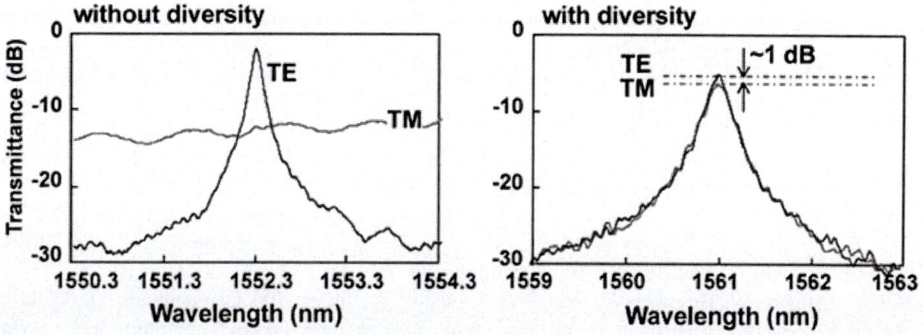

Fig. 1.33 Measured transmission spectra of a ring resonator with and without polarization diversity

as a fine wavelength filter for TE polarization, but it does not work at all for TM polarization. On the other hand, with polarization diversity, the circuit shows almost the same filter characteristics for both polarizations. The measured PDL of 1 dB is suitable for practical use.

1.6 Summary

Silicon photonic wire waveguides, featuring very strong optical confinement and compatibility with silicon electronics, provide us with a highly integrated platform for electronic–photonic convergence. The fundamental propagation performance of the waveguide has already become a practical standard. Some passive devices with sophisticated designs and dynamic devices with electronic structures have also been developed by using such waveguides. Moreover, the waveguide offers an efficient media for nonlinear optical functions. The polarization dependence, which is a serious obstacle to telecommunications applications, can be eliminated by using a monolithically integrated polarization diversity system.

In the next development stage, efforts will mainly shift to a large scale photonic integration in which passive, dynamic, and active devices will be integrated on a chip. The ultimate goal is, of course, a large-scale electronic–photonic integration, which is the most important advantage of the silicon photonic wire waveguide.

References

1. G.-L. Bona, R. Germann, B.J. Offrein, SiON high-refractive-index waveguide and planar light-wave circuits. IBM J. Res. Dev. **47**, 239–249 (2003) 2
2. M.J. Kobrinsky, B.A. Block, J.-F. Zheng, B.C. Barnett, R. Mohammed, M. Reshotko, F. Robertson, S. List, I. Young, K. Cadien, On-chip optical interconnects, Intel Technol. J. **8**, 129–141 (2004) 2
3. A.G. Rickman. G.T. Reed, F. Namavar, Silicon-on-insulator optical rib waveguide loss and mode characteristics. J. Lightwave Technol. **12**, 1771–1776 (1994) 2, 17
4. H. Rong, A. Liu, R. Jones, O. Cohen, D. Hak, R. Nicolaescu, A. Fang, M. Paniccia, An all-silicon Raman laser. Nature **433**, 292–294 (2005) 2, 20
5. K.K. Lee, D.R. Lim, H.-C. Luan, A. Agarwal, J. Foresi, L.C. Kimerling, Effect of size and roughness on light transmission in a Si/SiO2 waveguide: experiments and model. Appl. Phys. Lett. **77**, 1617–1619 (2000) 3
6. A. Sakai, G. Hara, T. Baba, Propagation characteristics of ultrahigh-Δ optical waveguide on silicon-on-insulator substrate. Jpn. J. Appl. Phys. **40**, L383–L385 (2001) 3, 14
7. K. Yamada, T. Tsuchizawa, T. Watanabe, J. Takahashi, E. Tamechika, M. Takahashi, S. Uchiyama, H. Fukuda, T. Shoji, S. Itabashi, H. Morita, Microphotonics devices based on silicon wire waveguiding system. IEICE Trans. Electron. **E87-C**, 351–358 (2004) 3, 14
8. T. Tsuchizawa, K. Yamada, H. Fukuda, T. Watanabe, J. Takahashi, M. Takahashi, T. Shoji, E. Tamechika, S. Itabashi, H. Morita, Microphotonics devices based on silicon microfabrication technology. IEEE J. Sel. Top. Quantum. Electron. **11**, 232–240 (2005) 3, 10
9. P. Dumon, W. Bogaerts, V. Wiaux, J. Wouters, S. Beckx, J.V. Campenhout, D. Taillaert, B. Luyssaert, P. Bienstman, D. Van Thourhout, R. Baets, Low-loss SOI photonic wires and ring resonators fabricated with deep UV lithography. Photon. Technol. Lett. **16**, 1328–1330 (2004) 3, 9

10. Y.A. Vlasov, S.J. McNab, Losses in single-mode silicon-on-insulator strip waveguides and bends. Opt. Express **12**, 1622–1631 (2004) 3, 13
11. T. Tsuchizawa, K. Yamada, H. Fukuda, T. Watanabe, S. Uchiyama, S. Itabashi, Low-loss Si wire waveguides and their application to thermooptic switches. Jpn. J. Appl. Phys. **45**, 6658–6662 (2006) 3, 10, 16
12. P. Lusse, P. Stuwe, J. Schule, H.-G Unger, Analysis of vectorial mode fields in optical waveguides by new finite difference method. J. Lightwave Technol. **LT-12**, 487–493 (1994) 4
13. M. Koshiba, S. Maruyama, K. Hirayama, A vector finite element method with the high-order mixed-interpolation-type triangular elements for optical waveguiding problems. J. Lightwave Technol. **LT-12**, 495–502 (1994) 4
14. A.S. Sudbo, Film mode matching: a versatile numerical method for vector mode field calculations in dielectric waveguides. Pure Appl. Opt. **2**, 211–233 (1993) 4
15. K. Okamoto, *Fundamentals of Optical Waveguides* (Academic Press, New York, 2000), p. 29 4
16. W. Bogaerts, P. Dumon, D. Van Thourhout, D. Taillaert, P. Jaenen, J. Wouters, S. Beckx, V. Wiaux, R. Baets, Compact wavelength-selective functions in silicon-on-insulator photonic wires. IEEE J. Sel. Top. Quantum. Electron. **12**, 1394–1401 (2006) 6, 14
17. F.P. Payne, J.P.R. Lacey, A theoretical analysis of scattering loss from planar optical wavewguide. Opt. Quantum. Electron. **26**, 977–986 (1994) 7
18. S. Janz, in *Silicon-Based Waveguide Technology for Wavelength Division Multiplexing*, ed. by L. Pavesi, D. Lockwood. Silicon Photonics. Top. Appl. Phys. 94 (Springer-Verlag, Berlin, 2004), p. 323 7
19. FIMMWAVE Software, Photon Design [Online]. Available: www.photond.com 7
20. T. Shoji, T. Tsuchizawa, T. Watanabe, K. Yamada, H. Morita, Low loss mode size converter from 0.3 μm square Si wire waveguides to single mode fibres. Electron. Lett. **38**, 1669–1670 (2002) 8, 20
21. Y. Hibino, Recent advances in high-density and large-scale AWG multi/demultiplexers with higher index-contrast silica-based PLCs. IEEE J. Sel. Top. Quantum. Electron. **8**, 1090–1101 (2002) 9
22. D. Taillaert, F. van Laere, M. Ayre, W. Bogaerts, D. van Thourhout, P. Bienstman, R. Baets, Grating couplers for coupling between optical fibers and nanophotonic waveguides. Jpn. J. Appl. Phys. **45**, 6071–6077 (2006) 9
23. T. Watanabe, K. Yamada, T. Tsuchizawa, H. Fukuda, H. Shinojima, S. Itabashi, Si wire waveguide devices. Proc. SPIE **6775**, 67750K (2007) 9
24. K.K. Lee, D.R. Lim, L.C. Kimerling, J. Shin, F. Cerrina, Fabrication of ultralow-loss Si/SiO_2 waveguides by roughness reduction. Opt. Lett. **26**, 1888–1890 (2001) 11
25. K. Yamada, T. Tsuchizawa, T. Watanabe, H. Fukuda, H. Shinojima, S. Itabashi, Fabrication and other related technologies for silicon photonic wire waveguides. Rev. Laser Eng. (in Japanese) **35**, 550–555 (2007) 14
26. F. Xia, M. Rooks, L. Sekaric, Y. Vlasov, Ultra-compact high order ring resonator filters using submicron silicon photonic wires for on-chip optical interconnects. Opt. Express **15**, 11934–11941 (2007) 14
27. F. Ohno, K. Sasaki, A. Motegi, T. Baba, Reduction in sidelobe level in ultracompact arrayed waveguide grating demultiplexer based on Si wire waveguide. Jpn. J. Appl. Phys. **45**, 6126–6131 (2006) 14
28. K. Yamada, T. Shoji, T. Tsuchizawa, T. Watanabe, J. Takahashi, S. Itabashi, Silicon-wire-based ultrasmall lattice filters with wide fre spectral ranges. Opt. Lett. **28**, 1663–1664 (2004) 14
29. K. Yamada, T. Tsuchizawa, T. Watanabe, J. Takahashi, H. Fukuda, M. Takahashi, T. Shoji, S. Uchiyama, E. Tamechika, S. Itabashi, H. Morita, Silicon wire waveguiding system: fundamental characteristics and applications. Electron. Commun. Jpn Part 2, **89**, 42–55 (2006) 14, 15
30. G. Cocorullo, I Rendina, Thermo-optical modulation at 1.5 μm in silicon etalon. Electron. Lett. 28, 83–85 (1992) 15
31. R. Kasahara, K. Watanabe, M. Itoh, Y. Inoue, A. Kaneko, Extremely low power consumption thermooptic switch (0.6 mW) with suspended ridge and silicon-silica hybrid waveguide structures. Proc 34th Eur. Conf. Opt. Commun. **5**, 55–56 (2008), Brussels 16, 17

32. R.A. Soref, B.R. Bennett, Electrooptical effects in silicon. IEEE J. Quantum Electron. **QE-23**, 123–129 (1987) 17
33. Q. Xu, B. Schmidt, S. Pradhan, M. Lipson, Micrometre-scale silicon electro-optic modulator, Nature **435**, 325–327 (2005) 17, 18
34. K. Yamada, T. Tsuchizawa, T. Watanabe, H. Fukuda, H. Shinojima, S. Itabashi, Applications of low-loss silicon photonic wire waveguides with carrier injection structures. Proc. 4th Intern. Conf. Group IV Photon., 116–118 (2007), Tokyo 17, 19
35. W.M.J. Green, M.J. Rooks, L. Sekaric, Y.A. Vlasov, Ultra-compact, low RF power, 10 Gb/s silicon Mach-Zehnder modulator. Opt. Express **15**, 17106–17113 (2007) 17, 18
36. H. Fukuda, K. Yamada, T. Shoji, M. Takahashi, T. Tsuchizawa, T. Watanabe, J.Takahashi, S. Itabashi, Four-wave mixing in silicon wire waveguides. Opt. Express **13**, 4629–4637 (2005) 20, 21
37. T.K. Liang, L.R. Nunes, T. Sakamoto, K. Sasagawa, T. Kawanishi, M. Tsuchiya, G.R.A. Priem, D. Van Thourhout, P. Dumon, R. Baets, H.K. Tsang, Ultrafast all-optical switching by crossabsorption modulation in silicon wire waveguides. Opt. Express **13**, 7298–7303 (2005) 20
38. K. Yamada, H. Fukuda, T. Tsuchizawa, T. Watanabe, T. Shoji, S. Itabashi, All-optical efficient wavelength conversion using silicon photonic wire waveguide. IEEE Photon. Technol. Lett. **18**, 1046–1048 (2006) 20, 21
39. M.A. Foster, A.C. Turner, J.E. Sharping, B.S. Schmidt, M. Lipson, A.L. Gaeta, Broad-band optical parametric gain on a silicon photonic chip. Nature **441**, 960–963 (2006) 20, 21
40. Q. Xu, R. Almeida, M. Lipson, Demonstration of high Raman gain in a submicrometer-size silicon-on-insulator waveguide. Opt. Lett. **30**, 35–37 (2005) 20
41. H. Takesue, H. Fukuda, T. Tsuchizawa, T. Watanabe, K. Yamada, Y. Tokura, S. Itabashi, Generation of polarization entangled photon pairs using silicon wire waveguide. Appl. Phys. Lett. **91**, 20118 (2007) 20
42. H. Takesue, H. Fukuda, T. Tsuchizawa, T. Watanabe, K. Yamada, Y. Tokura, S. Itabashi, Generation of polarization entangled photon pairs using silicon wire waveguide. Opt. Express **16**, 5721–5727 (2008) 20
43. K. Harada, H. Takesue, H. Fukuda, T. Tsuchizawa, T. Watanabe, K. Yamada, Y. Tokura, S. Itabashi, Generation of high-purity entangled photon pairs using silicon wire waveguide. Opt. Express **16**, 20368–20373 (2008) 20
44. O. Tadanaga, M. Asobe, H. Miyazawa, Y. Nishida, H. Suzuki, Efficient 1.55 μm-band quasi-phase-matched ZnO-doped LiNbO wavelength converter with high damage resistance. Electron. Lett. **39**, 1525–1575 (2003) 21
45. W. Bogaerts, D. Taillaert, P. Dumon, D. Van Thourhout, R. Baets, E. Pluk, A polarization-diversity wavelength duplexer circuit in silicon-on-insulator photonic wires. Opt. Express **15**, 1567–1578 (2007) 22
46. H. Fukuda, K. Yamada, T. Tsuchizawa, T. Watanabe, H. Shinojima, S. Itabashi, Silicon photonic circuit with polarization diversity. Opt. Express **16**, 4872–4880 (2008) 22, 25
47. H. Fukuda, K. Yamada, T. Tsuchizawa, T. Watanabe, H. Shinojima, S. Itabashi, Ultrasmall polarization splitter based on siliconwire waveguides. Opt. Express **14**, 12401–12408 (2006) 22, 23
48. H. Fukuda, K. Yamada, T. Tsuchizawa, T. Watanabe, H. Shinojima, S. Itabashi, Polarization rotator based on silicon wire waveguides. Opt. Express **16**, 2628–2635 (2008) 23, 24

Index

Chapter 2
Polarization Control in Silicon Photonic Waveguide Components Using Cladding Stress Engineering

Dan-Xia Xu

Abstract This chapter reviews the characteristics of SOI ridge waveguide birefringence, as governed by the waveguide cross-section geometry, the cladding stress level, and cladding thickness. Typical stress levels in dielectric cladding films such as silicon dioxide and silicon nitride are such that the stress-induced birefringence is of comparable magnitude to the waveguide geometrical birefringence. Therefore the total waveguide birefringence can be precisely controlled by counter-balancing these two factors. The application of this technique for achieving polarization independence in a variety of photonic components is described, as well as an example of polarization splitting. Passive and active tuning of the stress-induced birefringence is discussed. The use of birefringence tuning to enhance the efficiency in optical parametric processes and stress-induced Pockels electro-optic effect are also briefly addressed.

2.1 Introduction

Polarization is a property of electromagnetic waves that describes the orientation of the electric and magnetic field oscillations. In free space, these oscillations are in the plane perpendicular to the wave's direction of propagation. A lightwave of arbitrary polarization can be represented as the linear superposition of two orthogonally polarized modes, usually selected as the transverse electric (TE, with the electric field parallel to the wafer plane in a planar waveguide) and transverse magnetic (TM) modes. The fundamental polarization effects are polarization-dependent loss (PDL) and birefringence [1, 2]. Intrinsic PDL in silicon waveguides is mainly due to scattering from the waveguide core-cladding interfaces [3]. In a photonic component, a PDL may also arise from other loss sources, for example, the polarization-dependent bend loss [4]. In this chapter we focus on the effects of waveguide birefringence.

D.-X. Xu (✉)
Institute for Microstructural Sciences, National Research Council, Canada,
Ottawa K1A 0R6, Canada
e-mail: Danxia.Xu@nrc-cnrc.gc.ca

D.J. Lockwood, L. Pavesi (Eds.): *Silicon Photonics II*.
Topics in Applied Physics **119**, 31–70 (2011)
DOI 10.1007/978-3-642-10506-7_2

When a lightwave travels in a dielectric waveguide such as a glass fiber or a planar waveguide, the phase velocity v is decreased to $v = c/n_{\mathrm{eff}}$, where n_{eff} is the phase effective index of the waveguide and c is the speed of light in free space. Waveguide birefringence describes the anisotropy with respect to polarization in the propagation constant (or the effective index) of the optical modes. In other words, light waves with different polarization states travel at different speeds when a waveguide is birefringent. Waveguide birefringence may arise from intrinsic material properties such as in a garnet crystal, or due to geometrical anisotropies such as in polarization maintaining fibers and silicon ridge waveguides, or induced by external (mechanical) distortion such as often observed by bending a glass optical fiber. In structures with strong optical confinement, such as silicon photonic wire waveguides of sub-micron dimensions, the electric and magnetic oscillations are no longer perpendicular to the direction of travel, but have an appreciable component in that direction.

The polarization dependence of waveguides presents one of the most difficult challenges in the design of planar waveguide components. Light sources used in modern communication systems are often semiconductor lasers, which emit linearly polarized light. This light is often regenerated using erbium doped fiber amplifiers, which can decrease the degree of polarization due to the amplified spontaneous emission (ASE) background of the amplifiers [5]. Fibers used in optical links are generally not polarization maintaining, either due to accidental loss of circular symmetry in the fiber poling process or due to mechanical distortions in the fiber cable. The polarization state fluctuates randomly on a millisecond scale after travelling a distance of several kilometers [1, 2]. In spectroscopy, the excitation source may come from natural light sources such as the sun light, which may be partially polarized or unpolarized [6]. Even if the excitation comes from a laser with linear polarization, the state of polarization is often modified after transmission/reflection from the medium of interest. These phenomena are the root causes of impediments in optical systems including polarization-dependent wavelength shift (PDW), polarization-dependent mode dispersion (PMD), and polarization-dependent loss. The preferred approach to handle this issue is to produce devices with polarization insensitive performance. When this approach is not practical, a polarization diversity scheme may be adopted where the signal is divided into orthogonal polarization states using a polarization splitter and processed separately. Due to the increased complexity, the question of whether the polarization diversity method is commercially viable is still a subject of debate.

Silicon photonics has seen rapid progress as evidenced by this book sequel, and many building blocks have been developed for applications in optical telecommunication, interconnect, spectroscopy, and optical sensing. Control and utilization of polarization-dependent properties remains a challenge, but new solutions and functionalities also have been demonstrated [7–17]. The waveguides employed in commercial devices are mainly medium sized ridge waveguides for better fabrication tolerance and fiber-to-waveguide coupling. A considerable fraction of the silicon photonics research is concerned with sub-micron photonic wire waveguides to achieve a smaller device footprint and higher optical intensity, but accompanied by stringent fabrication requirements. Components using sub-micron waveguides demand a different strategy to deal with polarization controls and are beyond the scope of this work. In this chapter, we review the characteristics of polarization dependence, particularly birefringence, in silicon-on-insulator (SOI) waveguides of 1–5 μm ridge height, with regard to the two main sources of polarization anisotropy, namely the geometrical and the cladding stress-induced birefringence [11, 12, 18].

Since the early theoretical analysis [19, 20] and demonstration of using cladding stress-induced birefringence to produce a polarization-independent (PI) arrayed waveguide grating (AWG) [11, 21], the importance of stress-induced birefringence has increasingly been recognized [18, 22–26]. Cladding stress engineering has been applied to different components, including PI ring resonators [13, 14], PI Mach-Zehnder interferometers for differential phase shift keying (DPSK) [16, 23], low loss PI triplexers for fiber-to-the-home (FTTH) applications [15], phase matching in Raman wavelength conversion [24, 25], as well as broadband polarization splitters [26]. In this chapter we review the main characteristics of cladding stress-induced effects, design methodologies, and experimental demonstrations of their applications. Numerical analysis is focused on waveguides with 1.5–2.2 μm ridge heights, while scaling and the relative importance of stress-induced effects with core size are also discussed. The intent is to give a pictorial understanding of the pertinent characteristics in order to assist the readers to use this technique in their own applications.

This chapter is organized as follows. Section 2.2 discusses the general considerations in SOI waveguide design. Sections 2.3 and 2.4 describe the main characteristics of the geometrical and cladding stress-induced birefringence, respectively. Applications of cladding stress engineering to achieve different photonic functionalities are reviewed in Sect. 2.5, and finally some conclusions are drawn in Sect. 2.6.

2.2 SOI Waveguides: General Considerations

In planar waveguide components, two types of cross-section geometries are most frequently encountered, namely slab and ridge waveguides, as illustrated in Fig. 2.1a, b. Respective examples are the combiner sections in an arrayed waveguide

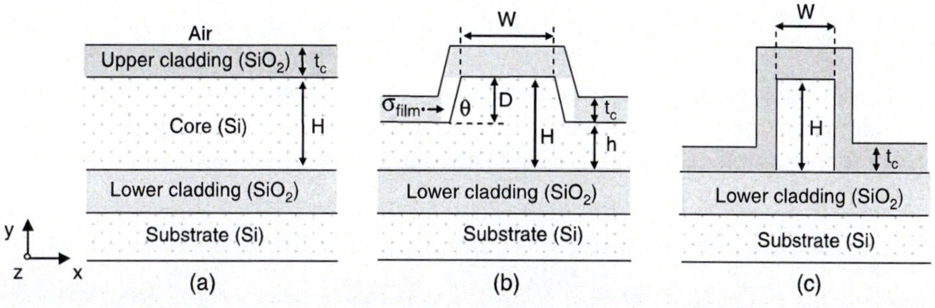

Fig. 2.1 Cross-section of an SOI (**a**) slab, (**b**) ridge, and (**c**) channel waveguide

grating (AWG) demultiplexer for the slab, and the input/output waveguides for the ridge. Channel waveguides (Fig. 2.1c), with the side slab etched down to the buried oxide layer, are also used for special purposes [27], but these are less common.

In the overall component design, it is often required that certain waveguide sections are both single mode and polarization independent. In applications such as dense wavelength division multiplexing (DWDM) and high resolution spectroscopy, the birefringence must be below 10^{-4} for acceptable polarization insensitive performance [28]. An appropriate mode shape to assist the smooth transition between different waveguide sections and a small bend radius to minimize device footprint are also essential. Another important consideration is the fabrication tolerance, which is the ability to achieve and maintain the pre-designed values of the effective index and birefringence using available fabrication technologies. All these properties depend on the waveguide cross-section geometry.

2.2.1 Single-Mode Condition

Due to the high-index contrast in the SOI material platform, waveguides with dimensions larger than a few hundred nanometers in cross-section usually support multiple modes. Nevertheless, large ridge waveguides can be designed to exhibit single-mode-like behavior provided the waveguide width (W) and etch depth (D) fulfil certain conditions relative to the ridge height H (see Fig. 2.1b). Soref et al. first proposed the following single-mode condition for SOI waveguides [29]:

$$\frac{W}{H} \leq 0.3 + \frac{r}{\sqrt{1-r^2}}, r = \frac{h}{H}, \tag{2.1a}$$

$$0.5 \leq r \leq 1, \tag{2.1b}$$

where $h = H - D$ is the slab height in the lateral etched sections. The restriction of Eq. 2.1b is imposed based on the argument that for $r \geq 0.5$, the effective index of the fundamental "vertical mode" in the etched slab regions becomes larger than the effective index of all vertical higher order modes in the ridge. Therefore the higher order modes in the ridge become "leaky," and eventually dissipate after a sufficiently long propagation distance. In the original investigation, a propagation distance of 2 mm was used to verify the single-mode-like behavior in waveguides with a height of $H = 4\,\mu\text{m}$, using the beam propagation method (BPM) [29]. Since then, several groups have continued the study on this subject [30–34], with modified single-mode conditions extended to waveguides of smaller dimensions and deeper etch. For example, Chan et al. established the following condition for waveguides with ridge height of $H = 1$–$1.5\,\mu\text{m}$ for the TM polarization [34]:

$$\frac{W}{H} \leq 0.05 + \frac{(0.94 + 0.25H)r}{\sqrt{1-r^2}}, \tag{2.2a}$$

$$0.3 < r < 0.5 \text{ and } 1.0 \leq H \leq 1.5, \tag{2.2b}$$

In this study, the single-mode boundary was established also by using the BPM algorithm, with a propagation length of 4 mm. Lousteau et al. have observed that near the single-mode condition, certain so called "sticky modes" can exist with rather low radiation loss, and therefore persist after a considerable propagation distance [33].

2.2.2 *Higher Order Mode Filtering*

The single-mode conditions described in Sect. 2.2.1 only apply to straight waveguides. It is well known that higher order modes suffer larger radiation losses at waveguide bends, and this mechanism can be used as an efficient method to filter out unwanted higher order modes. Both the fundamental and higher order mode losses, including the bend loss and scattering loss, are polarization sensitive. For the intrinsic bend loss (when the sidewall roughness is not considered), it has been shown that the TM modes can sustain a much smaller bend radius than the TE mode [4, 14]. Scattering loss due to sidewall roughness, on the other hand, depends on the mode-sidewall overlap, which in turn depends on the specifics of the waveguide geometry and interface condition. It is therefore more difficult to make a general statement, though this loss is certainly polarization dependent [3]. Waveguide bends have been used in various cases to allow a deeper etch depth than predicted by Eq. (2.1) or (2.2), with the final devices exhibiting single-mode behavior and having a reduced device size [11, 14, 35, 36]. In the sections to follow, we will discuss some waveguide designs that may not satisfy the single-mode conditions described previously, but have been validated through experiments to function satisfactorily. On the other hand, the effectiveness of higher order mode filtering using waveguide bends depends not only on the bend radius used but also the propagation distance and device requirements for the suppression of higher order modes. In determining the waveguide cross-section to achieve single-mode like behavior, all these aspects need to be taken into account to ensure acceptable device performance.

2.3 Waveguide Birefringence: Geometrical Effects

For ridge waveguides with a height of 1–5 μm, the minor components in the optical modes are sufficiently small, and the modes can usually be well described as TE and TM polarized modes. Modal phase birefringence can then be defined as the difference in the effective indices of the TM and TE polarizations, $\Delta n_{\mathrm{eff}} = n_{\mathrm{eff}}^{\mathrm{TM}} - n_{\mathrm{eff}}^{\mathrm{TE}}$, and it is a combined consequence of the geometrical and stress-induced anisotropy. The effective index can be expressed as $n_{\mathrm{eff}}^{\mathrm{TE,\,TM}} = n_{\mathrm{geo}}^{\mathrm{TE,\,TM}} + \delta n_{\mathrm{stress}}^{\mathrm{TE,\,TM}}$ for TE and TM polarizations, and the total birefringence is $\Delta n_{\mathrm{eff}} = \Delta n_{\mathrm{geo}} + \Delta n_{\mathrm{stress}}$. Here $n_{\mathrm{geo}}^{\mathrm{TE,\,TM}}$ and Δn_{geo} are the effective indices and birefringence solely determined by the waveguide geometry in the absence of stress, and $\delta n_{\mathrm{stress}}^{\mathrm{TE,\,TM}}$ and $\Delta n_{\mathrm{stress}}$ are the stress-induced effective index and birefringence changes, respectively. The geometrical birefringence originates from the different boundary conditions for the perpendicular and transverse electric field components of the optical modes at the waveguide core and cladding interfaces. Its magnitude depends on the strength of the field components at these interfaces, which is controlled by the refractive index contrast and the core size. In low index contrast platforms such as glass waveguides, the geometrical effect is negligible and the birefringence is primarily due to the residual stress in the waveguide layers. A large body of research has been devoted to this subject, where the main goal is to reduce the stress anisotropy in the waveguide core region, for example, by matching the thermal expansion coefficients of the cladding and core layers or creating stress relief features [37–42].

In high-index contrast waveguides such as SOI where the light is strongly confined, the cross-section geometry has the largest influence on the waveguide effective index. The cross-section of ridge waveguides is typically selected to fulfil the single-mode condition first, as discussed in Sect. 2.2. The resulting lack of axial symmetry generally causes a strong geometrical birefringence. Being a centro-symmetric crystal, silicon is optically isotropic in its natural state. When subjected to mechanical distortion, however, there is a stress-induced anisotropy in its polarizability and the associated material birefringence. In this section, we analyze the influence of the geometrical factors on the waveguide birefringence, while the stress-induced effects will be discussed in Sect. 2.4.

2.3.1 Waveguide Cross-Section

Using a full-vectorial finite element differential equation solver [43], mode solutions are calculated with the following parameters for ridge structures shown in Fig. 2.1b: a refractive index of silicon and silicon dioxide (SiO_2), in the absence of stress, of $n_{Si} = 3.476$ and $n_{oxide} = 1.444$, and a free-space wavelength of $\lambda = 1,550$ nm. The upper cladding is also assumed to be oxide in order to include the contribution of the oxide refractive index, but the presence of stress is excluded at this point.

Figure 2.2a shows the dependence of the geometrical birefringence Δn_{geo} on the etch depth D for waveguides with different widths, with vertical sidewalls ($\theta = 90°$) and a ridge height of $H = 2.2\,\mu$m. For a given waveguide width W, the polarization degeneracy ($\Delta n_{geo} = 0$) may only be achieved at a specific value of D. At the birefringence-free point for $W = 1.6\,\mu$m, Δn_{geo} changes by $\sim 1.5 \times 10^{-4}$ for an etch depth change of $\Delta D = \pm 10$ nm. When $W = 2.5\,\mu$m, Δn_{geo} is much less sensitive to the fluctuations in D ($\Delta n / \Delta D \sim 2 \times 10^{-5}$ for $\Delta D = \pm 10$ nm); however, the birefringence-free condition cannot be achieved. To reduce the birefringence below the level required for the state-of-the-art photonic devices ($\Delta n_{eff} < 10^{-4}$) by solely controlling the ridge aspect ratio W/D, dimensional control of less than 10 nm is required for these waveguides [11, 13, 35], which is difficult to attain in fabrication.

There has been extensive work on waveguides with vertical sidewalls that shows both polarization degeneracy and single-mode conditions can be simultaneously

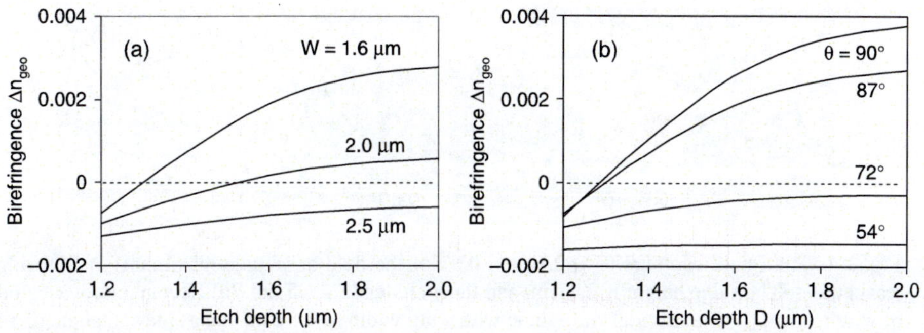

Fig. 2.2 Calculated modal geometrical birefringence Δn_{geo} for waveguides with $H = 2.2\,\mu\text{m}$. (**a**) Δn_{geo} in waveguides with vertical sidewalls ($\theta = 90°$) as a function of etch depth D. Waveguide width $W = 1.6$, 2, and $2.5\,\mu\text{m}$. (**b**) Δn_{geo} as a function of the etch depth for waveguides with sidewall angles of 54, 72, 87, and 90°. The top ridge width $W = 1.5\,\mu\text{m}$, and the bottom ridge width varies according to the sidewall angle θ

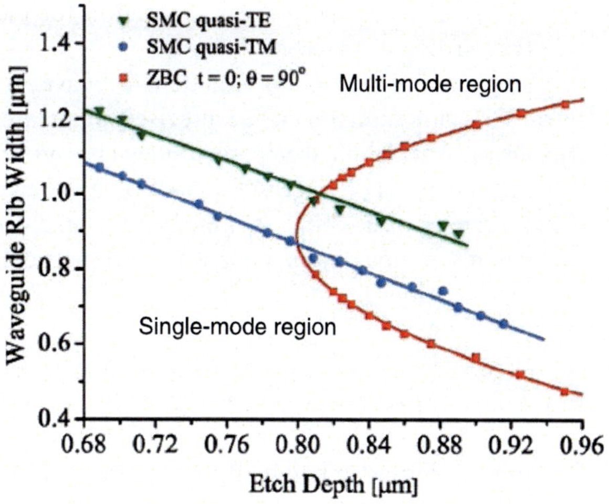

Fig. 2.3 Single-mode condition (SMC) and zero birefringence condition (ZBC) as a function of waveguide rib width and etch depth, for waveguides with vertical sidewalls, $H = 1.35\,\mu\text{m}$ and with no upper cladding (after Milosevic et al. [22])

fulfilled at specific waveguide aspect ratios [7, 8, 34]. Figure 2.3 gives such an example, which depicts the zero birefringence condition (ZBC) together with the single-mode boundaries for the quasi TE and TM modes [22]. The drawback of this exclusively geometrical approach is that the conditions are rather restrictive for both polarizations to be single mode as well as being degenerate. These designs also require narrow waveguide widths and deep etch (to $W < 0.9\,\mu\text{m}$ and $D > 0.8\,\mu\text{m}$ in this case), resulting in high sensitivity to dimensional fluctuations.

The effect of sidewall angle on Δn_{geo} is shown in Fig. 2.2b. For waveguides with near vertical sidewalls, an etch depth can be found to satisfy the birefringence-free condition, but a small change in θ or D causes a large variation in Δn_{geo}. The birefringence of a trapezoidal ridge waveguide is less susceptible to changes in the ridge geometry, including width and etch depth. However, the geometrical birefringence remains negative for the entire waveguide etch depth range when $\theta < 72°$. Figure 2.4 shows the fundamental (TE) mode equal-field contours in examples of

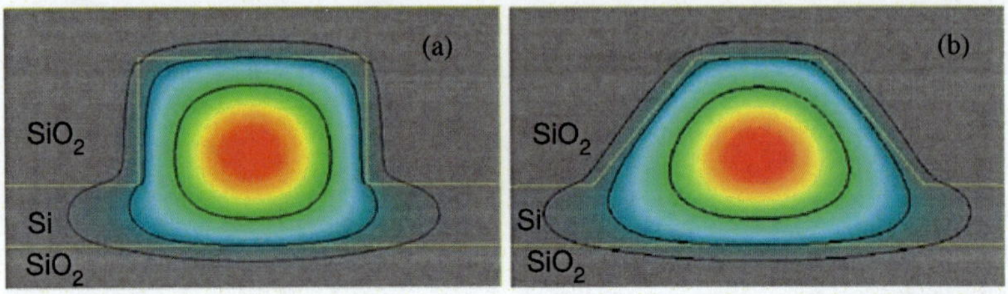

Fig. 2.4 Equal-field contours (0.5, 0.1, and 0.01 of the field amplitude maximum) of SOI ridge waveguides. The ridge height is 2.2 μm, and the etch depth is 1.5 μm. (**a**) Rectangular waveguide of width 2 μm. (**b**) Trapezoidal waveguide with a top width of 1.5 μm, and a sidewall angle of 54°

rectangular and trapezoidal waveguides. With the dimensions as indicated in the figure, these two waveguides have similar mode size and can support similar bend radii. On the other hand, the modal field for the trapezoidal waveguide has a smaller overlap with the Si/oxide boundaries as compared to that in the rectangular waveguide. This basic feature is the main reason for the different behaviors of the two types of waveguides presented above.

Using reactive ion etching to pattern the waveguides, near vertical sidewall angles ($\theta > 75°$) are common, while precisely maintaining $\theta = 90°$ is difficult. Using anisotropic wet chemical etching, a sidewall angle of 54° is expected for waveguides parallel to the < 110 > crystal planes on a wafer with (100) orientation. Indeed, a sidewall angle of $\theta < 85°$ is preferred for less stringent fabrication requirements. In-depth discussion on the influence of geometrical parameters can be found in previous publications [12, 44].

2.3.2 *Scaling of the Geometrical Birefringence with Core Size*

With decreasing ridge height, the dependence of Δn_{geo} on the structure parameters increases drastically. Figure 2.5a shows Δn_{geo} as a function of the waveguide aspect ratio $r = h/H$ for waveguides with ridge height from 4 to 1 μm and $W = 0.8\,H$. For a large core size such as the example of $H = 4\,\mu\text{m}$, the birefringence is generally low. Within a considerable range of r values, Δn_{geo} is at acceptable levels ($< 10^{-4}$). For small core sizes, however, only a narrow range of r values yields a low birefringence. Figure 2.5b shows the change in Δn_{geo} for a fluctuation in the etch depth of $D = \pm 2.5\% H$, with data extracted from Fig. 2.5a near the birefringence-free points. For $H = 1\,\mu\text{m}$, Δn_{geo} changes with a slope of $\sim 2 \times 10^{-3}$ for $\Delta D = \pm 10$ nm, which is prohibitively high for many applications.

Using contact-print lithography, waveguide width variations on the order of 100 nm can be expected. Using state-of-the-art high resolution lithography methods such as e-beam direct write or deep-UV steppers, dimensional control on the order of 10 nm is possible but is still very challenging. Because it is difficult to precisely control the waveguide dimensions due to limitations of the fabrication techniques, methods of controlling birefringence other than modifying waveguide geometries are desirable.

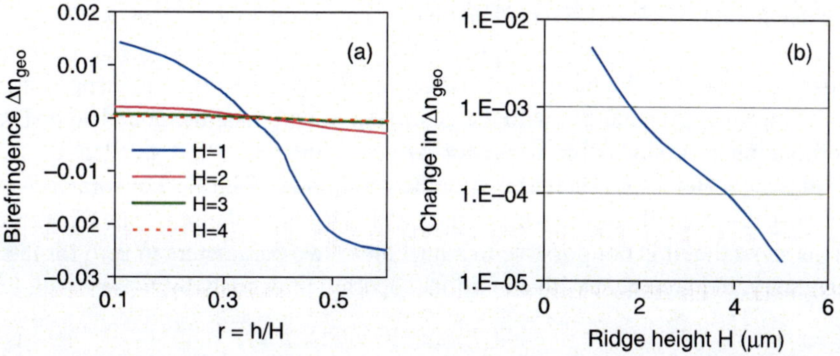

Fig. 2.5 (**a**) Geometrical birefringence Δn_{geo} as a function of the waveguide aspect ratio $r = h/H$, for ridge height of $H = 1, 2, 3$, and 4 μm, respectively. Waveguide width $W = 0.8$ H. (**b**) Change in Δn_{geo} for a change in the etch depth of $D = \pm 2.5\% H$, as a function of H, near the birefringence-free points shown in (**a**)

2.3.3 Dispersion and Group Index Birefringence

The properties of most filters, resonators, and spectrometers are dependent on the waveguide effective index n_{eff} and the group index $n_{\mathrm{G}} = n_{\mathrm{eff}} - \lambda \partial n_{\mathrm{eff}}/\partial \lambda$. The effective index generally is a function of wavelength, which in part originates from the material dispersion and also is influenced by the waveguide cross-section. Due to the strong modal confinement by the high-index contrast SOI waveguide boundaries, the effective index dispersion is a strong function of the waveguide geometry. Since the confinements of the TE and TM modes are different, the phase birefringence also has dispersion. The group index n_{G} is thus both wavelength and polarization dependent.

Figure 2.6 shows the geometrical (phase) effective index and the group index birefringence as a function of wavelength λ for waveguides of different widths and

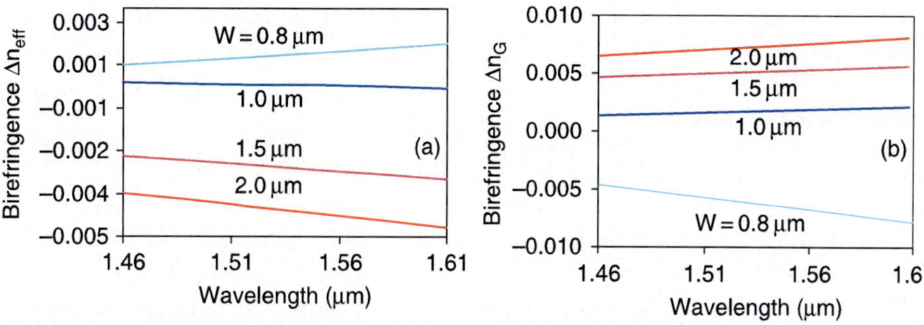

Fig. 2.6 Wavelength dependence of the waveguide geometrical birefringence (in the absence of cladding stress) for different waveguide widths. The ridge has a height of $H = 1.5$ μm and the etch depth is $D = 0.9$ μm. (**a**) Effective index birefringence $\Delta n_{\mathrm{eff}} = \Delta n_{\mathrm{geo}}$; (**b**) Group index birefringence Δn_{G}

a ridge height of $H = 1.5\,\mu\text{m}$. The phase index birefringence Δn_{geo} changes from negative values for wider waveguides (i.e., $n_{\text{geo}}^{\text{TM}} < n_{\text{geo}}^{\text{TE}}$, a slab like behavior) to positive values for the narrower waveguides ($W = 0.8\,\mu\text{m}$). The dispersion of the birefringence also changes with the geometry, in both magnitude and sign. The corresponding group index birefringence Δn_{G} as a function of the wavelength is shown in Fig. 2.6b. Here Δn_{G} increases with the waveguide width W. For some waveguide geometries (e.g., $W = 1.5\,\mu\text{m}$), Δn_{geo} and Δn_{G} have the opposite signs. Simulations show that it is not possible to adjust these two parameters to zero for the same geometry, in agreement with conditions reported previously by Vivien et al. [7].

2.4 Cladding Stress-Induced Birefringence: Theory and Modeling

The subject of photoelasticity deals with the birefringence developed in a solid under the application of mechanical stress. Discovered in 1815 by Sir David Brewster in jellies from the observation of artificial double refraction, this effect was subsequently analyzed in crystals by Pockels, who also discovered the electro-optic effect, now bearing his name [45]. The Pockels effect describes the generation of birefringence on the application of an electric field, and this is discussed further in Sect. 2.5.6. Today, photoelasticity is widely used in many areas of engineering, notably in identifying structural deformations.

Stress in planar waveguides can come from several sources. The most common is the mismatch in the thermal expansion coefficients between the waveguide core, cladding and substrate materials, or other thin films deposited on top of the waveguides such as metal electrical contacts. Cladding layers such as silicon dioxide are usually deposited or grown at elevated temperatures on the silicon substrate to provide passivation for the silicon surface and optical isolation from the external environment. A (compressive) stress is generally present in the SiO_2 films on Si substrates, since the latter has a much larger thermal expansion coefficient than that of the oxide ($\alpha_{\text{Si}} = 3.6 \times 10^{-6}\,\text{K}^{-1}$ and $\alpha_{\text{oxide}} = 5.4 \times 10^{-7}\,\text{K}^{-1}$). Intrinsic stress can also be significant in some thin films, particularly in metals. In this section we consider the effects of stress on a silicon waveguide, using a SiO_2 cladding as the primary example. Active tuning of the stress using a piezoelectric film cladding will be discussed in Sect. 2.5.5.

2.4.1 Photoelastic Effect

When a material is subjected to mechanical distortions, the resulting strain/stress modifies the crystal electronic structure on a microscopic scale by changing the bond lengths between the atoms. The crystal symmetry and therefore the polarizability of the material are altered. The change in the refractive index ellipsoid n_{ij} can be related to the strain ε_{ij} in terms of the photoelastic tensor p_{ij} as [19, 46]:

$$\Delta\left(\frac{1}{n^2}\right)_i = p_{ij}\varepsilon_j, \tag{2.3}$$

where the contracted tensor notation is used (i.e., 1 → xx, 2 → yy, 3 → zz, 4 → zy, 5 → zx, 6 → xy). In cubic crystal (hence optically isotropic) materials such as silicon, Eq. (2.3) can be simplified and expressed as the strain-induced change in the refractive index for light polarized along the i-axis ($i = 1, 2, 3$):

$$n_i - n_0 = -\frac{1}{2}n_0^3 p_{ij}\varepsilon_j. \tag{2.4}$$

Since the waveguide is usually very long in the propagation direction z (see Fig. 2.1), the shear strain in this direction can be ignored and the refractive index is considered unaltered. Using the stress-strain relation in an isotropic medium, the expression between the refractive index and stress σ_i is found to be [19, 46]

$$n_x - n_0 = -C_1\sigma_x - C_2(\sigma_y + \sigma_z), \tag{2.5a}$$

$$n_y - n_0 = -C_1\sigma_y - C_2(\sigma_z + \sigma_x). \tag{2.5b}$$

Here σ_i ($i = x, y, z$) are the principal stress tensor components along the x-, y-, and z-directions, respectively (see Fig. 2.1), n_i is the material refractive index in the corresponding direction, and n_0 is the (isotropic) refractive index in the absence of stress. Since stress levels in uniform thin films can be readily measured experimentally, the following discussions use the stress-related expressions. The stress-optic constants C_1 and C_2 are related to the Young's modulus (E), Poisson's ratio (υ), and the photoelastic tensor elements (p_{11} and p_{12}) as

$$C_1 = \frac{n_0^3}{2E}(p_{11} - 2\upsilon p_{12}), \tag{2.6a}$$

$$C_2 = \frac{n_0^3}{2E}(-\upsilon p_{11} + (1-\upsilon)p_{12}). \tag{2.6b}$$

When a stress (tensor) with axial anisotropy is imposed on an originally isotropic material, a material birefringence is induced:

$$n_y - n_x = (C_2 - C_1)(\sigma_y - \sigma_x). \tag{2.7}$$

The stress-optic constants and mechanical properties of silicon, glass and several other common optical materials are listed in Table 2.1. Stress-optic constants C_1 and C_2 for silicon and fused silica at the wavelength of 1.55 μm are calculated using Eq. (2.6) from the available values of p_{11} and p_{12} measured at $\lambda = 1.15$ μm as listed in the book by Xu and Stroud [46]. The refractive index values are from the handbook by Palik [47]. Stress optic constants for other materials are calculated

Table 2.1 Material constants of some materials: Young's modulus E, Poisson's ratio ν, thermal expansion coefficient α, wavelength λ, refractive index n_0, photoelastic tensor elements p_{11} and p_{12}, and stress optic constant C_1 and C_2 [46, 47]

Material	Si	Fused silica	Ge	GaAs	$LiNbO_3$
E (GPa)	130	76.7	103	85.5	166
ν	0.27	0.186	0.26	0.31	0.31
α (K^{-1})	3.6×10^{-6}	5.4×10^{-7}	5.8×10^{-6}	6.9×10^{-6}	6×10^{-6}
λ (μm)	1.55	1.55	2.0–2.2	1.15	0.633
n_0	3.476	1.444	4.1	3.44	2.28
p_{11}	−0.101	0.16	−0.063	−0.165	−0.026
p_{12}	0.0094	0.27	−0.0535	−0.140	0.090
C_1 (10^{-12} Pa^{-1})	−17.13	1.17	−11.77	−18.62	−2.92
C_2 (10^{-12} Pa^{-1})	5.51	3.73	−7.77	−10.82	2.50

for their respective wavelengths as listed. It has been shown that the silicon stress-optic constants have a weak dependence on the crystal orientation, but that is not considered here [48–50]. It is of interest to note that silicon is similar to $LiNbO_3$ in that the stress-optic constants C_1 and C_2 have opposite signs. Equation (2.7) predicts that these materials have a large birefringence resulting from stress anisotropy.

2.4.2 Ordinary and Normalized Plane Strain Models

In this section we introduce the models used for calculating the stress distributions in the SOI waveguide system consisting of a silicon substrate, a buried SiO_2 lower cladding layer, and a SiO_2 upper cladding layer (Fig. 2.1b). For SOI wafers formed using the Unibond process, the buried oxide is thermally grown with a typical stress in the range of −300 MPa (compressive). The upper cladding oxide films are generally deposited using plasma enhanced chemical vapour depositions (PECVD) in the temperature range of 150–400 °C, with a typical stress level in the range of −100 to −400 MPa. Apart from the deposition temperature, the residual stress in a blanket oxide film on silicon also depends on other process parameters such as gas flow rate, plasma power density, and deposition pressure. Even for films deposited at the same temperature, changes in other process parameters can cause large differences in the film density and stress [51]. For the purpose of our simulation, the upper and lower oxide claddings are assumed to have the same material constants, resulting in the same stress levels. Even though this assumption may not always reflect the real situation, simulations reveal that the lower cladding makes a negligible contribution to the stress in the silicon core, and the resulting inaccuracy in the optical properties is minimal.

Stress in a waveguide is usually non-uniformly distributed and has different values in different directions (anisotropic), determined by both the stress in the cladding film and the waveguide geometry. In our calculations, we assume that all material layers are at equilibrium, and therefore stress-free, at a reference temperature T_{ref}, which corresponds to the deposition temperature for an idealized film on a silicon substrate. Deformations occur and stresses build up upon cooling to the operating temperature T_0 as different layers contract at different rates. In other words, thermal stress is used to account for the total stress in the films for simulation, even though in real devices some intrinsic stress may be present in the cladding film.

The strain and stress components are related as follows, given the difference between the operating and reference temperatures ($\Delta T = T_0 - T_{\text{ref}}$):

$$\begin{bmatrix} \sigma_x \\ \sigma_y \\ \sigma_z \end{bmatrix} = \frac{E}{(1+\nu)(1-2\nu)} \begin{bmatrix} 1-\nu & \nu & \nu \\ \nu & 1-\nu & \nu \\ \nu & \nu & 1-\nu \end{bmatrix} \begin{bmatrix} \varepsilon_x \\ \varepsilon_y \\ \varepsilon_z \end{bmatrix} - \frac{\alpha E \Delta T}{1-2\nu}, \tag{2.8}$$

where σ_i and ε_i are the principal stress and strain components along the x-, y- and z-directions, respectively. A commonly used approximation is the "plane strain" model, which assumes that the strain component along the propagation direction z is negligible compared to the strain in the x- and y-directions. Under this plane-strain assumption where $\varepsilon_z = 0$, the three-dimensional (3D) stress analysis in (8) is reduced to a 2D problem with σ_x and σ_y decoupled from σ_z:

$$\sigma_x = \frac{E}{(1+\nu)(1-2\nu)} \left[(1-\nu)\varepsilon_x + \nu\varepsilon_y\right] - \frac{\alpha E \Delta T}{1-2\nu}, \tag{2.9a}$$

$$\sigma_y = \frac{E}{(1+\nu)(1-2\nu)} \left[\nu\varepsilon_x + (1-\nu)\varepsilon_y\right] - \frac{\alpha E \Delta T}{1-2\nu}, \tag{2.9b}$$

$$\sigma_z = \nu(\sigma_x + \sigma_y) - \alpha E \Delta T. \tag{2.9c}$$

Although the plane strain model has been widely used to calculate stress in waveguides [52–54], it introduces a strong artificial tensile stress σ_z in the silicon waveguide core and the substrate, but lower compressive stress in the cladding layers along the z axis. These changes in stress cause a uniform shift in the calculated material refractive indices n_x and n_y, which in turn changes the effective indices of the waveguide modes by approximately 10%, and therefore the stress-induced birefringence by a similar amount. This can be important in modelling interference based polarization controlled devices employing stress [12].

We have developed an alternative "normalized plane strain" model to circumvent this limitation [12]. Rather than assuming no distortion along the z-axis, we assume that the top layers conform to the substrate during cooling. The strain component in the z-direction then takes the form of $\varepsilon_z = \alpha_s \Delta T$, where α_s is the thermal expansion coefficient of the silicon substrate. Applying this relation to Eq. (2.8), the stress-strain relation can be expressed in a form similar to the 2D equations in Eq. (2.9):

$$\sigma_x' = \frac{E}{(1+\nu)(1-2\nu)}\left[(1-\nu)\varepsilon_x + \nu\varepsilon_y\right] - \left(\alpha - \frac{\nu}{1+\nu}\alpha_s\right)\frac{\alpha E \Delta T}{1-2\nu}, \quad (2.10a)$$

$$\sigma_y' = \frac{E}{(1+\nu)(1-2\nu)}\left[\nu\varepsilon_x + (1-\nu)\varepsilon_y\right] - \left(\alpha - \frac{\nu}{1+\nu}\alpha_s\right)\frac{\alpha E \Delta T}{1-2\nu}, \quad (2.10b)$$

$$\sigma_z' = \nu(\sigma_x' + \sigma_y') - \left(\alpha - \frac{\nu}{1+\nu}\alpha_s\right) E\Delta T + \frac{1}{1+\nu}\alpha_s E \Delta T, \quad (2.10c)$$

if "normalized" thermal expansion coefficients are adopted:

$$\alpha' = \alpha - \frac{\nu}{1+\nu}\alpha_s. \quad (2.11)$$

The final expression for $\sigma_z{}'$ is obtained by adding a constant correction term:

$$\sigma_z' = \sigma_z + \frac{1}{1+\nu}\alpha_s E \Delta T, \quad (2.12)$$

where the stress field σ_z is calculated as in Eq. (2.9), with the modification for the thermal expansion coefficients as in Eq. (2.11). This means that the ordinary plane-strain algorithm, commonly available in commercial multi-physics software such as COMSOL Physics, can be used as before to evaluate the x- and y-components of the stress distribution in the waveguides, using the normalized thermal expansion coefficients. Using this 2D "normalized plane-strain model", accuracies comparable to full 3D simulations can be achieved, but with much reduced computation time and memory requirement. Further details on this simulation method can be found in [12].

We define a parameter σ_{film} as the in-plane stress component (in the x direction, σ_x) that is present in the uniform upper cladding film, far away from the ridge. Allowing unconstrained thermal expansion in all directions, it can be expressed as

$$\sigma_{\text{film}} = \sigma_x = (\alpha_s - \alpha_{\text{oxide}})\frac{E}{1-\nu}\Delta T, \quad (2.13)$$

where α_s and α_{oxide} are the thermal expansion coefficients of the Si substrate and SiO_2 cladding film, respectively. The value of σ_{film} is also readily available from the simulation output. Experimentally, the value of the stress in a cladding film can be measured from the change in bowing radius of a blanket wafer before and after deposition of the stressed film [55, 56]. We use σ_{film} to characterize the stress in the cladding film to emulate the experimental conditions. Using the mechanical properties listed in Table 2.1, the temperature difference ΔT between the thermal equilibrium (deposition) and operating temperatures is used as a fitting parameter to match the experimentally measured σ_{film}. For example, $\Delta T = 400\,°\text{C}$ yields

$\sigma_{\text{film}} = -105$ MPa, whereas $\Delta T = 1,120\,°\text{C}$ yields $\sigma_{\text{film}} = -300$ MPa. The small stress gradient in the film thickness direction (y-direction) is neglected. As we have discussed at the beginning of this section, generally $\Delta T \leq 400\,°\text{C}$ for PECVD depositions and the measured film stress is in the range of -100 to -400 MPa. The discrepancy between experimental stress levels and the calculated values for PECVD deposition temperatures indicate that thermal stress alone does not account for the stress levels commonly observed in the oxide films, and intrinsic stress likely is present.

2.4.3 Cladding Stress-Induced Birefringence in Waveguides

The stress distribution in a waveguide depends on the core and cladding material constants, and the waveguide cross-section geometry, including the cladding film thickness and its coverage on the ridge sidewalls. Using the relations described in Eqs. (2.10–2.12), the stress components σ_x and σ_y, and the corresponding modified material refractive indices n_x and n_y can be computed. A full-vectorial mode solver is then used to calculate the modified effective index.

The stress distributions and the material birefringence $(n_y - n_x)$ in SOI ridge waveguides are discussed below for an SiO_2 lower cladding of thickness 0.37 μm, an SiO_2 upper cladding of thickness 1 μm and a step coverage of 70%, under a compressive stress of $\sigma_{\text{film}} = -300$ MPa. The ridges have a height of $H = 1.5$ μm and an etch depth of $D = 0.9$ μm. Two examples are shown, with Fig. 2.7 for a vertical sidewall waveguide with $W = 1.5$ μm and $\theta = 90°$, and Fig. 2.8 for a trapezoidal waveguide with $W = 1$ μm and $\theta = 54°$. These two waveguides have similar widths at the optical mode center and provide similar light confinement. In both cases, the primarily in-plane stress in the oxide film compresses the Si ridge, resulting in a compressive stress in the x-direction (Figs. 2.7a and 2.8a) and a higher tensile stress in the y-direction (Figs. 2.7b and 2.8b) in the silicon core. At the mode centre, $\sigma_x \sim -53$ MPa and $\sigma_y \sim 130$ MPa in Fig. 2.7, while $\sigma_x \sim -60$ MPa and $\sigma_y \sim 97$ MPa in Fig. 2.8. This stress anisotropy is the fundamental cause of the stress-induced birefringence. The corresponding local material birefringence $(n_y - n_x)$ is shown in Figs. 2.7c and 2.8c, with a value of $\sim 5 \times 10^{-3}$ at the mode centre for the vertical sidewall waveguide.

These modifications in the material index cause the modal effective index $n_{\text{eff}}^{\text{TM}}$ to increase and $n_{\text{eff}}^{\text{TE}}$ to decrease, for a cladding film with a compressive stress. In other words, the stress-induced changes in the effective index $\delta n_{\text{stress}}^{\text{TM}} < 0$ and $\delta n_{\text{stress}}^{\text{TE}} < 0$ for $\sigma_{\text{film}} < 0$. Figure 2.9 illustrates the dependence of the effective indices on the cladding thickness for the vertical waveguide with a material birefringence distribution shown in Fig. 2.7. These effective index changes in the opposite direction for TE and TM are due to the opposite signs of the stress-optic constants C_1 and C_2 for silicon (see Table 2.1). This fact makes stress engineering a particularly efficient method for changing the silicon waveguide birefringence, as compared to that for other materials where C_1 and C_2 are of the same sign.

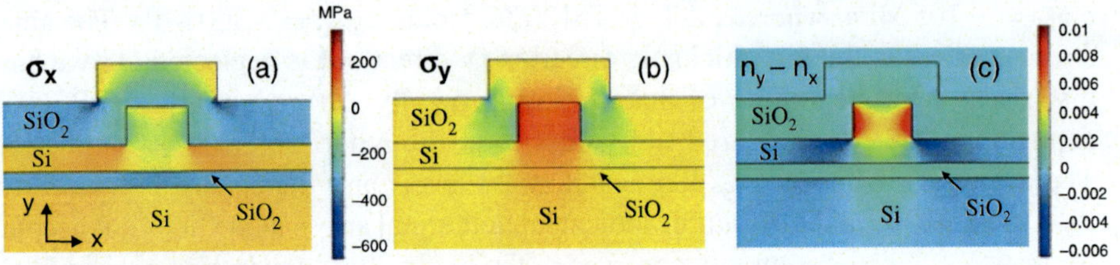

Fig. 2.7 Stress distributions in the (**a**) x-direction and (**b**) y-direction; and (**c**) stress-induced material birefringence $(n_y - n_x)$, for a waveguide with $H = 1.5\,\mu\text{m}$, $W = 1.5\,\mu\text{m}$, $D = 0.9\,\mu\text{m}$, and $\theta = 90°$. The cladding oxide is 1 μm thick with a compressive stress of $\sigma_{\text{film}} = -300\,\text{MPa}$

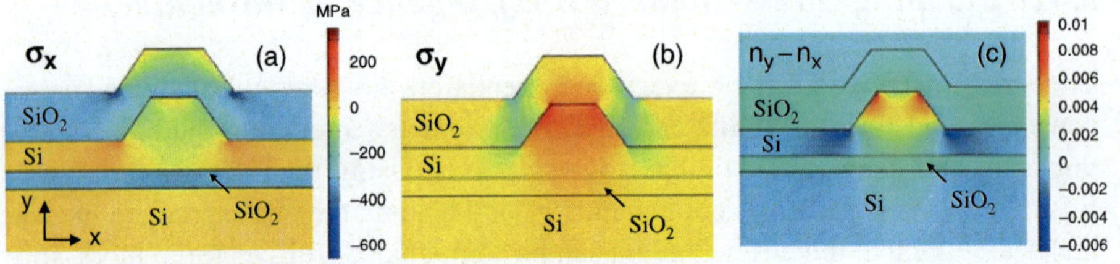

Fig. 2.8 Stress distributions in the (**a**) x-direction and (**b**) y-direction; (**c**) stress-induced material birefringence $(n_y - n_x)$, for a waveguide with $H = 1.5\,\mu\text{m}$, $W = 1\,\mu\text{m}$, $D = 0.9\,\mu\text{m}$, and $\theta = 54°$. The cladding parameters are the same as in Fig. 2.7

Fig. 2.9 (**a**) Effective indices for the TE and TM polarizations as a function of the oxide upper cladding thickness t_c, for different cladding stress levels of $\sigma_{\text{film}} = 0, -100$, and $-200\,\text{MPa}$, respectively. The waveguide dimensions are $H = 1.5\,\mu\text{m}$, $W = 1.5\,\mu\text{m}$, $D = 0.9$, μm and $\theta = 90°$

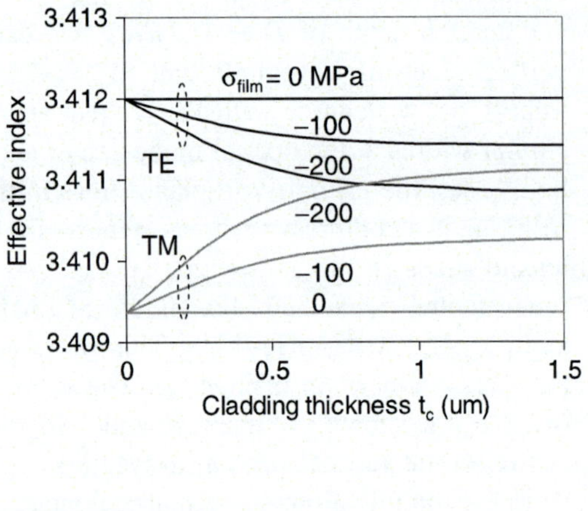

The upper cladding thickness t_c is another important parameter governing the stress-induced changes $\delta n_{\text{stress}}^{\text{TE}}$, $\delta n_{\text{stress}}^{\text{TM}}$, and thus $\Delta n_{\text{stress}} \left(= \delta n_{\text{stress}}^{\text{TM}} - \delta n_{\text{stress}}^{\text{TM}}\right)$, since the force exerted on the silicon core in the x-direction scales with both σ_{film} and film thickness. With increasing upper cladding thickness, $\left|\delta n_{\text{stress}}^{\text{TM}}\right|$ and $\left|\delta n_{\text{stress}}^{\text{TM}}\right|$ increase at first and eventually reach a plateau for large t_c, when the silicon ridge is entirely buried under the oxide and the force saturates. It is evident that the impact of stress increases with the etch depth D and the sidewall angle θ, but decreases with the waveguide width W. As can be observed in Figs. 2.7 and 2.8, $\sigma_x < 0$ and $\sigma_y > 0$ in the regions where the optical mode has significant intensity. On examining

Eq. (2.5), one can conclude that $|\delta n^{TM}| > |\delta n^{TM}|$ for most waveguide geometries, as also shown in Fig. 2.9.

The corresponding total modal birefringence Δn_{eff}, including the geometrical and stress-induced contributions, is illustrated as a function of the upper oxide thickness for the waveguides with vertical and slanted sidewalls in Figs. 2.10a and b, respectively. A cladding with a compressive stress induces a positive shift in Δn_{eff} that increases with the amplitude of σ_{film}, whereas a cladding material with a tensile stress (for example, some silicon nitride films) would induce a negative shift in Δn_{eff}. For the waveguide dimensions presented here, the optical field is sufficiently confined within a 0.3 μm thickness of the oxide cladding owing to the high-index contrast. A thickness range of 0.3–1.5 μm is available for birefringence tuning, giving a tuning range of $\Delta n_{stress} \sim 5 \times 10^{-3}$ for $\sigma_{film} = -300$ MPa. For thicknesses larger than 1.5 μm, little additional stress-induced changes can be obtained for these waveguides. As expected from the material birefringence distribution shown earlier, the stress-induced birefringence is slightly smaller in the trapezoidal waveguide.

The total birefringence Δn_{eff} can therefore be controlled with two parameters: the oxide cladding thickness t_c and the film stress level σ_{film}. The stress-induced refractive index variations resulting from commonly used oxide cladding films are of comparable magnitude to the geometrical birefringence found in typical SOI ridge waveguides. Depending on the specific value of the geometrical birefringence Δn_{geo}, Δn_{stress} can be used to adjust the total modal birefringence to be zero or some other desired values to fulfil the particular required functions. For example, as will be discussed in Sect. 2.5.2, the condition of $\Delta n_{eff} = 0$ is not always mandatory for achieving polarization-independent performance of a device. For components such as ring resonators and Mach-Zehnder interferometers (MZI) which have a cyclic spectral response, shifting the TE and TM spectra by one or even several free spectral ranges (FSRs) can also give apparent polarization independence.

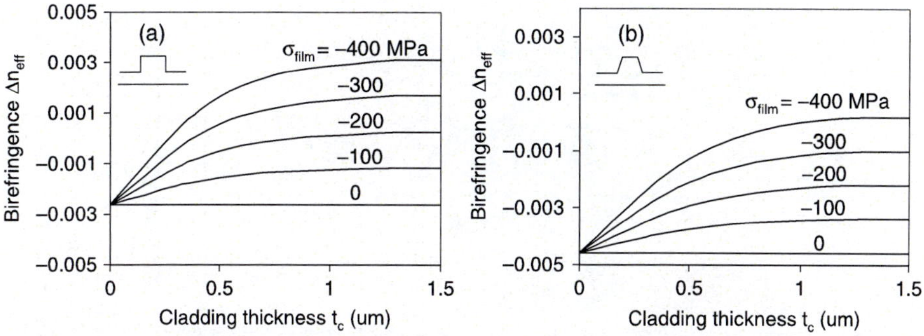

Fig. 2.10 Total modal birefringence Δn_{eff} as a function of the oxide upper cladding thickness t_c, for different cladding stress levels σ_{film} of 0, −100, −200, −300, and −400 MPa, respectively. (**a**) The vertical sidewall waveguide as shown in Fig. 2.7; (**b**) the trapezoidal waveguide as shown in Fig. 2.8

Stress-induced modifications in the material refractive indices n_x and n_y are linearly proportional to the stress tensor components (Eq. 2.5). From consulting Maxwell's equations and Eq. (2.5), it can be shown that the stress-induced changes in the mode parameters $n_{\mathrm{eff}}^{\mathrm{TE}}$, $n_{\mathrm{eff}}^{\mathrm{TM}}$ and Δn_{eff} are also linear functions of the stress components. For a given waveguide core geometry, we can first obtain the values of $\delta n_0^{\mathrm{TE,\,TM}}$ for a particular stress level σ_0, over a specific range of the cladding thickness t_{c}. Once the curves at σ_0 are determined, the values for other cladding stress levels σ_{film} can be readily calculated. It is verified through simulations that the following relations apply:

$$\delta n_{\mathrm{stress}}^{TM} = \frac{\sigma_{\mathrm{film}}}{\sigma_0}\delta n_0^{TM}, \tag{2.14a}$$

$$\delta n_{\mathrm{stress}}^{TE} = \frac{\sigma_{\mathrm{film}}}{\sigma_0}\delta n_0^{TE}, \tag{2.14b}$$

$$\Delta n_{\mathrm{stress}} = \frac{\sigma_{\mathrm{film}}}{\sigma_0}(\delta n_0^{TM} - \delta n_0^{TE}). \tag{2.14c}$$

The results presented in Fig. 2.10 illustrate these relations. For example, the curves for $\sigma_{\mathrm{film}} = -100\,\mathrm{MPa}$ can be used as baselines $\delta n_0^{\mathrm{TE,TM}}$ for each waveguide, determined by the waveguide core geometry and cladding thickness, and the values at other stress levels scale linearly by a factor $\sigma_{\mathrm{film}}/\sigma_0$.

Figure 2.11 shows the calculated results for the maximum $\Delta n_{\mathrm{stress}}$ that can be induced with $\sigma_{\mathrm{film}} = -300\,\mathrm{MPa}$ (with cladding thickness $t_{\mathrm{c}} = 2\,\mu\mathrm{m}$) in waveguides with $H = 2\,\mu\mathrm{m}$ and different aspect ratios r [57]. This illustrates the range, sign and general characteristics of the stress-induced birefringence $\Delta n_{\mathrm{stress}}$ available in SOI waveguides that can be advantageously employed for different applications.

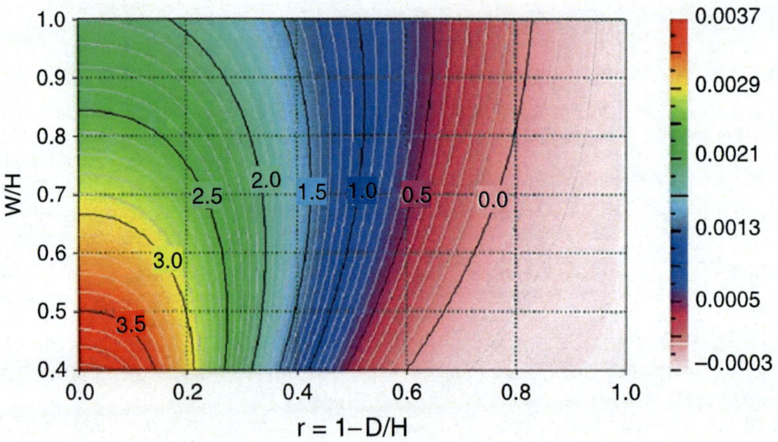

Fig. 2.11 Stress-induced modal birefringence $\Delta n_{\mathrm{stress}}$ for ridge waveguides with different aspect ratios, with $H = 2\,\mu\mathrm{m}$, $t_{\mathrm{c}} = 2\,\mu\mathrm{m}$ and $\sigma_{\mathrm{film}} = -300\,\mathrm{MPa}$

2.4.4 Stress-Induced Mode Mismatch

To fulfil some photonic functions, as in the example of a polarization splitter to be discussed in Sect. 2.5.3, it is required that the upper cladding layer only be applied to a selected section of the components. The mode mismatch at the junction between the bare silicon and oxide clad core, as shown in Fig. 2.12a, is a potential source of propagation loss.

From previous discussions, we observe that the stress-induced changes are $\left|\delta n_{\text{stress}}^{\text{TE, TM}}\right| \leq 5 \times 10^{-3}$, for oxide stress levels that are commonly available ($|\sigma_{\text{film}}| < 400\,\text{MPa}$). These changes are orders of magnitude smaller than the cladding-core index contrast of ~2 in SOI waveguides. The stress-induced modification to the mode profile is thus negligible, leading to minimal mode mismatch loss. Figure 2.12b shows the corresponding mode mismatch loss, calculated as the optical mode overlap integral between the two sections, for waveguides of widths $W = 1.5$ and $2.5\,\mu\text{m}$, a ridge height of $H = 2.2\,\mu\text{m}$ and an etch depth of $D = 1.5\,\mu\text{m}$. The loss is slightly larger for the narrower waveguide, but still practically negligible (less than 0.001 dB). The loss for TM is slightly higher than that for TE, due to the fact that $|\delta n_{\text{stress}}^{\text{TM}}| > |\delta n_{\text{stress}}^{\text{TE}}|$ as shown in Fig. 2.9. For waveguides with even smaller core dimensions, this loss increases, but is expected to be within acceptable levels for values of H down to $1\,\mu\text{m}$.

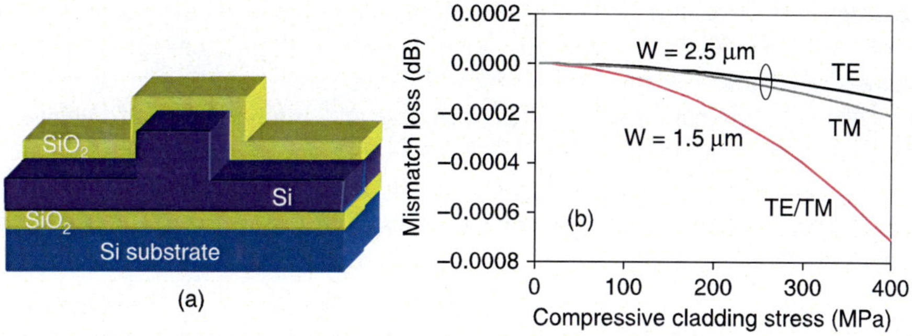

Fig. 2.12 **(a)** Schematic of a junction between the bare silicon and oxide clad waveguide core; **(b)** stress-induced mode mismatch loss at the junction as a function of the cladding stress level, for TE and TM polarizations. The waveguide dimensions are $H = 2.2\,\mu\text{m}$, $D = 1.5\,\mu\text{m}$, $W = 1.5$ and $2.5\,\mu\text{m}$, with a cladding of $1\,\mu\text{m}$ thick. For $W = 1.5\,\mu\text{m}$, the curves for TE and TM nearly overlap

2.4.5 Stress-Induced Effect on the Group Index Birefringence

As we have discussed in Sect. 2.3.3, due to the high confinement in SOI waveguides, there is a significant geometrical contribution to the waveguide dispersion in both the effective index and birefringence. It was noted from Fig. 2.6 that for some waveguide geometries (e.g., $W = 1.5\,\mu\text{m}$), the effective index geometrical birefringence Δn_{geo} and the group index birefringence Δn_{G} have opposite signs, and that

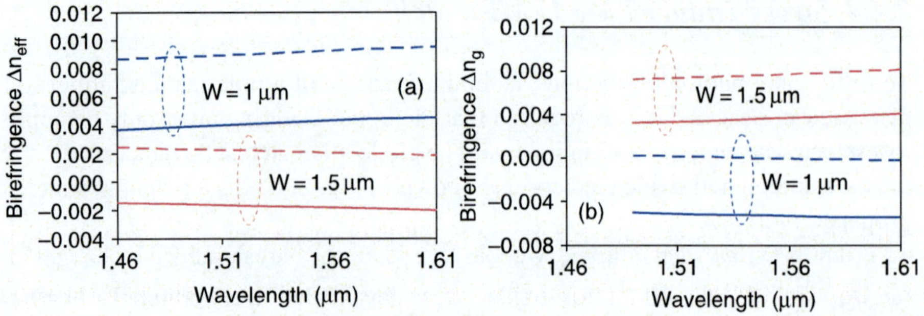

Fig. 2.13 The effect of cladding stress on the effective index and group index birefringence, as a function of wavelength. Here $H = 1.5\,\mu\text{m}$, $W = 1$ and $1.5\,\mu\text{m}$, $D = 0.9\,\mu\text{m}$, $t_c = 0.8\,\mu\text{m}$ and $\sigma_{\text{film}} = -250\,\text{MPa}$. *Solid lines*: in the absence of stress; *dashed lines*: including stress effects. (**a**) Effective index birefringence Δn_{eff}; (**b**) group index birefringence Δn_G

it is not possible to adjust both Δn_{geo} and Δn_G to zero for the same geometry when only the geometrical contribution is considered.

The effects of stress on the total birefringence Δn_{eff} and Δn_G as a function of the wavelength are shown in Fig. 2.13. The cladding film with a compressive stress induces a positive shift of $\Delta n_{\text{stress}} \sim 5 \times 10^{-3}$ in both Δn_{eff} and Δn_G, for $t_c = 0.8\,\mu\text{m}$ and $\sigma_{\text{film}} = -250\,\text{MPa}$. These shifts do not significantly vary with the wavelength. This follows similar arguments, as presented in Sect. 2.4.4, that due to the high refractive index contrast of SOI, the stress-induced changes in the mode profiles are minimal. Consequently, there is only negligible change in the mode dispersion. Therefore, the stress-induced birefringence Δn_{stress} can be used as a broadband tuning parameter to adjust the phase and group index birefringence. On examining the relation $n_G = n_{\text{eff}} - \lambda \partial n_{\text{eff}}/\partial \lambda$ and since the stress only introduces constant shifts in $n_{\text{eff}}^{\text{TE,TM}}$, it is clear that the following statement still holds even in the presence of stress: it is not possible to adjust both Δn_{eff} and Δn_G to zero for the same geometry.

2.4.6 *Scaling of Stress-Induced Birefringence with the Core Size*

Up to this point we have mainly examined the characteristics of stress-induced effects for a ridge height of $H = 1.5\,\mu\text{m}$. In this section, we discuss how these effects scale for different core sizes. Figure 2.14 shows the distribution of the material birefringence $(n_y - n_x)$ in ridges having heights of 1, 3, and $5\,\mu\text{m}$, but of the same aspect ratio $(r = 0.375)$ and with an oxide upper cladding of $1\,\mu\text{m}$ thickness and $\sigma_{\text{film}} = -105\,\text{MPa}$. When the ridge is buried under the oxide $(t_c \geq D)$, as for the case of $H = 1\,\mu\text{m}$ shown in Fig. 2.14a, the entire Si core is dominated by the tensile stress in the y-direction and $n_y - n_x > 0$ for the whole ridge area. As the ridge height H increases relative to t_c to the point that the top of the ridge is above the cladding/air boundary of the slab regions $(t_c < D)$, this fraction of the core is actually stretched in the x-direction and compressed in the y-direction. At the same

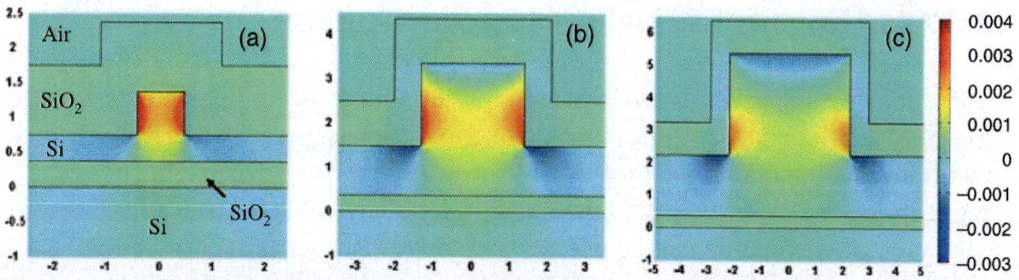

Fig. 2.14 Material birefringence $(n_y - n_x)$ in SOI waveguides with a width of 0.9 H, and an aspect ratio r of 0.375, for ridge height H of (**a**) 1 μm, (**b**) 3 μm, and (**c**) 5 μm. The oxide thickness is 1 μm, with a stress $\sigma_{\text{film}} = -105$ MPa

time, the lower fraction of the core "sandwiched" between the in-plane cladding film can only extend in the y-direction (i.e., under a tensile stress), as shown for the case of $H = 3$ and 5 μm in Figs. 2.14b, c. As long as the tensile-stressed fraction of the core has a significant overlap with the optical mode field, a positive shift is induced in the birefringence.

Figure 2.15 shows the dependence of Δn_{stress} on the cladding thickness t_c for core sizes ranging from 1 to 5 μm. For a given ridge height H and etch depth D, with increasing cladding thickness, there is an increasing fraction of the core under the in-plane force ($\sigma_{\text{film}} t_c$) of the cladding, and the stress-induced birefringence Δn_{stress} increases. When t_c approaches D, Δn_{stress} approaches a maximum and saturates. For the same cladding thickness, a smaller Δn_{stress} is induced for a larger core. The respective Δn_{stress} saturation values, however, are of the same order of magnitude for the ridge heights considered here, $\Delta n_{\text{stress}} \sim 10^{-3}$ for $\sigma_{\text{film}} = -105$ MPa. Values of Δn_{stress} for other σ_{film} levels can be calculated using the relations in Eq. (2.14). The stress-induced contribution to the total modal birefringence is significant for a wide range of core sizes, even for cladding films with a relatively low stress. These effects have to be taken into consideration in the design of SOI waveguide components, and can be advantageously used for precise control of birefringence, as will be discussed next.

Fig. 2.15 Stress-induced birefringence Δn_{stress} as a function of the cladding thickness t_c, in SOI waveguides with $W = 0.9\,H$ and an aspect ratio $r = 0.375$, for ridge heights $H = 1, 2, 3, 4$, and 5 μm, respectively, as indicated in the figure. The cladding stress $\sigma_{\text{film}} = -105$ MPa

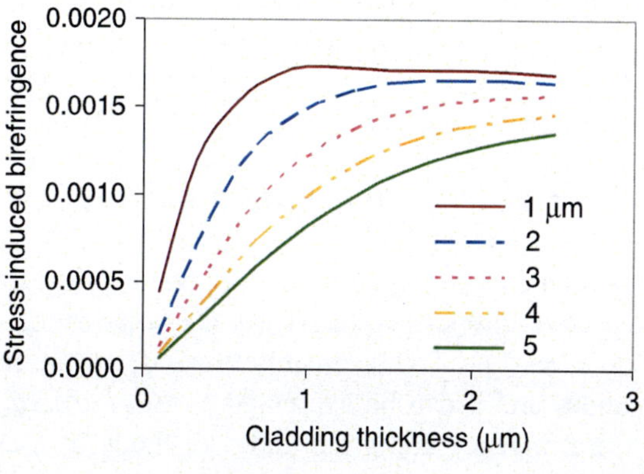

2.5 Cladding Stress Engineering: Applications

In the mid-1990s, several studies were reported that utilized the photoelastic effect to make strip-loaded waveguides in silicon [58, 59]. Using a strip or a gap of $\sim 6\,\mu m$ width in a WNi metal film on top of a Si substrate, it was shown that the corresponding strip-loaded waveguide only guided the TE or TM mode, respectively. Therefore each type of waveguide can be used as a polarizer and an extinction ratio of 14 dB was achieved [59]. In a later work, periodic oxide cladding sections were used to make polarization rotators [60]. In these cases, stress was the only mechanism for producing the desired functionalities. In the early 2000s, we proposed and demonstrated that cladding stress introduces a birefringence in silicon ridge waveguides that is of comparable magnitude to its geometrical birefringence [61, 62] and that it can serve as an effective means to counter the effect of geometrical birefringence, yielding birefringence-free waveguides [11].

In this section, we present several examples of using cladding stress engineering to control polarization-dependent properties in both passive and active SOI waveguide devices. In Sect. 2.5.1 we present polarization-independent (PI) AWG demultiplexers that we have developed, where the birefringence in the waveguide array is eliminated by the judicious selection of the oxide cladding stress and thickness [11, 35]. Examples of PI ring resonators [13, 14], MZI differential phase shift keying (DPSK) demodulators [16] and a low loss PI triplexer [15] are introduced in Sect. 2.5.2. Benefiting from the cyclic response of these devices, oxide (compressive) or nitride (tensile stress) claddings were used to modify the birefringence so that the TE and TM spectra were shifted by integer multiples of the free-spectral-range (FSR). As the last example of stress-engineered passive devices, we discuss in Sect. 2.5.3 a polarization splitter in a zero-order AWG configuration, where a prism-shaped oxide patch is placed in the arrayed waveguide section to separate the TE and TM modes into different output waveguides. Since this AWG operates at near zero-order, broadband performance is achieved [26]. In Sect. 2.5.4, methods of birefringence trimming in passive waveguide components are presented. Cladding stress engineering has also been used in active photonic components. Active birefringence tuning and its use for phase matching in Raman amplification is described in Sect. 2.5.5. Finally, the use of stress to alter the centro-symmetric crystal structure of silicon, thereby introducing a Pockels electro-optic effect, is discussed in Sect. 2.5.6.

2.5.1 Polarization-Independent AWGs

Arrayed waveguide gratings are arguably the most commonly used wavelength dispersive filters [28]. In AWGs the polarization-dependent wavelength shift (PDW) $\Delta\lambda = \lambda_{TM} - \lambda_{TE}$ mainly arises from the birefringence of the arrayed waveguides and it can be expressed as $\Delta\lambda = \lambda\Delta n_{eff}/n_G$, where Δn_{eff} and n_G are the birefringence and group index of the array waveguides. In glass waveguide-based

AWGs, a frequently used approach for obtaining PI performance is to minimize the residual stress to make the arrayed waveguides birefringence-free. Another common method is to insert a half-wave plate in the middle of the phased array to convert TE polarization to TM and vice versa [63]. The insertion loss penalty, reflections at the interfaces, and higher order mode excitation make this method impractical for high-index contrast platforms like SOI. Introducing a triangular-shaped region in the combiner or the array section, which has a modified birefringence and acts as a prism, can be used to compensate the PDW as originally proposed and demonstrated by He et al. on InP AWGs [64]. Cheben et al. have shown that this method is effective in eliminating PDW in AWG devices made of SOI ridge waveguides; however, it is also limited by the associated loss penalty [65]. We have subsequently shown that the use of a uniform cladding oxide with the appropriate stress and thickness to produce birefringence-free waveguide arrays is effective and simple to implement, as discussed below.

The AWGs used for the demonstration of polarization compensation using stress engineering were made on SOI substrates with a silicon thickness of 2.2 μm, with nine output channels spaced at 200 GHz and centred at 1,550 nm [11, 66] (see Fig. 2.16a). The arrayed waveguide gratings of order 40 were formed by 100 waveguides of nominal width of 2 μm. Waveguides were produced using either wet

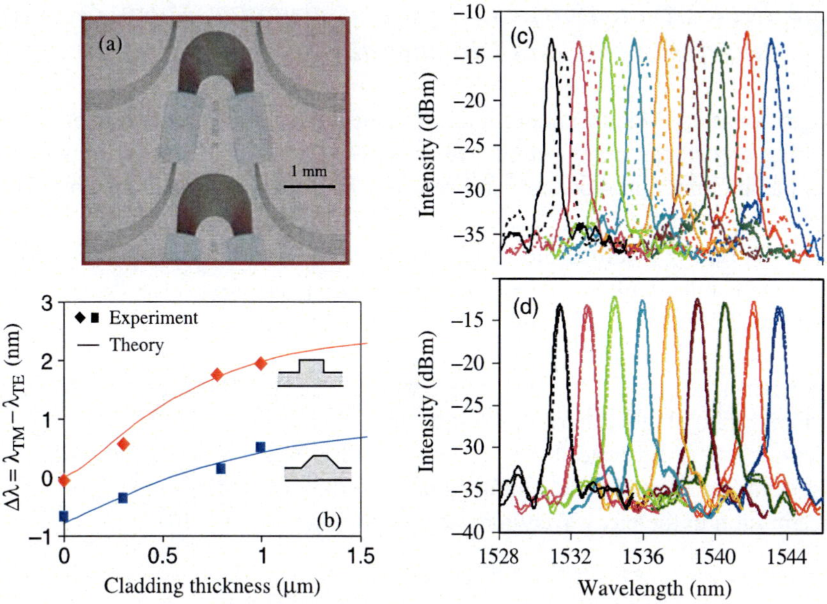

Fig. 2.16 (**a**) Optical image of a fabricated AWG. (**b**) Polarization-dependent wavelength shift $\Delta\lambda$ as a function of the cladding thickness t_c, for AWGs fabricated using wet (average waveguide top width $W = 1.1\,\mu$m and $\theta = 47°$) and dry ($W = 1.85\,\mu$m and $\theta = 87°$) etching to a depth of $D = 1.47\,\mu$m. The cladding stress is $\sigma_{film} = -320$ MPa. Measured TM (*solid*) and TE (*dashed*) spectra are shown for an SOI AWG (**c**) without the oxide upper cladding, and (**d**) compensated using a 0.6-μm-thick SiO_2 claddings with $\sigma_{film} = -320$ MPa

or dry etching, yielding waveguide sidewall angles of ~54 and 90°, respectively. To validate the simulation results, the waveguide birefringence was calculated for average measured waveguide dimensions. Upper cladding oxide films of different thicknesses were deposited at ~ 400 °C using PECVD, with a film stress of $\sigma_{\mathrm{film}} = -320$ MPa as measured from wafer bowing using films deposited on blanket substrates. Figure 2.16b shows the calculated and measured changes in $\Delta\lambda$ as a function of the cladding thickness t_c, for both trapezoidal and rectangular waveguides. Calculated results agree well with experiments for both types of waveguides, showing that $\Delta\lambda$ can be modified over an ~2-nm wavelength range [35].

Figure 2.16c, d show the TE and TM spectra of a wet-etched AWG before and after polarization compensation. The adjacent channel cross-talk is less than −25 dB. The non-compensated AWG device has a polarization-dependent wavelength shift of $\Delta\lambda \sim 0.6$ nm (Fig. 2.16c), arising from the geometrical birefringence of $\Delta n_{\mathrm{geo}} \sim -1.3 \times 10^{-3}$ in the arrayed waveguide. With an oxide upper cladding film having a stress of −320 MPa and thickness of 0.6 μm, $\Delta\lambda$ was then reduced to below 0.04 nm for all channels, corresponding to $\Delta n_{\mathrm{eff}} < 1 \times 10^{-4}$ (Fig. 2.16d). The polarization-dependent loss (PDL) was also found to be negligible in these devices after the compensation.

2.5.2 Polarization-Independent Ring Resonator, Mach-Zehnder Interferometer, and Directional Coupler

Ring resonators, Mach-Zehnder interferometers, and directional couplers are basic building blocks frequently employed in a diverse range of applications including communications, interconnect, and sensing. However, polarization sensitivity has often been an obstacle that demands considerable effort to mitigate.

In SOI ring resonators, the strong polarization sensitivity in both the resonance wavelength and quality factor has long been noted [9, 10, 14, 67]. A PI ring resonator was reported in 2004 achieved solely by adjusting the device geometries [9], but it was later realized that the oxide cladding stress present in those devices had a significant role [68]. In the following we will discuss the use of cladding stress engineering to obtain polarization-insensitive ring resonators [13, 14].

Figure 2.17a illustrates a ring resonator with a single access waveguide, and the ring cavity is excited with a multi-mode-interference (MMI) coupler. The power transmission in the bus waveguide can be expressed as [69, 70]:

$$|T|^2 = \alpha_C{}^2 \left[\frac{\alpha^2 - 2\alpha t \cos\phi + t^2}{1 - 2\alpha t \cos\phi + \alpha^2 t^2} \right], \tag{2.15}$$

where $\phi = 2\pi n_{\mathrm{eff}} L/\lambda$ is the total round trip phase accumulation, L is the cavity length, and $\alpha^2 = \alpha_C^2 \cdot \alpha_{\mathrm{ring}}^2$ is the combined power loss factor including both the coupler and the ring loss. The self-coupling coefficient t^2 is the ratio of power transfer from the input to the output of the coupler. A polarization dependence in t

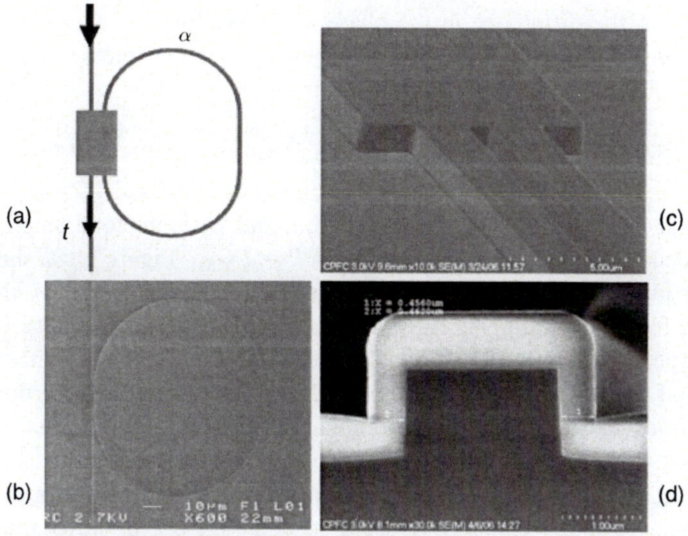

Fig. 2.17 (**a**) Schematics of an MMI-coupled ring resonator; (**b**) a fabricated resonator; (**c**) close-up of the MMI coupler structure prior to oxide cladding deposition; and (**d**) cross-section of the waveguide with the oxide cladding

and/or α lead to different resonance quality factors and spectral widths for TE and TM polarizations, and a birefringence in the effective and group indices result in polarization sensitive resonance wavelengths and free spectral range, respectively.

The example discussed here is a PI ring resonator using an MMI coupler designed in SOI waveguides having a $1.5\,\mu\text{m}$ core thickness. It is well known that MMIs are less polarization sensitive compared to directional couplers (DC) commonly used in ring resonators [14, 71, 72]. PI ring resonator designs using DC have been investigated [9, 73], but required long coupler length and very stringent fabrication control. The resonators here are designed with waveguide widths of $W = 1.5\,\mu\text{m}$ and $1\,\mu\text{m}$ and an etch depth of $D = 0.9\,\mu\text{m}$. A 2×2 restricted interference coupler with dimensions of $7.5\,\mu\text{m} \times 84\,\mu\text{m}$ and a 50:50 power splitting ratio was used [71], where a similar excess loss of $\sim -0.2\,\text{dB}$ was found for both polarizations. The choice of this MMI length is based on the assumption that losses from all other mechanisms are similar for TE and TM, which is not necessarily the case. For example, the waveguide propagation loss due to scattering depends on the optical mode overlap with the roughness at the waveguide boundaries and is generally polarization dependent. This loss factor is proportional to the propagation distance and therefore to the cavity length of the ring resonator. Another source is the radiation loss in waveguide bends [4, 14], as well as possible losses at discontinuities such as the junctions between the straight and curved waveguides [74]. These losses are also highly polarization dependent. These additional loss contributions need to be considered in the device design, and detailed discussions can be found in [14]

Since the stress-induced birefringence in the ring waveguide Δn_s^{ring} is much larger than that in the wider MMI section [12] and the ring waveguide section is much longer, the difference in the phase accumulation $\Delta\phi$ between TM and TE polarizations for one round-trip of the ring cavity can be expressed as [13]:

$$\Delta\phi = \frac{2\pi}{\lambda}\left[\Delta n_{\text{geo}}^{\text{MMI}} L_{\text{MMI}} + \Delta n_{\text{geo}}^{\text{ring}} L_1 + \Delta n_s^{\text{ring}} L_1\right]. \tag{2.16}$$

Here L_{MMI} is the length of the MMI section, and L_1 is the total length of the ring waveguide sections, $L_1 = L - L_{\text{MMI}} = 2\pi R + L_{\text{MMI}}$. Figure 2.18a shows the ring resonator phase difference, $\Delta\phi = \phi^{\text{TM}} - \phi^{\text{TE}}$, as a function of the oxide cladding thickness for two cladding stress levels. Since a ring resonator has a cyclic transmission spectrum, apparent polarization independence in the resonance wavelength is achieved for $\Delta\phi = 2\pi j (j = 0, \pm1, \pm2 \ldots)$. As indicated in the figure, various combinations of cladding stress and thickness can be found for $\Delta\phi$ to satisfy such conditions.

Figure 2.17b, c show the SEM images of a fabricated MMI-coupled ring resonator prior to oxide cladding deposition, and Fig. 2.17d shows the waveguide cross-section with the oxide cladding. Measured transmission spectra for TE and TM polarizations for a ring resonator, having $R = 200\,\mu\text{m}$ and a 0.8-μm oxide cladding having $\sigma_{\text{film}} = -250\,\text{MPa}$, are shown in Fig. 2.18b. Over the 4-nm scan range, the polarization-dependent wavelength shift is measured to be below 2 pm, which is the laser scan step used in these measurements. The measured free spectral range (FSR) is ~0.46 nm, and the quality factor of $Q \sim 15,000$ [14, 75].

Similar to ring resonators, a MZI is another example of devices that have a periodic spectral response. An MMI-coupled Mach-Zehnder delay interferometer (DI) realized in 4-μm SOI waveguide technology has been used for 40-GB/s differential phase shift keying (DPSK) demodulation [16], with the device layout shown in Fig. 2.19a. In these SOI waveguides the polarizationdependent wavelength (or frequency) shift comes approximately in equal fraction from the geometrical and

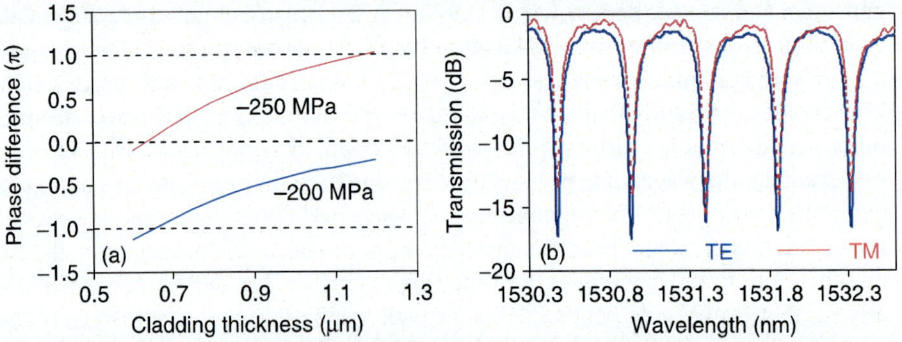

Fig. 2.18 (**a**) Phase difference $\Delta\phi = \phi^{\text{TM}} - \phi^{\text{TE}}$ as a function of the oxide cladding thickness for a MMI-coupled ring resonator with a radius R of 200 μm and an oxide cladding stress of −200 and −250 MPa, respectively; (**b**) measured TE and TM transmission spectra of the polarization-compensated ring resonator, with an upper SiO_2 cladding of 0.8 μm in thickness and a film stress of $\sigma_{\text{film}} = -250\,\text{MPa}$

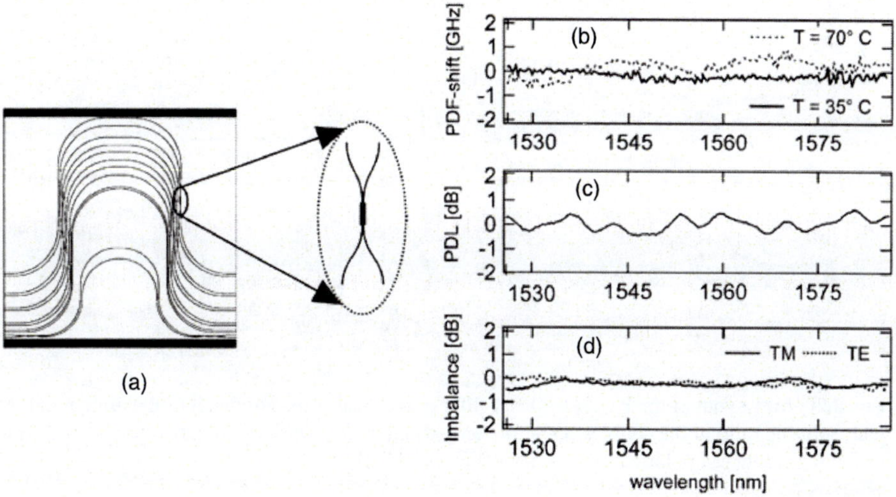

Fig. 2.19 (**a**) Layout of the Mach-Zehnder delay interferometers, showing five DIs and test structures. The insert shows a 2 × 2 MMI coupler as a part of one DI. (**b**) PDF shift for two temperatures ($T = 35$ and 70 °C). (**c**) PDL of the filter curve maxima for TE and TM mode across the C-band. (**d**) Port imbalance of TE and TM light (After Voigt et al. [16])

stress-induced birefringence, and very precise birefringence control ($\sim 10^{-5}$), which is a key requirement for DPSK receivers, can be achieved by adjusting the stress-induced birefringence. A silicon nitride film was used as the cladding layer, so that the tensile stress in the cladding shifts the TE response by one FSR relative to that of TM, therefore achieving apparent polarization independence. The measured performance is shown in Figs. 2.19b–d. These devices also showed a very low temperature dependence (polarization-dependent frequency shift PDF < 1 GHz) for chip temperatures up to 70 °C. Although the governing mechanisms are not yet fully understood, these results may originate from a combination of material thermo-optic effect and cladding stress-induced changes, similar to that suggested in the work on a porous silicon filter [76].

A triplexer for routing voice, data and video signals is an important component for fiber-to-the-home (FTTH) networks. Feng et al. have demonstrated a low loss triplexer design, shown in Figs. 2.20a, b, consisting of two-stage cascaded transmitter/receiver filters using Fourier-transform-based MZIs [15]. The device is made insensitive to fabrication variations by employing appropriate directional couplers and optical delay lines. The key design parameters are the coupling properties of the directional couplers and the optical path length difference of the phase shift sections ΔL_s, and both are functions of the waveguide effective index. The triplexer was fabricated in 3-μm-thick SOI. Two layers of thin films are applied on top of the device to compensate the polarization-dependent phase difference. The coupler splitting ratio before and after the film deposition is shown in Figs. 2.20c, d. The coupling length difference between TE and TM was reduced from 25 to 2%, and the polarization-dependent loss (PDL) was reduced to less than 0.1 dB.

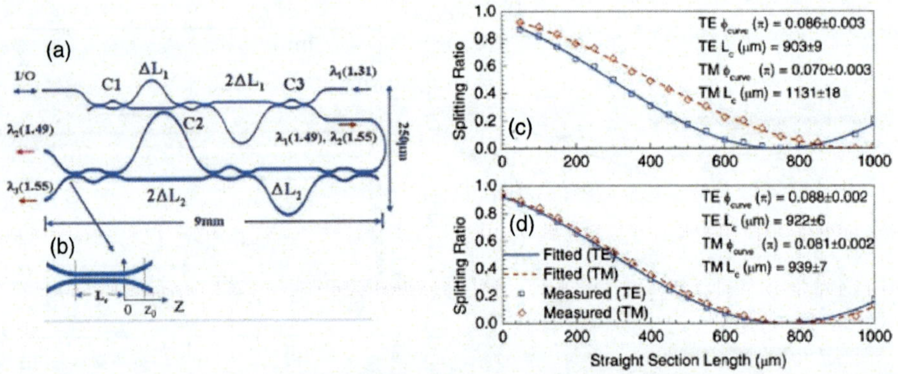

Fig. 2.20 (**a**) Layout of the FTTH triplexer filter (not to scale) and (**b**) detail of the directional coupler; splitting ratio of the coupler (**c**) before and (**d**) after stress-induced polarization compensation for both TE and TM polarizations (After Feng et al. [15])

2.5.3 Broadband Polarization Splitter in a Zero-Order AWG Configuration

Polarization splitters are key elements in devices based on the polarization diversity scheme, which can find applications in signal processing, network monitoring, polarimetric sensing, imaging, and data storage. Different configurations of waveguide polarization splitters such as the Y-branch [77], multimode interference coupler [78], directional coupler [79], Mach–Zehnder interferometer [80–82], and AWG [83] have been reported. These designs employ modifications to waveguide geometries in selected sections and have shown limitations in the fabrication tolerances, in the operating bandwidth, or in the density of integration.

Ye et al. have demonstrated a zero-order AWG polarization splitter that utilizes the stress-induced effect to produce a high level of birefringence in selected waveguide areas, as shown in Fig. 2.21. A triangular patch of oxide cladding is placed in the arrayed waveguide section with a constant length increment ΔL between the adjacent waveguides. The stress in the cladding induces a polarization-dependent phase difference for the light propagating in the waveguide array (see Fig. 2.21a), causing a polarization-dependent tilt in the phase front of the light in the slab waveguide combiner region (see Fig. 2.21a). Since all waveguides have the same physical length, the phase difference for light propagating in the adjacent waveguides of the arrayed section depends solely on the cladding stress-induced index change and the patch-length increment. Since the stress-induced index change in the TE and TM polarization modes have opposite signs for silicon waveguides, the two modes are spatially displaced in the opposite directions in the focal region relative to the combiner centreline, as shown in Fig. 2.21b. With proper placement of the two receiver waveguides along the combiner focal plane, the two polarizations are spatially separated.

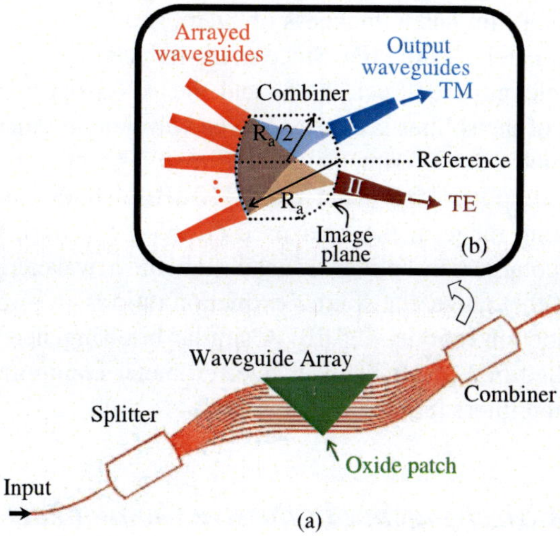

Fig. 2.21 (**a**) Schematic layout of a broadband zero-order AWG polarization splitter; (**b**) geometry of the output slab waveguide combiner with a rowland configuration (After Ye et al. [26])

As we have discussed in Sect. 2.4.4, the stress-induced index modifications are several orders of magnitude smaller than the core-cladding refractive index contrast in SOI, so that the effect of a SiO_2 cladding on the mode shape is negligible. Thus there is negligible mode mismatch loss or polarization-dependent loss at the junctions between waveguide sections with and without the claddings.

Polarization splitters fabricated on SOI wafers having a 2.2-μm-thick silicon layer are shown in Fig. 2.22a, with an overall device size of ~12 mm × 4 mm. The

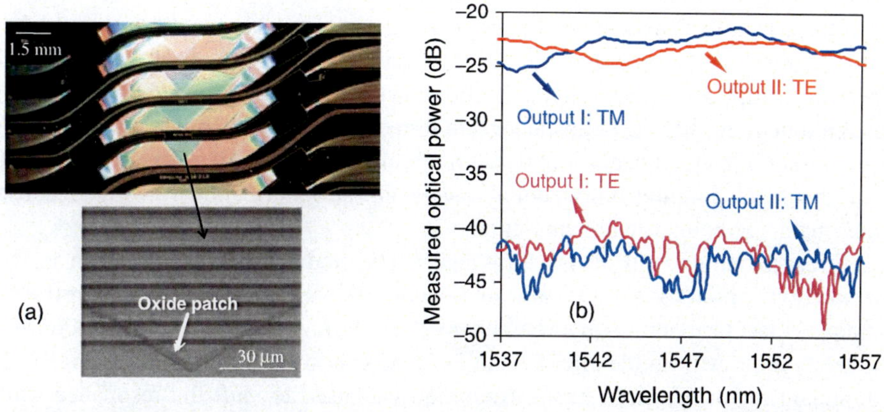

Fig. 2.22 (**a**) A top view of the fabricated wavelength-independent zero-order AWG polarization splitters (several devices shown). Insert: SEM image of the oxide cladding triangular patch in the waveguide array; (**b**) measured transmission as a function of wavelength of a zero-order AWG polarization splitter (After Ye et al. [26])

deposited oxide cladding film had a thickness of 1 μm, and the measured cladding film stress was approximately −340 MPa. Since stress-induced effects do not appreciably vary with wavelength (see Sect. 2.4.5) and the AWG operates near order zero, the performance of the splitter is expected to be wavelength independent over a large bandwidth, which was experimentally confirmed. The measured extinction ratio of the polarization splitter was better than −10 dB for both output polarizations over the full tuning range of the laser used (1,465–1,580 nm). Figure 2.22b shows the measured polarization splitter transmission for a wavelength range of 1,537–1,557 nm. Over this range, the splitter extinction ratio is < −15 dB, and the highest measured extinction ratio is −20 dB. A similar birefringence modification method was also applied in a Mach-Zehnder interferometer configuration to form polarization splitters and filters [82].

2.5.4 Trimming of Birefringence in Passive Components

Even with advanced processing tools available today, variations in critical dimensions of devices fabricated with either an electron beam or deep-UV optical lithography are still inevitable. These variations may arise from silicon thickness non-uniformity from wafer to wafer, or from deviations on the wafer caused by lithography imperfection or etch non-uniformity. From discussions in Sect. 2.3.2, we noted that the effective index and birefringence are sensitive to dimensional variations, particularly as the core size is reduced. These changes lead to device central wavelength shifts and polarization dependence. Trimming of the waveguide effective index is still necessary in certain situations. Active thermal tuning has been used for this purpose, but this approach adds to design complexity and power consumption. Another approach is trimming of the devices after fabrication, by adjusting the cladding stress level or thickness, which may be performed on a wafer level or on a device-by-device basis.

Stress in thin films can be modified by thermal treatments. The extent of the modification depends on the film microscopic properties, thermal annealing temperatures, and durations. Figure 2.23a shows such an example where the stress magnitude in a PECVD deposited oxide film decreases with annealing temperature and duration [61]. Anneals at 600 °C for 2 and 17 min were performed on an AWG, and the corresponding polarization wavelength shifts are shown in Fig. 2.23b, in reasonable agreement with simulations.

Stress in oxide or nitride thin films can also be modified by electron beam and UV irradiation, which causes a volume compaction as well as a change in the refractive index, as has been shown by Schrauwen et al. [84]. Although both effects have an impact on the waveguide effective index, it was shown that the stress change is the dominant factor. Electron beam irradiation was used to shift the resonance wavelength of a SOI wire waveguide ($220 \times 500\,\text{nm}^2$) ring resonator with a 2-μm-thick oxide cladding, and a red shift of ~4.9 nm was demonstrated. Figure 2.24 shows the evolution of the resonator transmission spectrum during one stage of the trimming

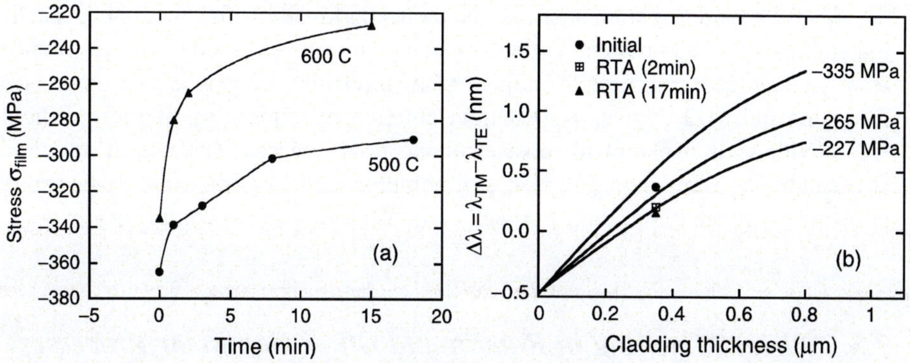

Fig. 2.23 (**a**) Changes in the stress level as a function of anneal time for a blanket PECVD oxide film; (**b**) calculated (*line*) and measured (*symbols*) polarization-dependent wavelength shift for the corresponding anneals (After Xu et al. [61])

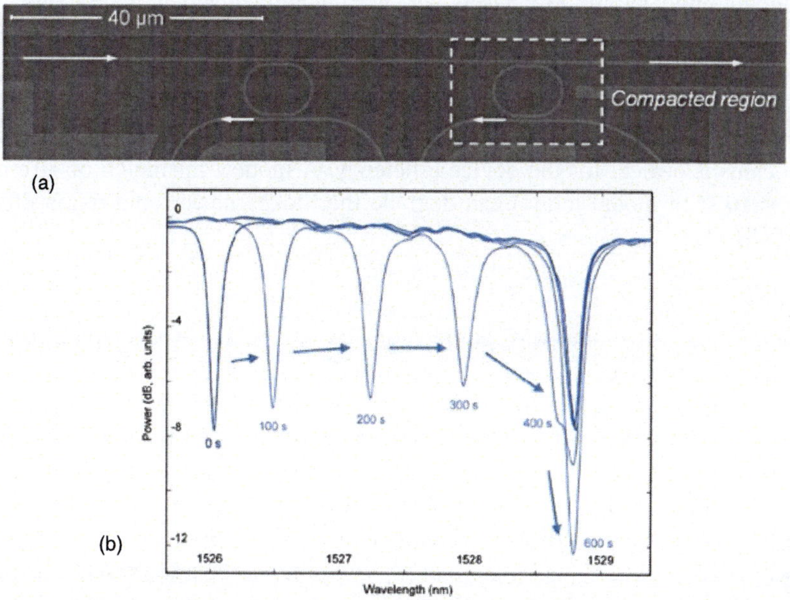

Fig. 2.24 (**a**) Overview of the device used for an electron beam trimming experiment: the *right*-hand ring is trimmed by electron beam compaction; the *left* one is kept intact as a reference to exclude temperature or ambient variations. (**b**) Evolution of the resonance during the scanning by a 0.84-nA electron beam (After Schrauwen et al. [84])

process, when the ring resonator on the right side was scanned by a 0.84-nA electron beam. Although the contribution of the different microscopic mechanisms and the long term stability of this method still require further investigation, this method can potentially be used for trimming individual devices in an integrated photonic circuit.

The oxide cladding thickness may be reduced globally with chemical or reactive ion etching. The thickness of a stress-inducing amorphous silicon film was reduced locally using laser trimming for adjusting the birefringence in glass waveguide ring resonator and MZI devices [85]. Direct ablation of oxide films is more difficult, since oxide is transparent to most available laser sources. The use of additional stress inducing films on oxide clad SOI components is also possible. Even though allowing local trimming, this approach certainly adds process complexity.

2.5.5 *Phase Matching in Raman and Other Nonlinear Processes and Active Birefringence Tuning*

Silicon as a nonlinear medium has received a surge of interest in recent years [25, 86]. Raman scattering was proposed and demonstrated as a mechanism for producing optical amplifiers, lasers, and wavelength converters. One motivation for this approach is the fact that the stimulated Raman gain coefficient in silicon is 10^3–10^4 times larger than that in glass fibers. The third order nonlinear susceptibility in silicon, arising from the Raman and non-resonant susceptibilities, is dominated by the Raman contribution. In these devices, phase matching of the interacting beams is critical for the device efficiency. A modest mismatch of $10\,\text{cm}^{-1}$ (or $\sim 2 \times 10^{-4}$ in modal index) can degrade the Stokes power conversion efficiency by 20 dB [87].

In four-wave-mixing, the phase mismatch is mainly caused by the waveguide and material dispersion, and the birefringence between the pump and (anti-) Stokes waves. It was shown that waveguide geometry can be adjusted so that Δn_{geo} compensates for the dispersion terms, thereby satisfying the phase matching condition [25, 87]. Maintaining the mismatch to below $1.3\,\text{cm}^{-1}$, however, requires width and etch depth control of 70 and 10 nm, respectively, even for the relatively large waveguide of $H = W = 2.3\,\mu\text{m}$ at $h = 1.38\,\mu\text{m}$. In those analyses, only the geometrical contributions were considered. Stress-induced birefringence can be utilized to mitigate these stringent requirements [24]. With reference to discussions in Sects. 2.3.1 and 2.4.3, one can observe that the waveguide cross-section can be designed to geometries less sensitive to dimensional fluctuations (e.g., $W > H$ and shallower etch), and employ stress-induced birefringence to reach phase matching.

Recently, active tuning of the cladding stress was demonstrated using a piezoelectric cladding layer [17, 88]. Lead zirconate titanate (PZT) sandwiched in Pt/Ti electrodes were deposited on the SOI waveguides with an oxide spacer, as shown in Figs. 2.25a, b. A linear dependence of the birefringence on the bias voltage was observed, and a tuning range of 1×10^{-4} was obtained by varying the bias voltage from -10 to 5 V [17], as shown in Fig. 2.25c. A corresponding efficiency tuning of 6 dB was demonstrated in the coherent anti-Stokes Raman scattering (CARS) wavelength conversion process. Optimization in waveguide designs and fabrication processes are predicted to further improve the efficiency and offer adaptive control of waveguide nonlinear processes.

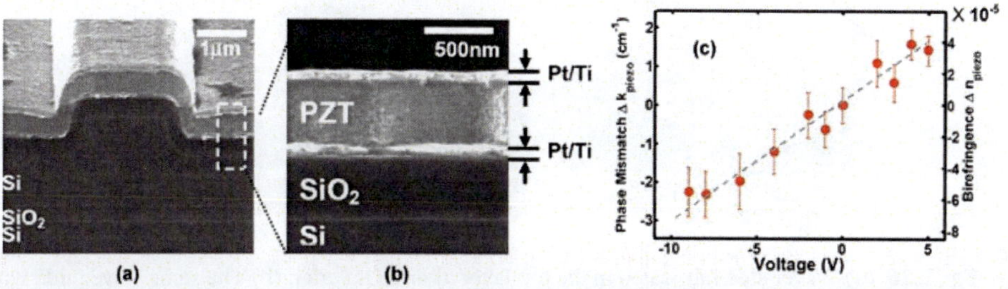

Fig. 2.25 (**a**) Scanning electron microscope image of the SOI waveguide with the PZT capacitor on top; (**b**) the enlarged cross-section of the PZT capacitor. The waveguide has a width of 1.5 μm, rib height of 2 μm, and slab height of 1.1 μm. The oxide-cladding is 500 nm thick. The PZT thickness is 500 nm. Both top and bottom Pt/Ti electrodes are 100/10 nm thick. (**c**) Phase mismatch (*left* axis) and the corresponding birefringence (*right* axis) of the SOI waveguide due to the piezoelectric effect in PZT as a function of DC voltage (After Tsai et al. [17])

2.5.6 Stress-Induced Pockels Electro-optic Effect in Silicon

The last example in this section introduces a strain-induced active phenomenon in silicon, i.e., the Pockels electro-optic effect which prescribes that the material refractive index varies linearly as a function of the external applied electric field. Although it is not a direct use of stress for polarization control, the Pockels effect and photoelasticity are two intimately related phenomena. The Pockels effect, as found in materials such as $LiNbO_3$ and GaAs, is the basis for most commercially successful high-speed electro-optic modulators. Being a centro-symmetric cubic crystal in its natural state, silicon lacks the inversion asymmetry which is necessary for the Pockels effect.

It was discovered in 2006 by Jacobsen et al. that the inversion symmetry in silicon can be broken by applying asymmetric strains to the crystal. Depositing a straining Si_3N_4 layer of 1 GPa compressive stress on top of a silicon photonic crystal waveguide, as shown in Fig. 2.26, introduced a second order nonlinearity $\chi^{(2)}$ [89]. The phenomenon was observed in conjunction with the enhancement of apparent nonlinearity by the high group index in the photonic crystal waveguides used in the experiments. An enhanced nonlinearity of ~830 pm V^{-1} was observed, and a material nonlinearity of $\chi^{(2)} \approx 15$ pm V^{-1} was deduced. This value is still much lower than that of $LiNbO_3$ (~360 pm V^{-1}), nonetheless the experiment constitutes a decisive step towards making silicon a versatile electro-optic material.

Figure 2.27 shows the simulated stress and strain in a sub-micron silicon waveguide by applying a cladding film with a high compressive stress ($\sigma_{\text{film}} = -1$ GPa) on half of the waveguide. The stress level is the same as in the work shown in Fig. 2.26. The waveguide deformed non-uniformly and anisotropically, and the strain level in the silicon core can reach more than ~1%. This is another possible configuration to introduce inversion asymmetry in the silicon core.

Stress levels on the order of −1GPa are very high values for dielectric layers deposited on silicon. Using epitaxially grown SiGe material, a stress of several GPa is common due to a 4% lattice mismatch between the silicon and germanium

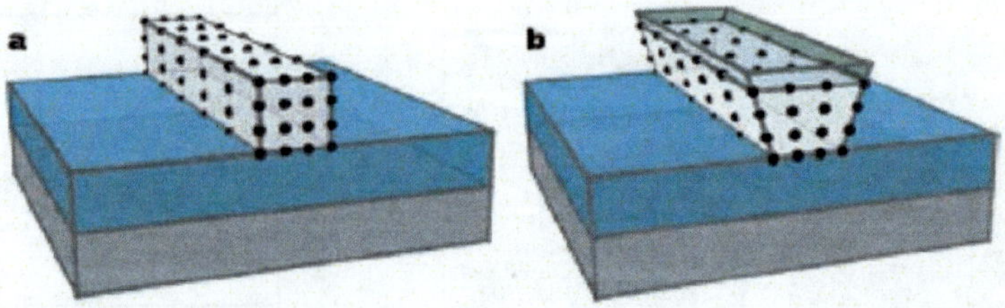

Fig. 2.26 (**a**) Waveguide fabricated in the top layer of an SOI wafer. (**b**) The same waveguide with a straining layer on *top*. The straining layer breaks the inversion symmetry and induces a linear electro-optic effect (After Jacobsen et al. [89])

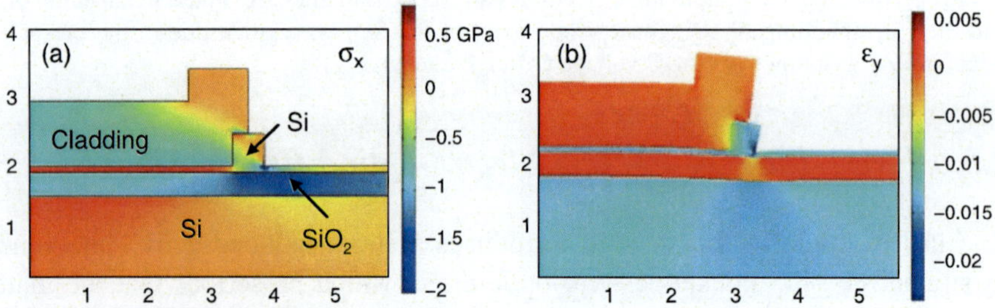

Fig. 2.27 Stress and strain distributions in an SOI submicron waveguide with asymmetric cladding stress. (**a**) Stress component σ_x distribution; (**b**) Strain component ε_y distribution. The deformation is enlarged by ten times relative to the waveguide dimensions. $H = 500\,\text{nm}$, $W = 500\,\text{nm}$, and $D = 400\,\text{nm}$. The stress applying film has $\sigma_{\text{film}} = -1\,\text{GPa}$

crystals. This high strain is widely used in the microelectronic industry to enhance the electron or hole mobilities in CMOS devices, by applying strain in different crystal orientations through selective growth of SiGe in localized areas. Even though not yet experimentally investigated, applying SiGe to the SOI waveguide and exploring the effect for waveguides along different crystal orientations may be a promising direction to enhance the electro-optic coefficient in silicon.

2.6 Conclusions

Birefringence in SOI waveguides is governed by the waveguide cross-section and the stress in the silicon core. A stress in the core can be introduced on purpose or inadvertently by the waveguide upper cladding films. The associated birefringence is important for a wide range of commonly employed SOI waveguide sizes and geometries. The significance of these stress-induced effects is now being recognized in the research community, and the technique is also being applied by industry in product development. If not taken into consideration, these effects can lead to large deviations in device characteristics from the designed specifications. Fortuitously, cladding stress-induced birefringence is of comparable order of magnitude to the waveguide geometrical birefringence, allowing precise compensation of the total birefringence. This affords a considerable degree of freedom in designing SOI waveguides to meet other performance criteria such as relaxed dimensional tolerance, reduced loss at waveguide bends, and overall improved device performance. The technique of cladding stress engineering is simple to implement and requires no additional process step other than the conventional cladding layer deposition. The stress-induced modifications to the effective indices and birefringence are readily controlled by the stress level and the thickness of the upper cladding layer. We have reviewed the application of cladding stress engineering to produce polarization-independent performance in several types of photonic components, including AWGs, ring resonators, DPSK demodulators, and low loss triplexers. Since the effect of the cladding stress on the mode profile is negligible, there is little mode mismatch loss or polarization-dependent loss at the junctions between waveguide sections with and without the upper cladding. Therefore, tailored cladding patches can be applied at discrete locations in a planar waveguide circuit with negligible insertion loss and PDL penalty, as demonstrated in the example of a broadband polarization splitter.

Active tuning of birefringence has also been demonstrated using a piezoelectric cladding film to enhance the conversion efficiency in optical parametric processes. On a related subject, cladding stress can be applied to break the inversion symmetry of the silicon crystal and introduce a Pockels electro-optic effect. This area holds much promise in making silicon a truly versatile photonic material. Cladding stress engineering can be envisioned in a variety of situations to enhance device functionalities, simplify fabrication, and improve operation tolerance.

Acknowledgments The author gratefully acknowledges Dr. Siegfried Janz, Dr. Winnie Ye, Dr. Pavel Cheben, Dr. André Delâge, and Dr. Philip Waldron for their collaboration through out this work, and for many insightful discussions and suggestions in preparing this manuscript. The contribution of Dr. Winnie Ye during her Ph.D. thesis work is particularly recognized. Acknowledgments are also due to Dr. Juan Caballero, Dr. Dan Dalacu, Dr. Boris Lamontagne, Ms. Marie-Josée Picard, and Ms. Edith Post for their contribution in device fabrication and testing. The author also acknowledges the Institute for Microstructual Sciences, National Research Council Canada, for providing the opportunity and facilities to carry out this research.

References

1. C. Poole, J. Nagel, Polarization effects in lightwave systems. Opt. Fiber Telecommun. **IIIA** (1997) 31, 32
2. J. Damask, *Polarization Optics in Telecommunications* (Springer, New York, 2005) 31, 32
3. K.P. Yap, A. Delâge, B. Lamontagne, S. Janz, D.-X. Xu, L. Lapointe, P. Waldron, J. Schmid, P. Chow-Chong, E. Post, B. Syrett, Scattering loss measurement of SOI waveguides using 5X17 integrated optical star coupler, in *Photonics West* (SPIE, San Jose, 2007), p. 64770J. 31, 35
4. M. Krause, H. Renner, E. Brinkmeyer, Polarization-dependent curvature loss in silicon rib waveguides. IEEE J. Sel. Top. Quantum Electron. **12**, 1359–1362 (2006) 31, 35, 55
5. M. Quintela, F. Madruga, C. Jauregui, J. Lopez-Higuera, S-EDFA and R-EDFA polarization properties comparison. Opt Commun. **255**, 72–80 (2005) 32
6. G. Konnen, *Polarized Light in Nature* (Cambridge University Press, Cambridge, 1985) 32

7. L. Vivien, S. Laval, B. Dumont, S. Lardenois, A. Koster, E. Cassan, Polarization-independent single-mode rib waveguides on silicon-on-insulator for telecommunication wavelengths. Opt. Commun. **210**, 43–49 (2002) 32, 37, 40
8. D. Dai, S. He, Analysis of the birefringence of a silicon-on-insulator rib waveguide. Appl. Opt. **43**, 1156–1161 (2004) 32, 37
9. W. Headley, G. Reed, S. Howe, A. Liu, M. Paniccia, Polarization-independent optical racetrack resonators using rib waveguides on silicon-on-insulator. Appl. Phys. Lett. **85**, 5523–5525 (2004) 32, 54, 55
10. I. Kiyat, A. Aydinli, N. Dagli, Polarization characteristics of compact SOI rib waveguide racetrack resonators. IEEE Photon. Technol. Lett. **17**, 2098–2100 (2005) 32, 54
11. D.-X. Xu, P. Cheben, D. Dalacu, A. Delage, S. Janz, B. Lamontagne, M.J. Picard, W.N. Ye, Eliminating the birefringence in silicon-on-insulator ridge waveguides by use of cladding stress. Opt. Lett. **29**, 2384–2386 (2004) 32, 33, 35, 36, 52, 53
12. W.N. Ye, D.-X. Xu, S. Janz, P. Cheben, M.J. Picard, B. Lamontagne, N.G. Tarr, Birefringence Control using stress engineering in silicon-on-insulator (SOI) waveguides. J. Lightwave Technol. **23**, 1308–1318 (2005) 32, 33, 38, 43, 44, 55
13. D.-X. Xu, S. Janz, P. Cheben, Design of polarization-insensitive ring resonators in silicon-on-insulator using MMI couplers and cladding stress engineering. IEEE Photon. Technol. Lett. **18**, 343–345 (2006) 32, 33, 36, 52, 54, 56
14. D.-X. Xu, P. Cheben, A. Delage, S. Janz, B. Lamontagne, E. Post, W.N. Ye, Polarization-insensitive MMI-coupled ring resonators in silicon-on-insulator using cladding stress engineering, in *Photonics West*(SPIE, San Jose, 2007), p. 64770D 32, 33, 35, 52, 54, 55, 56
15. N. Feng, D. Feng, H. Liang, W. Qian, C. Kung, J. Fong, M. Asghari, Low-loss polarization-insensitive silicon-on-insulator-based WDM filter for triplexer applications. IEEE Photon. Technol. Lett. **20**, 1968–1970 (2008) 32, 33, 52, 57, 58
16. K. Voigt, L. Zimmermann, G. Winzer, T. Mitze, J. Bruns, K. Petermann, B. Huttl, C. Schubert, Performance of 40-Gb/s DPSK demodulator in SOI-technology. IEEE Photon. Technol. Lett. **20**, 614–616 (2008) 32, 33, 52, 56, 57
17. K. Tsia, S. Fathpour, B. Jalali, Electrical control of parametric processes in silicon waveguides. Opt. Express **16**, 9838–9843 (2008) 32, 62, 63
18. D.-X. Xu, W.N. Ye, S. Janz, A. Delage, P. Cheben, B. Lamontagne, E. Post, P. Waldron, Stress induced effects for advanced polarization control in silicon photonics components. Adv. Opt. Technol. **2008**, Article ID 689715 (2008) 33
19. M. Huang, Stress effects on the performance of optical waveguides. Int. J. Solids Struct. **40**, 1615–1632 (2003) 33, 40, 41
20. W.N. Ye, D.-X. Xu, S. Janz, P. Cheben, A. Delage, M.J. Picard, B. Lamontagne, N.G. Tarr, Stress-induced birefringence in silicon-on-insulator (SOI) waveguides, in *Optoelectronic Integration on Silicon* (San Jose, 2004), p. 57 33
21. D.-X. Xu, P. Cheben, D. Dalacu, S. Janz, M. Picard, N.G. Tarr, W.N. Ye, Control and compensation of birefringence in SOI waveguides, in *IEEE LEOS 16th Annuanl Meeting* (Tucson, 2003) 33
22. M. Milosevic, P. Matavulj, B. Timotijevic, G. Reed, G. Mashanovich, Design rules for single-mode and polarization-independent silicon-on-insulator rib waveguides using stress engineering. J. Lightwave Technol. **26**, 1840–1846 (2008) 33, 37
23. K. Voigt, L. Zimmermann, G. Winzer, T. Mitze, J. Bruns, K. Petermann, C. Schubert, SOI delay interferometer with tuned polarization dependent wavelength shift for 40 Gbit/s DPSK demodulation, Group IV Photonics, Proceedings of the 4th IEEE International Conference, Tokyo (2007), pp. 1–3 33
24. V. Raghunathan, B. Jalali, Stress-induced phase matching in silicon waveguides, in *Conference of Lasers and Electro-Optics (CLEO)* (Long Beach, 2006), p. CMK5 33, 62
25. B. Jalali, V. Raghunathan, D. Dimitropoulos, O. Boyraz, Raman-based silicon photonics, IEEE J. Sel. Top. Quantum Electron. **12**, 412–421 (2006) 33, 62

26. W.N. Ye, D.-X. Xu, S. Janz, P. Waldron, P. Cheben, N.G. Tarr, Passive broadband silicon-on-insulator polarization splitter. Opt. Lett. **32**, 1492–1494 (2007) 33, 52, 59
27. P. Cheben, J. Schmid, A. Delage, A. Densmore, S. Janz, B. Lamontagne, J. Lapointe, E. Post, P. Waldron, D.-X. Xu, A high-resolution silicon-on-insulator arrayed waveguide grating microspectrometer with sub-micrometer aperture waveguides. Opt. Express **15**, 2299–2306 (2007) 34
28. P. Cheben, A. Delage, S. Janz, D.-X. Xu, *Echelle Gratings and Arrayed Waveguide Gratings for WDM and Spectral Analysis* (SPIE, Billingham, 2008) 34, 52
29. R. Soref, J. Schmidtchen, K. Petermann, Large single-mode rib waveguides in GeSi-Si and Si-on-SiO 2. IEEE J. Quantum Electron. **27**, 1971–1974 (1991) 34
30. S. Pogossian, L. Vescan, A. Vonsovici, The single-mode condition for semiconductor rib waveguides with large cross section. J. Lightwave Technol. **16**, 1851 (1998) 34
31. O. Powell, Single-mode condition for silicon rib waveguides. J. Lightwave Technol. **20**, 1851 (2002) 34
32. J. Xia, J. Yu, Single-mode condition for silicon rib waveguides with trapezoidal cross-section. Opt. Commun. **230**, 253–257 (2004) 34
33. J. Lousteau, D. Furniss, A. Seddon, T. Benson, A. Vukovic, P. Sewell, The single-mode condition for silicon-on-insulator optical rib waveguides with large cross section. J. Lightwave Technol. **22**, 1923 (2004) 34, 35
34. S. Chan, C. Png, S. Lim, G. Reed, V. Passaro, Single-mode and polarization-independent silicon-on-insulator waveguides with small cross section. J. Lightwave Technol. **23**, 2103 (2005) 34, 37
35. D.-X. Xu, W.N. Ye, A. Bogdanov, D. Dalacu, A. Delage, P. Cheben, S. Janz, B. Lamontagne, M.J. Picard, Design of polarization-insensitive components using geometrical and stress-induced birefringence in SOI waveguides, in *Photonics West* (SPIE, San Jose, 2005), p. 158 35, 36, 52, 54
36. B. Luff, D. Feng, D. Lee, W. Qian, H. Liang, M. Asghari, Hybrid silicon photonics for low-cost high-bandwidth link applications. Adv. Opt. Technol. **2008**, Article ID 245131 (2008) 35
37. J. Canning, Birefringence control in planar waveguides using doped top layers. Opt. Commun. **191**, 225–228 (2001) 36
38. L. de Peralta, A. Bernussi, H. Temkin, M. Borhani, D. Doucette, Silicon-dioxide waveguides with low birefringence. IEEE J. Quantum Electron. **39**, 874–879 (2003) 36
39. A. Kilian, J. Kirchhof, B. Kuhlow, G. Przyrembel, W. Wischmann, Birefringence free planar optical waveguide made by flame hydrolysis deposition (FHD) through tailoring of the overcladding. J. Lightwave Technol. **18**, 193 (2000) 36
40. C. Nadler, M. Lanker, E. Wildermuth, W. Hunziker, H. Melchior, Polarisation insensitive wavelength multiplexers using stress release grooves, *Proceedings of the ECOC 98*, Madrid (1998) 36
41. S. Ojha, C. Cureton, T. Bricheno, S. Day, D. Moule, A. Bell, J. Taylor, Simple method of fabricating polarisation-insensitive and very lowcrosstalk AWG grating devices. Electron. Lett. **34**, 78–79 (1998) 36
42. C. Doerr, L. Buhl, W. Lin, Simple method for mitigation of polarization crosstalk in silica planar lightwave circuit directional couplers. IEEE Photon. Technol. Lett. **18**, 1816–1818 (2006) 36
43. COMSOL Multiphysics, http://www.comsol.com/products/multiphysics/ 36
44. D.-X. Xu, J. Baribeau, P. Cheben, D. Dalacu, A. Delage, B. Lamontagne, S. Janz, M. Picard, W.N. Ye, Prospects and challenges for microphotonic waveguide components based on Si and SiGe, in *Electrochemical Society annual meeting* (ECS, Hawaii, 2004), pp. 619–633 38
45. T. Narasimhamurty, *Photoelastic and Electro-Optic Properties of Crystals* (Plenum Pub Corp, New York, 1981) 40
46. J. Xu, R. Stroud, *Acousto-Optic Devices: Principles, Design, and Applications* (Wiley, New York, 1992) 40, 41, 42
47. E. Palik, *Handbook of Optical Constants of Solids* (Academic press, New York, 1985) 41, 42
48. K. Gomi, Y. Niitsu, Effect of crystalline orientation on photoelastic constant of Si single crystal Proc. SPIE **2873** (1996), 238 42

49. M. Huang, S. Dev, T. Inc, T. Dallas, The influence of light propagation direction on the stress-induced polarization dependence of silicon waveguides. IEEE Photon. Technol. Lett. **18**, 1314–1316 (2006) 42
50. S. He, T. Zheng, S. Danyluk, Analysis and determination of the stress-optic coefficients of thin single crystal silicon samples. J. Appl. Phys. **96**, 3103 (2004) 42
51. V. Au, C. Charles, R. Boswell, Interface creation and stress dynamics in plasma-deposited silicon dioxide films. Appl. Phys. Lett. **88**, 234103 (2006) 42
52. K. Okamoto, *Fundamentals of Optical Waveguides* (Academic press, New York, 2006) 43
53. M. Okuno, A. Sugita, K. Jinguji, M. Kawachi, Birefringence control of silica waveguides on Si and itsapplication to a polarization-beam splitter/switch. J. Lightwave Technol. **12**, 625–633 (1994) 43
54. X. Zhao, Y. Xu, C. Li, Birefringence control in optical planar waveguides. J. Lightwave Technol. **21**, 2352–2357 (2003) 43
55. H. Liu, S. Murarka, Elastic and viscoelastic analysis of stress in thin films. J. Appl. Phys. **72**, 3458–3463 (1992) 44
56. S. Murarka, *Metallization Theory and Practice for VLSI and ULSI* (Butterworth-Heinemann, Northamptonshire, 1992), P. 270 44
57. A. Delage, P. Cheben, A. Densmore, S. Janz, B. Lamontagne, J. Lapointe, J. Schmid, P. Waldron, D.-X. Xu, Application of optical telecom planar technology to silicon-based microphotonics and sensors, in *ICTON* (Nottingham, 2006), p. Tu.D2.2 48
58. B. Almashary, H. Kim, Stress-induced vertical confinement of light in bulk GaAs and Si substrates. IEEE Photon. Technol. Lett. **8**, 1358–1360 (1996) 52
59. L. Yu, Z. Guan, Q. Liu, S. Lau, Silicon on insulator photoelastic optical waveguide and polarizer. Appl. Phys. Lett. **66**, 2016–2018 (1995) 52
60. O. Burke, I. Evans, H. Arnaud, B. Luff, Efficient polarization conversion in silicon-on-insulator waveguides, OSA trends in optics and photonics, Integrated Photonics Research Technical Digest **78**, paper IThA3 (2002) 52
61. D.-X. Xu, P. Cheben, D. Dalacu, S. Janz, M. Picard, N.G. Tarr, W.N. Ye, Control and compensation of birefringence in SOI waveguides, in *Proceedings of the 16th Annual Meeting of the IEEE Lasers & Electro-Optics Society* (*LEOS*) (LEOS, Tucson, 2003) 52, 60, 61
62. D.-X. Xu, P. Cheben, S. Janz, D. Dalacu, Control of SOI waveguide polarization properties for microphotonic applications, in *CLEO Pacific Rim* (OSA, Taipei, 2003), p. TU3I 52
63. Y. Inoue, H. Takahashi, S. Ando, T. Sawada, A. Himeno, M. Kawachi, Elimination of polarization sensitivity in silica-based wavelengthdivision multiplexer using a polyimide half waveplate. J. Lightwave Technol. **15**, 1947–1957 (1997) 53
64. J. He, E. Koteles, B. Lamontagne, L. Erickson, A. Delage, M. Davies, Integrated polarization compensator for WDM waveguidedemultiplexers. IEEE Photon. Technol. Lett. **11**, 224–226 (1999) 53
65. P. Cheben, A. Bezinger, A. Delage, L. Erickson, S. Janz, D.-X. Xu, Polarization compensation in silicon-on-insulator arrayed waveguide grating devices, in *Photonics West*(San Jose, 2001), p. 15 53
66. P. Cheben, *Wavelength Dispersive Planar Waveguide Devices: Echelle gratings and Arrayed Waveguide Gratings* (CRC Press, London, 2007)53
67. Y. Kokubun, S. Kubota, S. Chu, Polarisation-independent vertically coupled microring resonatorfilter. Electron. Lett. **37**, 90–92 (2001) 54
68. M. Milosevic, P. Matavulj, G. Mashanovich, Stress-induced characteristics of silicon-on-insulator rib waveguides, in *Proceedings of the 15th Telecommunications Forum TELFOR* (Serbia, Belgrade, 2007), pp. 401–404 54
69. A. Yariv, Universal relations for coupling of optical power betweenmicroresonators and dielectric waveguides. Electron. Lett. **36**, 321–322 (2000) 54
70. D.-X. Xu, A. Densmore, P. Waldron, J. Lapointe, E. Post, A. Delage, S. Janz, P. Cheben, J. H. Schmid, B. Lamontagne, High bandwidth SOI photonic wire ring resonators using MMI couplers. Opt. Express **15**, 3149–3155 (2007) 54

71. M. Bachmann, P. Besse, H. Melchoir, General self-imaging properties in N × N multimode interference couplers including phase relations. Appl. Optics **33**, 3905–3911 (1994) 55
72. L. Soldano, E. Pennings, Optical multi-mode interference devices based on self-imaging: principles and applications. J. Lightwave Technol. **13**, 615–627 (1995) 55
73. W. Headley, G. Reed, A. Liu, M. Paniccia, S. Howe, Polarization-independent optical ring resonators on silicon-on-insulator. Proc. SPIE **5451**, 276 (2004) 55
74. R. Halir, I. Molina-Fernandez, A. Ortega-Monux, J. Wanguemert-Perez, D.-X. Xu, P. Cheben, S. Janz, A design procedure for high-performance, rib-waveguide-based multimode interference couplers in silicon-on-insulator. J. Lightwave Technol. **26**, 2928–2936 (2008) 55
75. D.-X. Xu, A. Delage, S. Janz, P. Cheben, Broadband polarization compensation in silicon on insulator components, in *Photonics North* (SPIE, Montreal, 2008), p. 70991O 56
76. S.M. Weiss, M. Molinari, P. Fauchet, Temperature stability for silicon-based photonic band-gap structures. Appl. Phys. Lett. **83**, 1980–1982 (2003) 57
77. Y. Liao, R. Lu, C. Yang, W. Wang, Passive Ni: LiNbO 3 polarisation splitter at 1.3 μm wavelength. Electron. Lett. **32**, 1003–1005 (1996) 58
78. J. Hong, H. Ryu, S. Park, J. Jeong, S. Lee, E. Lee, S. Park, D. Woo, S. Kim, O. Beom-Hoan, Design and fabrication of a significantly shortened multimode interference coupler for polarization splitter application. IEEE Photon. Technol. Lett. **15**, 72–74 (2003) 58
79. I. Kiyat, A. Aydinli, N. Dagli, A compact silicon-on-insulator polarization splitter, IEEE Photon. Technol. Lett. **17**, 100–102 (2005) 58
80. L. Soldano, A. de Vreede, M. Smit, B. Verbeek, E. Metaal, F. Green, Mach-Zehnder interferometer polarization splitter in InGaAsP/InP. IEEE Photon. Technol. Lett. **6**, 402–405 (1994) 58
81. T. Liang, H. Tsang, Integrated polarization beam splitter in high index contrast silicon-on-insulator waveguides. IEEE Photon. Technol. Lett. **17**, 393–395 (2005) 58
82. W.N. Ye, D.-X. Xu, S. Janz, P. Waldron, N.G. Tarr, Stress-induced SOI polarization splitter based on Mach-Zehnder Interferometers (MZI), in *IEEE Group Four Photonics Conference* (IEEE, Ottawa, 2006), pp. 249–251 58, 60
83. A. Vellekoop, M. Smit, A small-size polarization splitter based on a planar optical phasedarray. J. Lightwave Technol. **8**, 118–124 (1990) 58
84. J. Schrauwen, D. Van Thourhout, R. Baets, Trimming of silicon ring resonator by electron beam induced compaction and strain. Opt. Express **16**, 3738–3743 (2008) 60, 61
85. A. Sugita, K. Jinguji, N. Takato, M. Kawachi, Laser-trimming adjustment of waveguide birefringence in optical FDMcomponents. IEEE J. Sel. Areas Commun. **8**, 1128–1131 (1990) 62
86. R. Dekker, N. Usechak, M. Forst, A. Driessen, Ultrafast nonlinear all-optical processes in silicon-on-insulator waveguides. J. Phys. D: Appl. Phys. **40**, R249–R271 (2007) 62
87. D. Dimitropoulos, V. Raghunathan, R. Claps, B. Jalali, Phase-matching and nonlinear optical processes in silicon waveguides. Opt. Express **12**, 149–160 (2004) 62
88. K. Tsia, S. Fathpour, B. Jalali, Electrical tuning of birefringence in silicon waveguides, Appl. Phys. Lett. **92**, 061109 (2008) 62
89. R. Jacobsen, K. Andersen, P. Borel, J. Fage-Pedersen, L. Frandsen, O. Hansen, M. Kristensen, A. Lavrinenko, G. Moulin, H. Ou, Strained silicon as a new electro-optic material. Nature **441**, 199–202 (2006) 63, 64

Index

Chapter 3
Interfacing Silicon Nanophotonic Integrated Circuits and Single-Mode Optical Fibers with Diffraction Gratings

Günther Roelkens and Dries Van Thourhout

Abstract Exploiting the high refractive index contrast achievable on the silicon-on-insulator (SOI) material platform enables a drastic scaling down of the footprint of integrated optical functions. While this allows for large-scale integration of optical functions, it severely complicates the interfacing with a single-mode optical fiber, due to the huge mismatch in mode size between the optical fiber and nanophotonic silicon waveguide. In this chapter we elaborate on the use of diffraction gratings to achieve an efficient, compact, alignment-tolerant, polarization-independent, and broadband optical coupling. Besides using these components as an interface with an optical fiber, its use to interface with opto-electronic components will be discussed. Finally, an optical probe will be presented, based on a diffraction grating integrated on the facet of a single-mode fiber. Such an optical probe allows testing of individual components in a silicon-on-insulator nanophotonic integrated circuit.

3.1 Nanophotonic SOI Waveguide Circuits

Silicon-on-insulator (SOI) is emerging as the most important platform for large-scale photonic integrated circuits. This is due to the high refractive index contrast between the silicon waveguide core and SiO_2 or air cladding. This huge refractive index contrast allows a drastic reduction in the size of individual optical functions, bringing large-scale integrated optical circuits closer to reality. Moreover, these photonic integrated circuits can be fabricated using standard CMOS technology on 200-mm or 300-mm wafers [1]. This allows for low-cost mass manufacturing and leverages from the billion dollar investments done in the development of stable

G. Roelkens (✉)
Photonics Research Group, Department of Information Technology, Ghent University/IMEC, B-9000 Ghent, Belgium
e-mail: Gunther.Roelkens@intec.ugent.be

D. Van Thourhout (✉)
Photonics Research Group, Department of Information Technology, Ghent University/IMEC, B-9000 Ghent, Belgium
e-mail: Dries.Vanthourhout@intec.ugent.be

D.J. Lockwood, L. Pavesi (Eds.): *Silicon Photonics II*.
Topics in Applied Physics **119**, 71–94 (2011)
DOI 10.1007/978-3-642-10506-7_3

and high-yield processes for CMOS manufacturing. While the degree of scaling is a clear merit of the use of high refractive index contrast waveguide systems like silicon-on-insulator, it severely complicates the interfacing with an optical fiber. This is clear from the huge mismatch between the modal area of a single-mode optical fiber ($\approx$100 μm^2) and that of a nanophotonic monomode waveguide ($\approx$0.1 μm^2). This is perhaps one of the most important issues that high-index contrast waveguide circuits face for practical applications.

3.2 Solutions to the Fiber-Chip Coupling Problem

Various solutions to the fiber-chip coupling problem have been proposed in the literature. The most common solution implies the use of a spot size converter to transform the size of the mode of the nanophotonic waveguide to that of the optical fiber. In principal, a three-dimensional adiabatic taper structure can be used for this purpose, both laterally and vertically transforming the size of the silicon waveguide [2]. The technology required to do this is, however, not CMOS compatible, thereby compromising its use in practical applications. A second approach to change the spot size on the photonic integrated circuit is to use a lateral inverted taper structure. In these structures, the width of the silicon nanophotonic waveguide is tapered down to sub 100-nm dimensions (which can be achieved using state-of-the-art deep UV lithography used in CMOS fabrication), thereby expanding the size of the optical mode. By realizing a low refractive index waveguide on top of this inverted taper structure, very efficient coupling can be obtained between the nanophotonic waveguide and the low refractive index waveguide [3]. However, to interface with a single-mode fiber, this requires a low refractive index waveguide with cross-sectional dimensions comparable to that of the single-mode fiber core. This approach therefore suffers from two important drawbacks: the integration of such a large core optical waveguide requires a huge topography (to CMOS standards) on the photonic integrated circuit, compromising further processing. Moreover, to adiabatically transform the mode from a nanophotonic silicon waveguide to such a large core waveguide would require taper structures of thousands of micrometers in length, thereby canceling out the advantages of using a high-index contrast waveguide system for increasing the density of optical functions on the SOI chip. This is why in current implementations, the use of inverted adiabatic taper structures is limited to interfacing with a lensed optical fiber, since this reduces the topography on the integrated circuit to about 3 μm (which is still large, but manageable) and reduces the taper length to hundreds of microns, a more typical length scale in high-index contrast waveguide circuits. This approach allows for a high efficiency (an insertion loss lower than 1 dB has been shown) and large optical bandwidth (typically several hundreds of nanometers) coupling between a lensed fiber and a nanophotonic integrated circuit [3]. It however requires the use of specialty optical fiber and the associated small alignment tolerances. Besides the requirement of a specialty optical fiber, maybe the most important issue with this approach is the requirement of a cleaved and polished facet to interface with the photonic integrated circuit. While this is the standard approach for current photonic integrated circuits (like planar lightwave circuits, edge-emitting laser diodes, etc.), the advent of large-scale photonic integrated circuits requires a fiber coupling approach that does not require singulated chips for optical interfacing, e.g., for optical testing on a wafer scale, comparable to the wafer scale electrical testing in electronic integrated circuits.

Therefore the use of a new type of optical interface between a standard single-mode fiber and nanophotonic waveguide circuit using diffractive grating structures was proposed [4]. In these coupling structures, the lateral adaptation of the mode size is achieved by a conventional in-plane taper structure. The interfacing with the optical fiber, however, is done by redirecting the light exiting the photonic integrated circuit out of the plane of the photonic integrated circuit, using a diffraction grating (vice versa for optical coupling to the photonic integrated circuit). In this way the optical fiber interfaces with the silicon photonic integrated circuit from the top, no longer requiring a singulated chip for coupling. This approach even enables packaging on the wafer level, by integrating fiber alignment structures on the wafer level. These two aspects, namely the possibility of wafer scale testing and wafer level packaging of photonic integrated circuits, could enable a dramatic reduction in the cost of the photonic integrated circuit, since these are two of the most cost-intensive operations in the whole fabrication cycle of the device. In the subsequent sections we will outline the operation principle, the design and the experimental realization of high efficiency, wide band and polarization independent grating coupler structures. We will show that, besides their use as fiber-chip couplers, they can also be used to interface a silicon photonic integrated circuit with a (flip-chipped) opto-electronic component. In Fig. 3.1, the three approaches discussed above for fiber-chip coupling are schematically depicted.

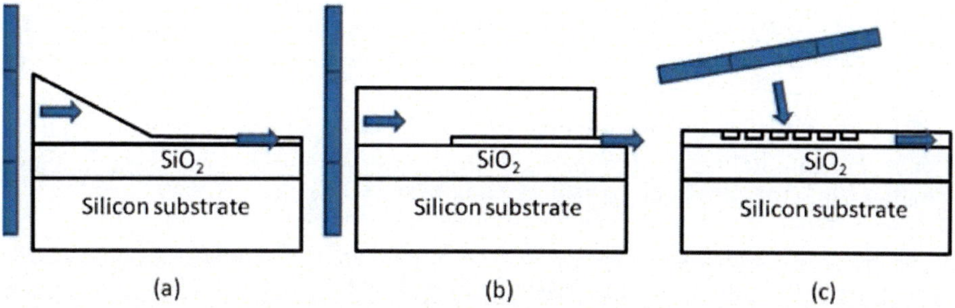

Fig. 3.1 Comparison of two spot-size converter approaches (3D adiabatic tapering in (**a**), lateral inverted tapers in (**b**)) to interface with an optical fiber and (**c**) the diffractive grating based coupling

3.3 Fundamentals of Fiber-Chip Diffraction Grating Couplers

In its most simple form, a diffractive grating coupler consists of a one-dimensional periodic structure defined on or in the silicon waveguide layer (later on we will show that also two-dimensional periodic structures can be used), as shown in Fig. 3.2a.

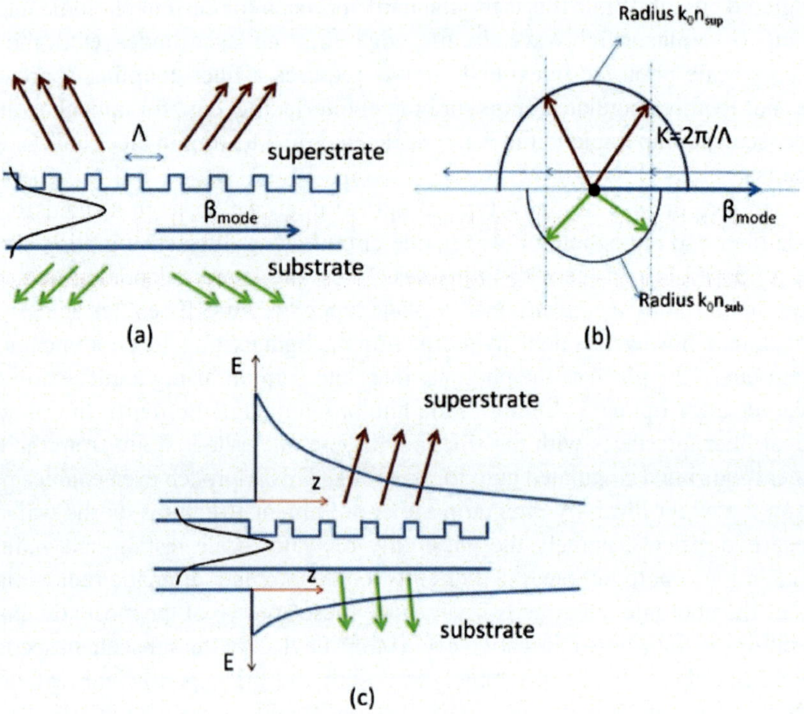

Fig. 3.2 Fundamental operation principle of a one-dimensional diffraction grating structure: (**a**) device layout, (**b**) geometric representation of the projected Bragg condition, and (**c**) the exponential decay of the diffracted field profile

The periodicity implies that light is diffracted from these structures in preferential directions. This can be understood by looking at the one-dimensional grating structure as a collection of scatter centers (located at the interface between the slits and teeth of the grating), which have a fixed phase relation due to the periodicity of the structure. This fixed phase relation results in a strongly angle-dependent interference. For infinitely extending periodic structures, this interference behavior can be mathematically described by the projected Bragg condition

$$\boldsymbol{k}_{z,m} = \boldsymbol{\beta}_{\mathbf{mode}} + m\boldsymbol{K} \tag{3.1}$$

with m an integer. In this formula $\boldsymbol{k}_{z,m}$ is the z-component of the wave vector of the diffracted wave of the mth order (which arises due to the fact that all scatter centers interfere constructively in a particular direction). $\boldsymbol{\beta}_{\mathbf{mode}}$ is the effective wave vector of the optical mode in the silicon grating. In low-index contrast grating structures this is approximately the propagation constant of the guided mode in the uncorrugated waveguide section. In a high refractive index contrast structure this perturbation analysis is no longer valid and the effective propagation constant is an average of the propagation constant of the guided mode in the unetched and etched waveguide region. $\boldsymbol{K}$ is the reciprocal lattice vector of the periodic structure. Its length is given by $K = \frac{2\pi}{\Lambda}$ with Λ the period of the diffraction grating.

The projected Bragg condition directly determines the angle of diffraction, since $k_z = \frac{2\pi n \sin\theta_0}{\lambda}$, with n the refractive index of the medium in which the diffracted wave is propagating. Therefore the diffraction angle θ_0 is determined by

$$\sin\theta_0 = \frac{n_{\text{eff}}(\lambda)}{n} + m\frac{\lambda}{n\Lambda} \tag{3.2}$$

since $\beta_{\text{mode}} = \frac{2\pi n_{\text{eff}}(\lambda)}{\lambda}$. The right hand side of this equation has to be in between -1 and 1 for diffraction to occur. By properly choosing the grating period Λ (for a given λ,$n_{\text{eff}}(\lambda)$ and n) one can realize a grating in which diffraction occurs into only one diffraction order, i.e., only the diffraction order $m = -1$ is present. This is the preferred situation for the fiber-chip grating couplers, which we will describe in this chapter, since it allows efficient collection of the light diffracted from the grating, by positioning the single-mode optical fiber on top of the grating structure at an angle corresponding to the diffraction angle of this single diffraction order. A geometrical representation of this projected Bragg condition is shown in Fig. 3.2b.

The choice of the diffraction angle depends on the particular application. Typically an angle between 8 and 15° off vertical is chosen. The reason why an off-surface normal coupling angle is chosen becomes clear from the Bragg diffraction. To obtain the condition of perfectly vertical coupling to the chip, the length of the reciprocal lattice vector has to obey

$$K = \beta_{\text{mode}} \tag{3.3}$$

since $m = -1$. Although this condition indeed produces only a single order of diffraction into the superstrate, a second-order reflection also occurs, back into the waveguide, since

$$k_{z,-2} = \beta_{\text{mode}} - 2K = -\beta_{\text{mode}} \tag{3.4}$$

The second-order reflection toward the photonic integrated circuit is unwanted and reduces the coupling efficiency to the optical fiber. Therefore a slightly off-normal operating grating coupler is typically used (a so-called detuned grating), since in this case no second-order Bragg reflection occurs. In Sect. 3.7, we will elaborate on how this performance reduction for vertical incidence can be avoided, for the integration of surface emitting opto-electronic components on the silicon photonics platform.

The Bragg formalism described above is only valid for infinitely extending structures. In the design of the fiber-chip grating couplers, one has to take into account the finite (or at least the semi-infinite) nature of the diffraction grating. When the optical mode in the silicon input waveguide is exciting the diffraction grating, optical power is coupled out of the waveguide layer during the propagation due to diffraction (besides some scattering at the interface between the access waveguide and the grating, due to a slight mismatch between the waveguide mode and the grating mode). This results in an exponentially decaying diffracted field profile (different from the plane wave considered in infinitely extending structures). This exponentially decaying field profile, as shown in Fig. 3.2c has two consequences: first of all, it can be considered as a superposition of plane waves that propagate at an angle θ centered around θ_0, with an amplitude distribution $A(\theta)$ determined by the Fourier transform of this exponentially decaying field profile. The extent of the angular distribution of these plane wave components is therefore inversely proportional to the decay length of the diffracted field. This implies that a grating with a strong refractive index contrast (resulting in a short decay length) generates a whole range of plane waves. Since the angle θ_0 only varies slowly with wavelength (typically 1°/10 nm), a large optical bandwidth coupling can be realized in high-index contrast grating structures, which will be discussed in more detail in Sect. 3.5. A second consequence of the exponentially decaying diffracted field profile is the fact that there is a mismatch with the Gaussian fiber mode in a single-mode optical fiber. This limits the achievable fiber coupling efficiency for a uniform grating structure. These aspects will be discussed in the following sections.

So far, the influence of the polarization of the incident light has not yet been considered. Due to the high refractive index contrast obtained in silicon-on-insulator waveguide circuits, these waveguides show a strong modal birefringence. This implies that the diffraction angle for transverse electric and transverse magnetic polarized light is substantially different, making these one-dimensional diffraction gratings very polarization dependent. In Sect. 3.6 we will describe, however, how we can utilize this strong polarization dependence to realize a polarization diversity approach based on two-dimensional diffraction gratings, effectively rendering the photonic integrated circuit polarization independent. For the one-dimensional grating structures described in the subsequent sections, transverse electric (TE) polarized light is always assumed, with the dominant electrical field component oriented along the grating lines.

3.4 High-Efficiency Fiber-Chip Grating Couplers

Simply calculating the overlap between the exponentially decaying field profile and the Gaussian mode profile of a single-mode fiber (with a $1/e^2$ intensity mode field diameter of 10.4 μm) shows that there is an optimal decay length in the diffracted field profile of $L = 7.6\,\mu$m, for which the overlap integral is maximal [4]. This results in a minimal insertion loss of about 1 dB for uniform grating couplers. While this theoretical minimal loss can be improved by the use of non-uniform diffraction gratings, where the coupling strength of the grating structure varies along the grating, this is not considered in this chapter, since it requires the definition of features which cannot be defined and/or well controlled by state-of-the-art deep UV lithography.

While there is a 1 dB theoretical limit in the achievable fiber coupling efficiency for one-dimensional uniform diffraction grating structures, there is another, more prominent, loss mechanism in these devices. So far, we have only considered the light that is diffracted toward the optical fiber. However, since the refractive index of the buried oxide layer, separating the silicon waveguide layer from the silicon substrate, has a comparable index to that of the superstrate (either air, SiO_2 or index matching glue), diffraction also occurs toward the silicon substrate, as is clear from Fig. 3.2. This limits the directionality of the grating structure (being defined as the ratio of the optical power diffracted toward the optical fiber to the total diffracted optical power). In a basic diffraction grating structure, where the one-dimensional grating is defined by etching in the silicon waveguide layer, typically a 50% directionality is obtained, implying that only a maximum coupling efficiency of −4 dB (−3 dB from the limited directionality and −1 dB from the mode profile mismatch) can be obtained. Therefore, tackling this problem is, in a first phase, more critical than trying to match the mode profiles.

To improve the directionality, we envisaged two types of solutions, schematically outlined in Fig. 3.3. In one type of solution, the downward diffracted light from the grating was redirected upward by integrating a mirror below the grating structure. Both a metallic mirror and distributed Bragg reflector mirror were considered. While in this case a very high directionality is obtained, care has to be taken to optimize the distance between this bottom mirror and the grating. This is due to the fact that the

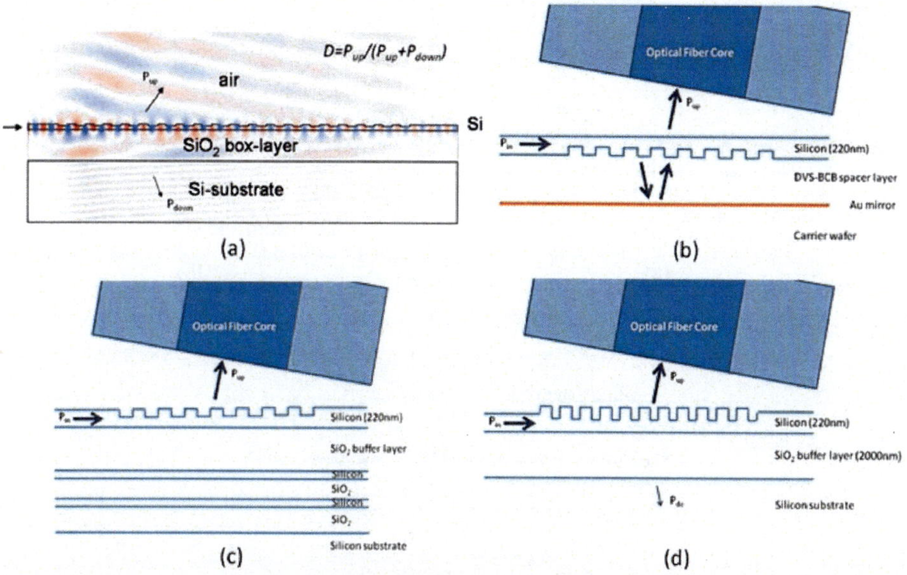

Fig. 3.3 Schematic outline of the different strategies pursued to realize high efficiency fiber-to-chip grating couplers: (**a**) the standard – moderate efficiency – grating coupler structure with the grating lines directly etched in the silicon waveguide layer, (**b**) the metallic and (**c**) DBR type bottom mirror approach, and (**d**) the silicon overlay approach

optical field reflected on the bottom mirror will interfere with the directly upward diffracted wave. This interference modifies the coupling strength of the grating (i.e., when there is destructive interference between both diffracted waves, little light will be coupled out of the grating and hence the coupling length will be very long). Since one still needs to obtain a grating coupling length matched to the dimensions of the single-mode optical fiber, care has to be taken in optimizing these structures.

Device optimization of the bottom mirror grating structures was performed using finite difference time domain (FDTD) and eigenmode expansion methods. To limit the required computational resources, simulations are limited to two-dimensional simulations. This is a good approximation for one-dimensional fiber-chip grating couplers, since the waveguides are in practice 12 μm wide (i.e., 24 material wavelengths) to interface with a single-mode fiber. While this shows that the diffracted field profile coming from a two-dimensional simulation corresponds well to the full three-dimensional simulation, still a correction factor of 0.97 (on a linear scale) in the fiber-chip coupling efficiency needs to be taken into account, due to the slight mismatch in lateral field profile between the silicon waveguide mode (approximately cosinusoidal) and the Gaussian fiber mode. In Fig. 3.4a and b, the electric field component (for TE polarized light) of the diffraction in each type of optimized grating structure is plotted.

Both a metallic bottom mirror-based fiber-chip grating coupler [5] and a DBR bottom mirror-based fiber-chip grating coupler [6] were experimentally realized. For the metallic mirror approach, a 50-nm-thick layer of gold was used. The bottom mirror diffraction grating was realized used a wafer bonding approach. In

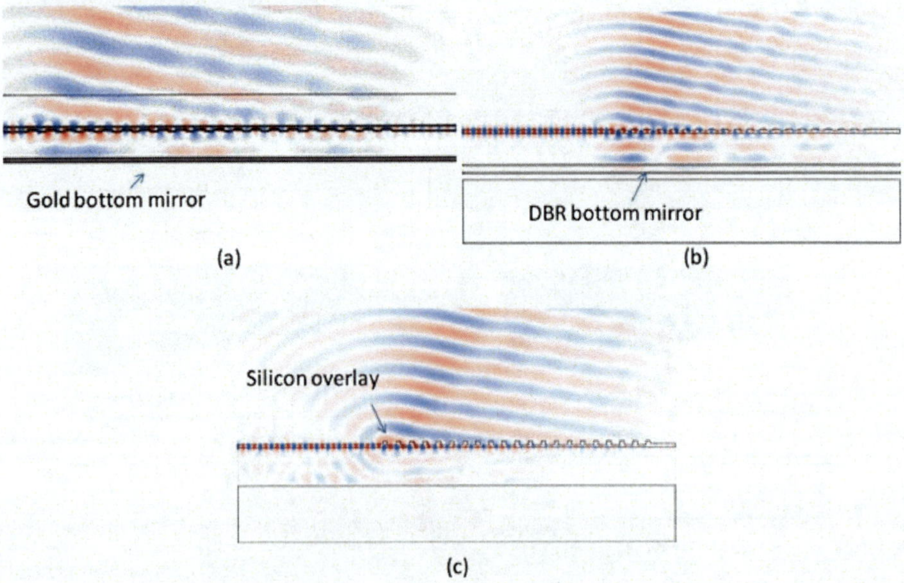

Fig. 3.4 Simulation results for the optimized grating structures for the different strategies pursued to realize high efficiency fiber-to-chip grating couplers: (**a**) gold bottom mirror approach, (**b**) DBR type bottom mirror approach, and (**c**) the silicon overlay approach

this approach a uniform one-dimensional grating structure was defined on an SOI waveguide wafer using the standard lithography and etching processes. Afterwards a polymer (DVS-BCB) spacer layer was applied, after which the gold mirror is defined on top of the grating coupler structure. Finally, the waveguide layer is transferred to a silicon carrier wafer substrate using a DVS-BCB bonding technique. After removal of the silicon substrate, using the buried oxide layer as an etch stop layer, the grating structure is accessible from the top by an optical fiber. While this approach allows the integration of a bottom mirror below a crystalline silicon waveguide layer, the layer transfer process required to achieve this is not standard CMOS technology. Therefore, a DBR bottom mirror approach was considered, building up the DBR/waveguide layer stack by means of chemical vapor deposition of silicon dioxide and amorphous silicon (a-Si) on a 200-mm wafer scale. In this case a two pair SiO_2/a-Si DBR mirror was used. The experimentally obtained fiber coupling efficiency was in both cases about -1.5 dB, approaching the theoretical limit of a uniform one-dimensional grating structure. The 1 dB optical bandwidth of these coupling structures is approximately 35 nm. In Fig. 3.5, microscope images and SEM images of the fabricated grating structures are shown, together with the experimentally obtained fiber coupling efficiency spectrum. As is clear from these results, a dramatic improvement in coupling efficiency can be obtained compared to a standard, directly etched grating coupler. The parameters of these fabricated devices are listed in Table 3.1 in the appendix to this chapter.

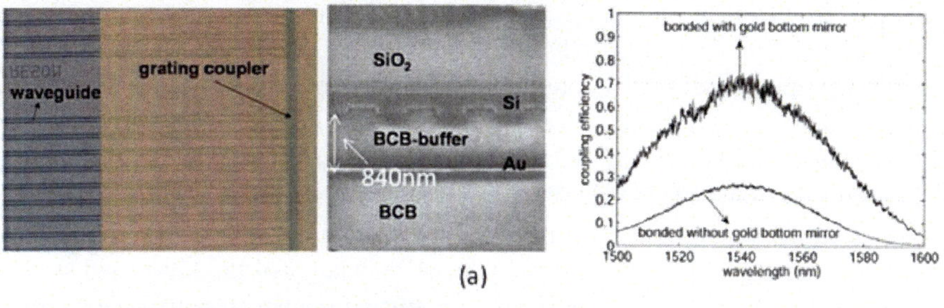

Fig. 3.5 Microscope image and SEM cross-section image of the fabricated high-efficiency grating coupler structures based on a bottom mirror approach, together with the experimentally obtained fiber coupling efficiency spectrum: (**a**) the metallic bottom mirror approach [5] and (**b**) the DBR bottom mirror approach [6]. The DBR consists of two 112-nm-thick a-Si layers with a 267-nm layer of SiO_2 in between

While the integration of a bottom mirror allows for high-efficiency couplers, the drawback is that either a wafer bonding approach is needed or that the silicon waveguide layer is an amorphous silicon layer (e.g., problematic for the integration of high speed modulators). Therefore, another approach was investigated allowing for high-efficiency coupling to a crystalline silicon waveguide layer, in which only CMOS technology is used for the fabrication. The mechanism here to realize high-efficiency coupling is to intrinsically modify the grating directionality by modifying its design. The proposed fabrication procedure consists of the definition of a silicon overlay mesa in the area where the grating will be defined, prior to etching the grating lines [7]. While the optimal design originated from a pure optimization process, the operation principle of the device can be understood as follows. Modifying the design of the grating by having this silicon overlay allows the grating to be designed in such a way that the different scattering centers (which have a fixed phase relation) interfere constructively in the direction of the optical fiber while destructive interference is obtained toward the substrate. Hence, an enhanced directionality can be obtained. Besides the high directionality, the coupling length also needs to be optimized to match with the dimensions of the Gaussian fiber mode (unless a lens system is used to transform the waist of the Gaussian beam as in [9]). Both requirements can be achieved by optimizing the silicon overlay thickness and grating etch depth. Simulations show that in this way efficiencies close to the −1 dB theoretical coupling limit can be obtained.

High-efficiency diffraction gratings based on the silicon overlay approach were fabricated using CMOS technology. The silicon overlay was defined using silicon epitaxial growth in an opened SiO_2 mask layer. Experimentally, a coupling efficiency of −2.6 dB was obtained with a 1 dB optical bandwidth of 50 nm [8]. This is below the expected −1 dB coupling efficiency due to the fact that the fabricated device dimensions deviated from the designed ones. Taking into account the dimensions of the actual fabricated structure, an excellent agreement is obtained between the two-dimensional FDTD simulations and the experimentally obtained fiber coupling spectrum, strengthening our belief that with an optimized fabrication process a −1 dB coupling efficiency is obtainable. SEM images of the fabricated structures and the experimentally obtained fiber coupling efficiency spectrum are shown in Fig. 3.6. The device parameters are listed in Table 3.1. While the optimal grating layer thickness is on the order of 350–400 nm, the optimal waveguide layer thickness still is lower (about 220 nm) in order to obtain a vertically single mode strip waveguide.

3.5 Multi-band Fiber-Chip Grating Couplers

The fiber-chip coupling structures that were outlined in the previous section all show a 1-dB (3 dB) optical bandwidth in the order of 40–50 nm (80–100 nm). While this is sufficient for applications where only a single wavelength band needs to interface with the photonic integrated circuit, a set of applications require a more broadband

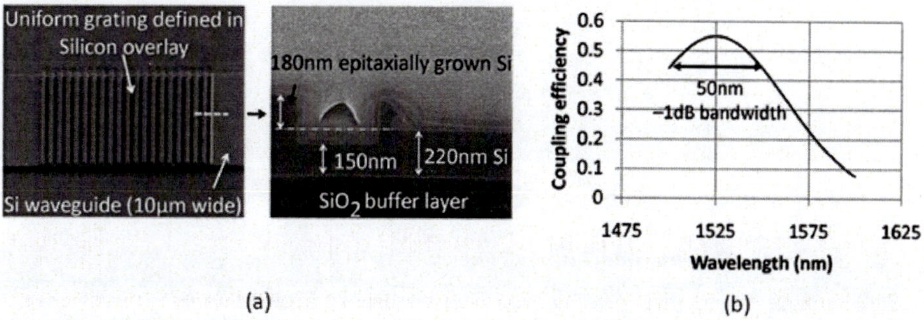

Fig. 3.6 (**a**) SEM images and (**b**) measured fiber-chip coupling efficiency spectrum of a silicon overlay-based grating coupler [8]

optical interface since optical signals in different wavelength bands need to be processed by the photonic integrated circuit. This is, for example, the case in integrated transceivers for optical access networks, where both a 1,310 nm optical signal (i.e., for upstream at the subscriber side) and a 1,490/1,550 nm optical signal (for the downstream analog and digital signal) needs to be coupled to the photonic integrated circuit.

This type of application can also benefit from the use of a diffraction grating to interface with an optical fiber. In this case both access waveguides to the one-dimensional grating structure are being used, as shown in Fig. 3.7a [10]. This allows the diffraction grating to be designed such that the Bragg condition is satisfied for both wavelength bands, by coupling one wavelength band in the forward direction and the second wavelength band in the opposite direction. This approach therefore allows not only to efficiently couple both wavelength bands to the photonic integrated circuit, it also realizes wavelength duplexing, which is required, for example, in the integrated transceiver application. For optimal performance, all device parameters need to be optimized: fiber tilt angle, grating period, grating etch depth, duty cycle, and number of grating periods. This can be achieved using dedicated optimization algorithms such as particle swarm or genetic optimization algorithms, written around the two-dimensional FDTD and eigenmode expansion simulations. The same methods as in the case of a single wavelength band grating coupler discussed in the previous section can be used to increase the fiber coupling efficiency. For example, using a silicon overlay approach to increase the directionality of the grating structures allows realizing a coupling efficiency of about −2.5 dB, for both a 1,310- and 1,490-nm wavelength as shown in Fig. 3.7b. The respective 3 dB optical bandwidth is 55 and 60 nm.

An important issue for these grating duplexer structures, especially when they are used in an upstream/downstream configuration, is the parasitic coupling of the (out-coupled) upstream wavelength channel into the downstream waveguide, resulting in cross-talk (see Fig. 3.7a). While in a grating structure that is optimized for optimal coupling efficiency for both wavelength bands a substantial fraction of the light

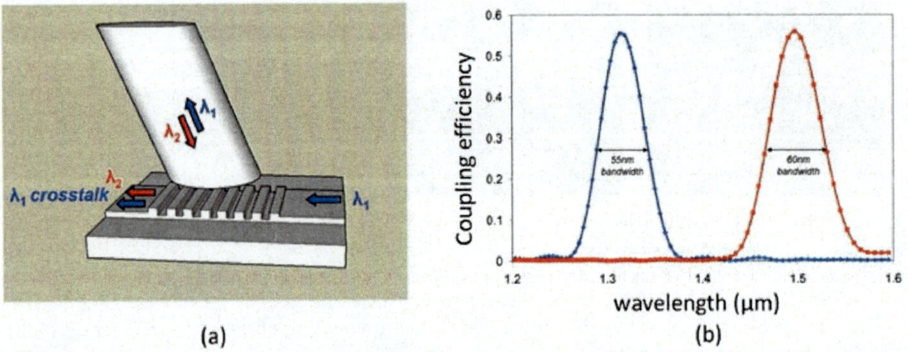

Fig. 3.7 (**a**) Concept of the one-dimensional grating duplexer structure and (**b**) simulated device performance in a silicon overlay grating configuration [10]

(typically −10 dB) is still coupled from the upstream waveguide channel into the downstream waveguide, this can be improved by incorporating a wavelength filter in the downstream wavelength path, in order to combine a high fiber-chip coupling efficiency and low crosstalk behavior. The combination of a one-dimensional grating duplexer and an optical filter to improve the crosstalk performance (and at the same

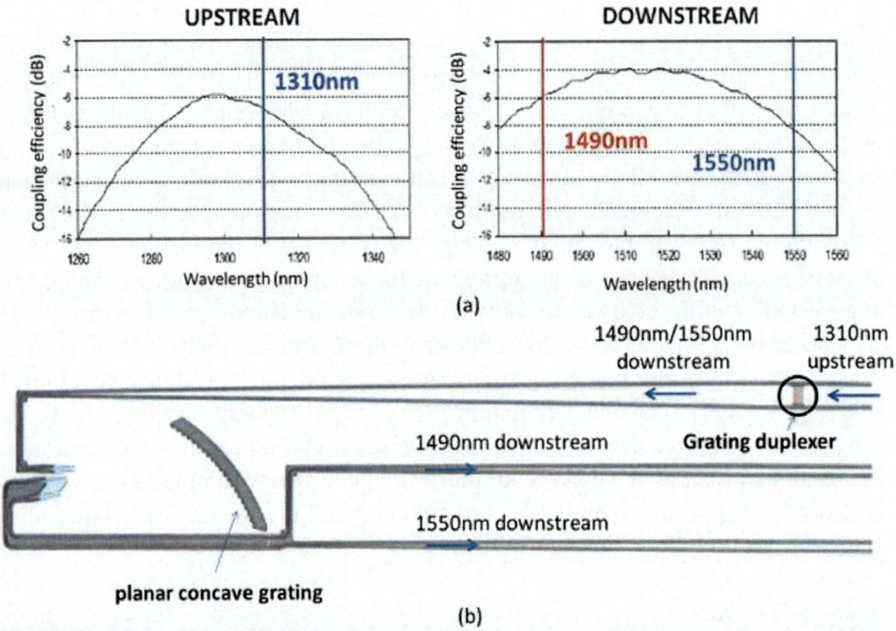

Fig. 3.8 (**a**) Fiber-chip coupling efficiency spectrum for the one-dimensional grating duplexer for the upstream and downstream wavelength channel and (**b**) a microscope image of the fabricated photonic integrated circuit including a one-dimensional grating duplexer and a planar concave grating to split the downstream wavelength channels, while at the same time suppressing the upstream wavelength channel crosstalk to below −30 dB [11]

time split the 1,490 and 1,550 nm wavelength channels) was experimentally realized [11]. In this case, a "standard" grating design was used by directly etching the grating lines into the silicon waveguide layer. The wavelength filter was implemented as a planar concave grating filter, resulting in more than 30 dB suppression of the upstream wavelength channel into the downstream wavelength path and demultiplexing of the 1,490 and 1,550 nm wavelength channels. The fabricated photonic integrated circuit, together with the experimentally obtained fiber-chip coupling spectrum is shown in Fig. 3.8. The device parameters of this device are again listed in Table 3.1.

3.6 Polarization Independent Fiber-Chip Coupling

As described in Sect. 3.3, silicon-on-insulator nanophotonic waveguides experience large modal birefringence, due to the high refractive index contrast of the waveguides and the typical rectangular shape of the photonic wire. This implies that one-dimensional diffraction gratings are inherently very polarization sensitive, since the diffraction angle for transverse electric and transverse magnetic polarized light is substantially different. In most practical applications, however, the polarization of the light in the input optical fiber is unknown and varying over time. While the lateral inverted taper discussed in Sect. 3.2 can be designed to allow for efficient optical coupling for both transverse electric and magnetic polarization, still the photonic components based on these silicon photonic wires behave very differently for both polarizations. This is a well-known issue in high-index contrast waveguide systems. The most popular solution to solve this problem is to use a polarization diversity approach. In this approach, both polarization states are spatially separated in the two arms of the polarization diversity circuit. In the case of a lateral inverted taper structure, first the TE and TM polarized light are spatially separated (e.g., using a directional coupler structure). Then the polarization in one of the arms of the polarization diversity circuit is first rotated to the orthogonal polarization state, thereby allowing the use of identical high index contrast waveguide structures in both arms. However, before combining both arms of the polarization diversity circuit (using an identical polarization combiner structure such as a directional coupler), the polarization needs to be rotated again in one arm [12]. While this approach enables polarization-independent operation of nanophotonic waveguide circuits (which are inherently very polarization dependent), it is clear that this requires several extra integrated optical components (besides the additional requirement of a cleaved and polished facet for optical interfacing).

Because of these drawbacks, we pursued an approach based on diffraction gratings to achieve polarization-independent operation of high-index contrast photonic integrated circuits [13]. The basic idea is to realize a grating structure that is the superposition of two one-dimensional diffraction gratings. Placing these two one-dimensional grating structures orthogonal to each other allows the coupling of both orthogonal polarization states in the optical fiber to the optical waveguide layer. Moreover, the optical modes that are excited by these two orthogonal polarization states in the orthogonal waveguides have identical polarization (i.e., transverse electric polarization), no longer requiring separate polarization splitter and polarization converter waveguide structures. The superposition of two identical one-dimensional grating structures leads to a square lattice grating for interfacing with the optical fiber. Therefore, a photonic integrated circuit can be made to work in a polarization independent way by incorporating two such square lattice grating structures on a photonic integrated circuit and duplicating the optical functions in both arms of the polarization diversity circuit. The concept of this polarization diversity approach and the operation principle of the square lattice grating structure is schematically shown in Fig. 3.9.

While this is conceptually an easy way to realize polarization-independent operation, two important issues complicate its actual realization. First of all there is the need for identical photonic integrated circuits in both arms of the polarization diversity configuration, which is also the case for the edge-coupling method. Although state-of-the-art deep UV lithography is used, this is far from trivial to achieve. It turns out that deep sub-nanometer dimensional control is required to match for example the resonance frequencies of a silicon ring resonator (a rule of thumb says that one nanometer variation in waveguide dimension results in a one nanometer wavelength shift). One way to tackle this problem is to thermo-optically tune the photonic integrated circuits, to compensate for the non-uniformity. In some applications, however, it is possible to use a single optical component for both arms of the polarization diversity circuit, by passing the light through the device in opposite directions. This mechanism is not limited to the diffraction grating approach and can be used in all polarization diversity approaches. This device concept was demonstrated for the case of polarization independent operation of an SOI arrayed waveguide grating [14] and ring resonator structure [12]. Another difficulty specific to the two-dimensional grating coupler implementation is the fact that the previous reasoning pointing to a square lattice grating structure as a polarization independent

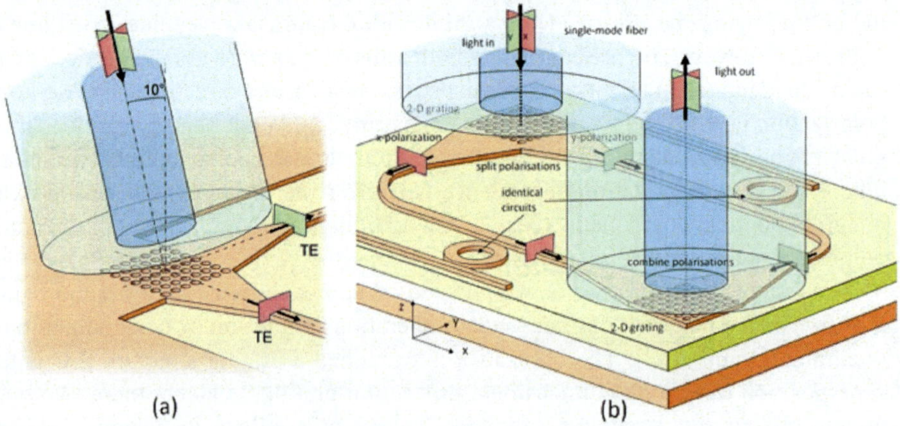

Fig. 3.9 Concept of the two-dimensional square lattice grating as (**a**) a polarization independent interface to the optical fiber and (**b**) the polarization diversity configuration

coupler is, in principle, only valid for a perfectly perpendicular positioned optical fiber. However, in Sect. 3.3 we pointed out that perfectly vertical positioning of the optical fiber results in strong second-order Bragg reflection and associated reduced fiber coupling efficiency. Therefore, a tilt of the optical fiber is still required to interface with a two-dimensional grating structure. To preserve the symmetry of the polarization diversity circuit, this tilt should be along the bisection line of the diffraction grating. This tilt has two implications. First, tilting the optical fiber along the bisection line of the two-dimensional grating implies that the waveguides connected to this grating structure also need to be tilted inwards, as is clear from the projected Bragg condition

$$\boldsymbol{k}_{z,i} = \boldsymbol{k}_{\mathbf{in,proj}} + \boldsymbol{K}_i \tag{3.5}$$

with $i = 1, 2$ and $\boldsymbol{K}_1 = \frac{2\pi}{\Lambda}\mathbf{1}_x$, $\boldsymbol{K}_2 = \frac{2\pi}{\Lambda}\mathbf{1}_y$. Since

$$\boldsymbol{k}_{\mathbf{in,proj}} = \frac{2\pi n}{\lambda} \sin(\theta_0) \left(\frac{\sqrt{2}}{2}\mathbf{1}_x + \frac{\sqrt{2}}{2}\mathbf{1}_y \right) \tag{3.6}$$

wave vectors $\boldsymbol{k}_{z,1}$ and $\boldsymbol{k}_{z,2}$ are no longer orthogonal to each other, as can be seen in Fig. 3.10.

This can however easily be implemented in the design of the integrated circuit by tilting the access waveguides inwards (i.e., 3.1° tilt for a 10° tilted fiber). A more important issue associated with this tilting is the fact that a perfect polarization independent coupling can no longer be obtained. This can be understood by looking at the polarization in the optical fiber along the bisection line of the grating and the orthogonal polarization. By tilting the optical fiber, the polarization along the bisection line is tilted out of the plane of the photonic integrated circuit, while the orthogonal polarization stays in this plane, as shown in Fig. 3.11a. This implies

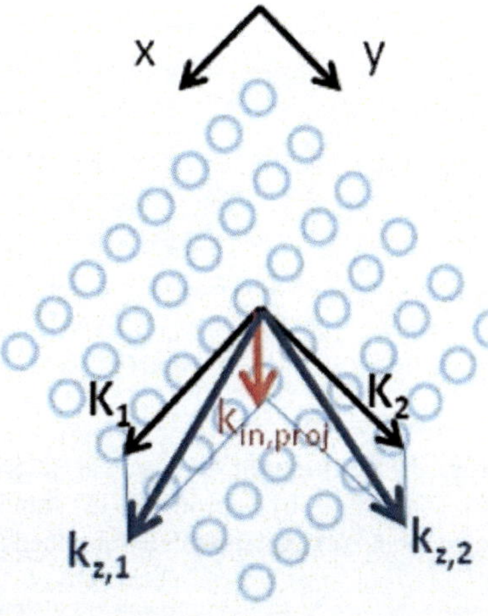

Fig. 3.10 Calculation of the non-orthogonal wave vectors in the silicon waveguide layer due to the finite projected wave vector of the incident light arising from the tilting of the optical fiber

that the coupling spectrum for both polarizations will be different when the fiber is tilted off vertical, resulting in an intrinsic polarization-dependent loss (PDL) of the grating structure. This is confirmed by rigorous three-dimensional FDTD simulations (see Fig. 3.11b), showing the presence of a wavelength-dependent PDL. This wavelength-dependent PDL is also observed in the experiment, as shown in Fig. 3.11c. The parameters of the grating coupler structure are listed in Table 3.1. While this wavelength-dependent PDL arises due to the tilting of the optical fiber, improvement can be expected by optimizing the design of the grating structure. This requires however extensive three-dimensional FDTD simulations and has not been studied in detail so far.

Recently we studied another approach to realize a polarization-independent interface to a silicon photonic integrated circuit, using a one-dimensional grating structure [15]. In this case, the same mechanism as described in Sect. 3.5 is used to interface with both polarizations on the photonic integrated circuit, as shown in Fig. 3.12a. Although this allows one to couple both polarizations to the photonic integrated circuit (and spatially separate them on the chip), the modes in the waveguides still are TE and TM polarized, respectively. This means that a polarization rotator is still required to realize a true polarization diversity scheme. In this case, the grating coupler fulfills the role of fiber-chip coupler and polarization splitter at the same time. In some applications, the need for a polarization converter is not

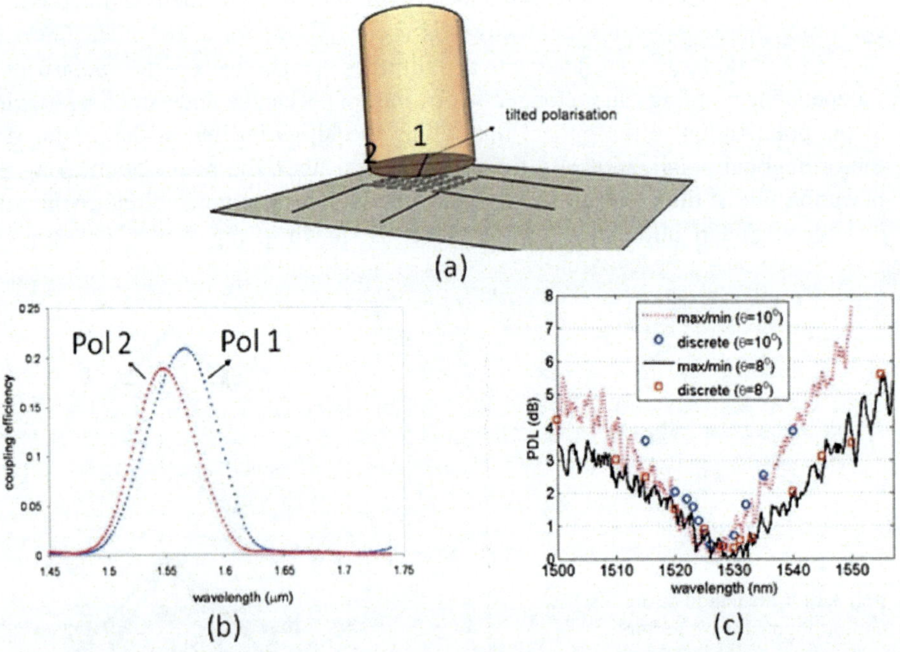

Fig. 3.11 Tilting of the optical fiber and its influence on the polarization-dependent loss: (**a**) schematic, (**b**) rigorous FDTD simulation of the polarization dependence of the coupling efficiency, and (**c**) experimentally obtained polarization dependent loss as a function of the wavelength

even required. This is, for example, the case for a point-to-point transceiver in which the downstream wavelength only needs to be coupled to the chip in a polarization-independent way, where it is immediately detected. By optimization of the grating structure, polarization-independent coupling of the downstream wavelength and single polarization coupling of the upstream wavelength using a one-dimensional grating structure can be realized at the same time. This was recently demonstrated using a standard diffraction grating approach for polarization-independent coupling of a 1,310 nm wavelength signal and single polarization coupling of a 1,610 nm wavelength signal. The operation wavelength of the device is defined both by the grating period and angle of the optical fiber, although the waveguide layer thickness and grating etch depth also play a significant role. Because the device geometry was fixed, the important wavelength combination of 1,310/1,550 nm could not be reached, although a slight modification in waveguide geometry and etch depth allows one to do so. The operation principle of this device was demonstrated using standard diffraction grating structures (a 220-nm-thick silicon waveguide layer and a 70-nm etch depth for the grating). A coupling efficiency of −5.9 dB for the 1,610 nm wavelength channel was obtained, while a −5.2 dB coupling efficiency was obtained for 1,310 nm, which is independent of the polarization of the incident light. The fiber-to-fiber polarization dependent loss measured using this device configuration is lower that 1 dB over a broad wavelength range of 70 nm. The fiber-to-chip coupling efficiency for both wavelength channels and the polarization dependent loss of the

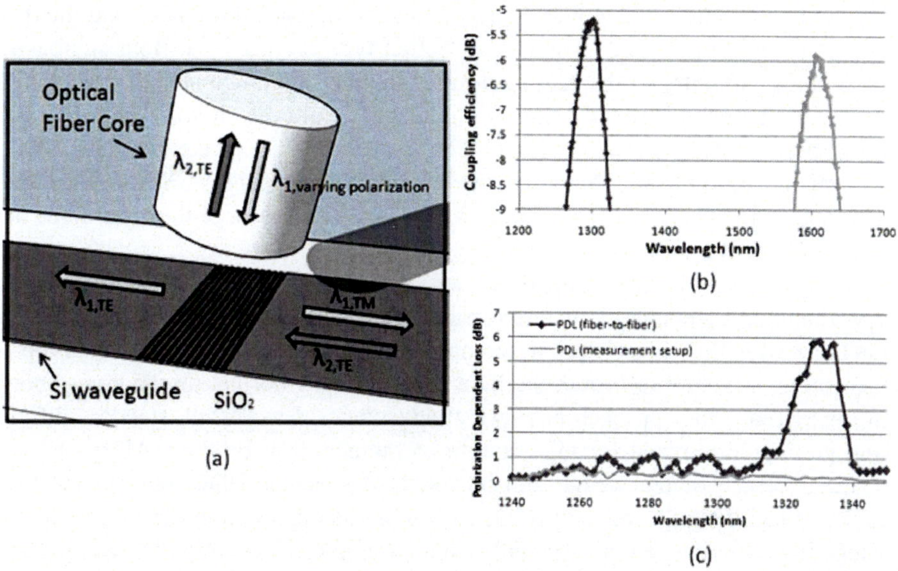

Fig. 3.12 A one-dimensional diffraction grating for polarization independent coupling of a λ_1 wavelength channel and single polarization coupling of an upstream wavelength channel λ_2: (**a**) device principle, (**b**) experimentally obtained fiber coupling spectrum, and (**c**) measured polarization dependent loss for the downstream wavelength channel (**c**)

downstream wavelength channel is shown in Fig. 3.12b and c. The parameters of the grating structure used in the experiment are listed in Table 3.1.

3.7 Integration of Opto-electronic Components

So far we have discussed the use of diffraction gratings as a means to interface with a single-mode optical fiber. The use of grating couplers is however not restricted to this application: interfacing with hybrid integrated III–V opto-electronic components can be envisioned as well. In this case, these opto-electronic components need to be flip-chipped on top of the diffraction gratings. However, one particular aspect of the diffraction gratings needs to be considered with care. In Sect. 3.3 we elaborated on the fact that diffractive grating structures only perform well for oblique diffraction. Otherwise, second-order reflection and reduced coupling efficiency occur. While this is not a problem for a flip-chipped photodetector (or at least can be taken into account in the design), this is a serious issue for vertically emitting laser diodes, like vertical cavity surface emitting lasers (VCSELs). Although there are ways to flip chip components in a tilted fashion, this is not the standard approach. Therefore, to interface a VCSEL with a silicon photonic integrated circuit, a modified grating layout is required.

One approach we have pursued to solve this problem is the modification of the grating structure by adding either a grating at the backside of the coupling grating functioning as a high-reflectivity grating [4], or adding an additional slit at the front side of the grating, acting as an anti-reflection type coating [16]. Both approaches create an asymmetry in the grating design allowing the efficient coupling of light from a flip-chipped vertically emitting light source into the photonic circuit. While this approach allows realizing a high VCSEL-chip coupling efficiency, there still is a parasitic reflection from the grating that couples back into the VCSEL light source. This is an unwanted situation, since it will increase the noise of the laser diode and can even result in polarization switching of the device.

Therefore, an alternative approach was investigated. In this case, an obliquely operating grating coupler structure is used, as discussed in the previous section. To interface with the opto-electronic component, a wedge can be defined on top of the grating structure to refract the light to the surface normal direction, as shown in Fig. 3.13a. This approach has several advantages. First of all, standard diffraction gratings optimized for high-coupling efficiency can be used. Moreover, due to the presence of the wedge, the influence of specular reflection into the laser cavity is significantly reduced, yielding a more stable solution. This approach of integrating a wedge on top of the diffraction grating has recently been demonstrated [17]. In this case, a polymer wedge was defined on top of the grating structures by means of a nanoimprint lithography technique, as shown in Fig. 3.13b. Standard diffraction gratings were used as proof-of-principle components. The performance of these grating coupler structures was assessed by probing the wedge/grating combination using a vertically positioned single-mode fiber and comparing the cou-

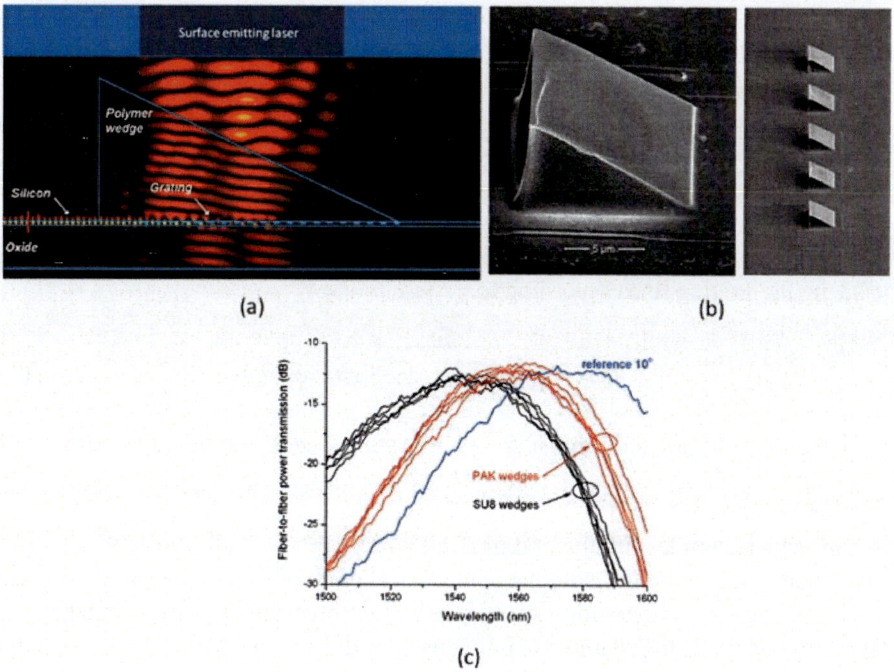

Fig. 3.13 Integration of a polymer wedge on top of diffractive grating couplers for interfacing with a VCSEL light source: (**a**) FDTD simulation, (**b**) SEM picture of the fabricated structure, and (**c**) measured fiber-to-fiber coupling efficiency after a thermal cycle at 300°C for 10 min [17]

pling efficiency thus obtained to the oblique coupling efficiency on a reference grating coupler without the wedge structure. Only a marginal penalty in device performance was observed, showing that wedges with good optical quality can be obtained. To emulate the flip-chip integration process and the influence this process has on the device performance (due to the relatively high temperatures used), the polymer wedge/grating coupler combination was exposed to a 300°C temperature cycle of 10 min. Both the configurations based on SU-8 (MicroChem) wedges and PAK polymer (Toyo Gosei) wedges showed no appreciable degradation in device performance, as is clear from Fig. 3.13c.

3.8 Small Footprint Fiber-Chip Coupling Structures

While the grating coupler structures presented in the previous sections are very compact (typically 12 by 12 μm in size), a relatively long adiabatic taper is still needed to laterally transform the optical mode size from that of a single-mode fiber (10 μm) to that of a single-mode nanophotonic waveguide (0.5 μm). This taper, on the order of 150 μm in length, determines the footprint of the fiber-chip coupling structure. Since one of the advantages of working on a high refractive index contrast waveguide platform is the density of integration that can be obtained, it would be interesting to be able to scale down the footprint of the fiber-chip coupling structure. This can be achieved by modifying the layout of the grating structure. Instead of using straight grating lines in a one-dimensional grating structure, a curved grating can be used to couple the light from the fiber to the waveguide plane and at the same time focus the light in that waveguide plane onto the aperture of the optical waveguide. When the waveguide plane is chosen to be the (Y, Z) plane of a right-handed Cartesian coordinate system, with Z along the waveguide axis and the origin chosen to be in the desired focal point, it can be shown that a focusing grating can be obtained by curving the grating lines according to

$$q\lambda_0 = n_{\text{eff}}\sqrt{y^2 + z^2} - zn_t\cos(\theta_c) \tag{3.7}$$

Here, q is an integer number for each grating line, θ_c is the angle between the fiber and the chip surface, n_t is the refractive index of the environment, λ_0 is the vacuum wavelength, and n_{eff} is the effective index felt by the cylindrical wave in the broad waveguide with the grating. The grating lines are ellipses with a common focal point, that coincides with the optical focal point of the coupler. While rigorous design of the focusing gratings would require full three-dimensional simulations, we approximate by extrapolating two-dimensional designs of standard linear gratings for TE-polarization (electric field parallel to the grating lines) and coupling at $\theta_c = 80°$. This way, the use of curved grating structures to reduce the device footprint was demonstrated. We showed that by implementing these curved grating structures an eight-fold reduction in the device footprint is obtained without a penalty in device performance [18].

Besides the demonstration of a one-dimensional curved grating structure, the footprint of a two-dimensional fiber-chip grating coupler can also be scaled down by changing the grating layout. Again the idea is to consider the two-dimensional grating as the superposition of two one-dimensional curved gratings. At the intersection of both one-dimensional curved grating lines, a hole is etched in the silicon

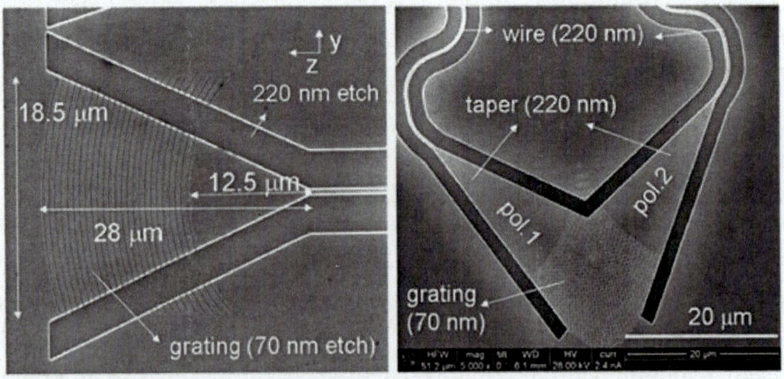

Fig. 3.14 The implementation of compact one-dimensional and two-dimensional curved gratings [18]

waveguide layer, acting as a scatter center. Also in this way, ultra-compact polarization independent coupling to a silicon-on-insulator waveguide circuit was demonstrated without penalty in device performance compared to a standard square lattice two-dimensional grating structure [18]. A one-dimensional and a two-dimensional curved grating structure is shown in Fig. 3.14.

3.9 Optical Probing of Nanophotonic Integrated Circuits

One of the major drivers to use a diffraction grating is the need for interfacing non-singulated chips with an optical fiber. This enables wafer-scale testing of large-scale photonic integrated circuits, comparable to the wafer-scale electronic testing for CMOS circuits. Since both electrical probing and optical probing can be done from the top of the wafer surface, both types of measurements can be combined in a single measurement system, in the case of electronic/photonic integrated circuits. This approach allows using a modified probing station to do a fully automated mapping of the device performance across the wafer. This, however, only allows the characterization of the input-output characteristics of the full photonic integrated circuit, since it is only at the optical input/output ports of the circuit that these grating couplers can be implemented. In the development stage of a photonic integrated circuit, device engineers typically would like to access the individual components in the actual integrated circuit. Therefore, there is the need for a true optical probe, which allows one to do so, without the need for a diffraction grating on the photonic integrated circuit.

The approach we followed for such an optical probe is, instead of defining the diffraction grating on the photonic integrated circuit, to implement a high refractive index contrast grating on the facet of a single-mode optical fiber [19]. By making physical contact between this probe and an input/output waveguide of an individual component of the integrated circuit, a grating coupler is formed allowing the coupling of light to and from the photonic integrated circuit. A prototype fiber probe was realized by defining a uniform metallic periodic structure on the core of a single-mode fiber, by means of an imprint technology. First, the fiber with UV-curable resist on the facet is aligned over a specially prepared mold carrying the 10 by 10 μm gold grating pattern in the mold trenches. The mold was obtained by starting off from an SOI sample containing silicon gratings of 630 nm period and an etch depth of 220 nm. This surface was treated with an antistiction coating. Then, gold was evaporated onto the mold. Finally, the gold on top of the mold grating lines was selectively removed by microcontact printing on another substrate. In this way, the original SOI mold becomes a carrier of the gold grating pattern by leaving the gold only in the grating trenches. After attachment of the fiber to the mold, the cavities are filled and the resist is UV cured. Finally, the mold is released. The metal grating is now attached to the fiber. The fabrication process is schematically outlined in Fig. 3.15a together with an SEM picture of a fabricated device, while Fig. 3.15b

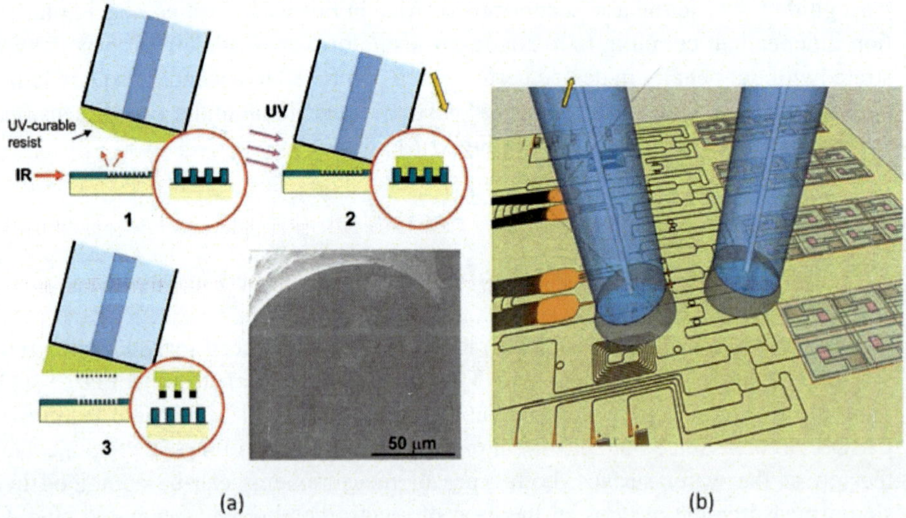

Fig. 3.15 The optical probe: (**a**) probe fabrication and SEM image of a transferred grating onto a fiber facet and (**b**) artist's impression of how the optical probe is used to probe individual integrated optical functions [19]

shows an artist's impression of how this optical probe is used to assess the properties of individual components on a photonic integrated circuit.

Two gold grating fiber probes were fabricated, both consisting of a gold grating of 20 nm thickness. Both probes were brought into contact with a straight 220 nm by 3 μm SOI waveguide. A 15% coupling efficiency and a 1-dB bandwidth of 38 nm was demonstrated in this way for transverse electric polarized light, which is sufficient for testing purposes.

3.10 Conclusions

Diffractive grating structures provide an elegant way of interfacing a high refractive index contrast single-mode fiber and a nanophotonic integrated circuit. Polarization-independent, large optical bandwidth, high-efficiency coupling can be obtained. Even duplexing of two wavelength bands can be realized using the same type of structure. As diffractive grating structures allow accessing a photonic integrated circuit, it paves the way to wafer scale testing of photonic integrated circuits and wafer level packaging of photonic integrated circuits. In this chapter the main focus was on the interfacing of a single mode fiber and a photonic integrated circuit. We showed, however, that it is possible to also interface hybrid integrated opto-electronic components such as flip-chipped photodetectors and VCSELs with the photonic integrated circuit, allowing the realization of complex active/passive photonic integrated circuits. Finally, an optical probe has been presented, creating the possibility to assess the performance of the individual subcomponents forming the photonic integrated circuit.

Acknowledgments We would like to thank Dr. Dirk Taillaert, Dr. Frederik Van Laere, Dr. Wim Bogaerts, Dr. Jonathan Schrauwen, Dr. Pieter Dumon, Ir. Diedrik Vermeulen, Ir. Shankar Selvaraja, and Prof. Roel Baets for the fruitful discussions.

References

1. W. Bogaerts, D. Taillaert, B. Luyssaert, P. Dumon, J. Van Campenhout, P. Bienstman, D. Van Thourhout, R. Baets, V. Wiaux, S. Beckx, Opt. Express **12**, 1583 (2004) 71
2. A. Sure, T. Dillon, J. Murakowski, C.C. Lin, D. Pustai, D.W. Prather, Opt. Express **11**, 3555 (2003) 72
3. S.J. McNab, N. Moll, Yu.A. Vlasov, Opt. Express **11**, 2927 (2003) 72
4. D. Taillaert, W. Bogaerts, P. Bienstman, T.F. Krauss, P. Van Daele, I. Moerman, S. Verstuyft, K. De Mesel, R. Baets, J. Quantum Electron **38**, 949 (2002) 73, 76, 88
5. F. Van Laere, G. Roelkens, M. Ayre, J. Schrauwen, D. Taillaert, D. Van Thourhout, T. Krauss, R. Baets, J. Lightwave Technol. **25**, 151 (2007) 78, 79
6. S. Selvaraja, D. Vermeulen, M. Schaekers, E. Sleeckx, W. Bogaerts, G. Roelkens, P. Dumone, D. Van Thourhout, R. Baets, Highly efficient grating coupler between optical fiber and silicon photonic circuit, p. CTuC6, CLEO, Baltimore, 2009 78, 79
7. G. Roelkens, D. Van Thourhout, R. Baets, Opt. Express **14**, 11622 (2006) 80
8. G. Roelkens, D. Vermeulen, D. Van Thourhout, R. Baets, S. Brision, P. Lyan, P. Gautier, J.M. Fedeli, Appl. Phys. Lett. **92**, 131101 (2008) 80, 81
9. L. Vivien, D. Pascal, S. Lardenois, D. Marris-Morini, E. Cassan, F. Grillot, S. Laval, J.M. Fedeli, L.El Melhaoui, J. Lightwave Technol. **24**, 3810 (2006) 80
10. G. Roelkens, D. Van Thourhout, R. Baets, Opt. Express **15**, 10091 (2007) 81, 82
11. D. Vermeulen, G. Roelkens, J. Brouckaert, D. Van Thourhout, R. Baets, R. Duijn, E. Pluk, G. Van den Hoven, Silicon-on-insulator nanophotonic waveguide circuit for Fiber-to-the-Home transceivers, p. Tu3C6, ECOC, Brussels, 2008 82, 83
12. H. Fukuda, K. Yamada, T. Tsuchizawa, T. Watanabe, H. Shiojima, S. Itabashi, Opt. Express **16**, 4872 (2008) 83, 84
13. D. Taillaert, H. Chong, P.I. Borel, L.H. Frandsen, R.M. De La Rue, R. Baets, Photon. Technol. Lett. **15**, 1249 (2003) 83
14. W. Bogaerts, D. Taillaert, P. Dumon, D. Van Thourhout, R. Baets, Opt. Express **15**, 1567 (2007) 84
15. D. Vermeulen, G. Roelkens, D. Van Thourhout, R. Baets, Silicon-on-insulator polarization independent one-dimensional grating duplexer for fiber-to-chip coupling, submitted for publication in J. Lightwave Technol. 86
16. G. Roelkens, D. Van Thourhout, R. Baets, Opt. Lett. **32**, 1495 (2007) 88
17. J. Schrauwen, S. Scheerlinck, D. Van Thourhout, R. Baets, Polymer wedge for coupling VCSELs to a silicon photonic integrated circuit, p. 72180B, PhotonicsWest, San Jose, 2009 88, 89
18. F. Van Laere, W. Bogaerts, P. Dumon, G. Roelkens, D. Van Thourhout, R. Baets, Group IV photonics **15**, 10091 (2007) 90, 91
19. S. Scheerlinck, D. Taillaert, D. Van Thourhout, R. Baets, Appl. Phys. Lett. **91**, 031104 (2008) 91, 92

Appendix

In this appendix the device parameters and measurement conditions for the various types of considered grating coupler structures are listed.

Table 3.1 Parameters of the various types of fiber-to-chip grating coupler structures (DC is the grating duty cycle, n_{sup} is the refractive index of the medium on top of the grating and N is the number of grating periods)

Coupler	Period (nm)	Etch (nm)	DC	Fiber tilt (Degrees)	n_{sup}	N
Standard	630	70	0.5	10	1.0	20
Gold mirror	630	70	0.5	10	1.0	20
DBR bottom mirror	630	70	0.5	10	1.45	20
Silicon overlay	610	250	0.5	10	1.45	20
1D duplexer	520	70	0.5	10	1.45	20
2D coupler	605	70	Diam = 390	10	1.0	Rows = 19
1D pol. div.	535	70	0.5	14	1.45	18

Index

Chapter 4
Development and Application of Er-Doped Silicon-Rich Silicon Nitrides and Er Silicates for On-Chip Light Sources

Jee Soo Chang, Kiseok Suh, Moon-Seung Yang, and Jung H. Shin

Abstract On-chip light sources are a critical part for an integrated Si photonic technology, yet they lag other photonic components in their level of development. In this chapter, Er used as an optical dopant and utilizing its intra-4f transition at 1.54 μm will be introduced as a viable means for on-chip light generation that has the advantage of being compatible with long-distance telecom as well. First, a general introduction to the topic of developing a Si-based/compatible light source will be presented, with emphasis on the need for a novel high-index material for Er-doping that can provide both a higher Er content and refractive index than has been reported so far. Second, Er-doped silicon-rich silicon nitride (SRSN) and Er silicates will be introduced as a promising host material for compact on-chip light sources. Finally, results of fabricating basic photonic components using Er-doped SRSN and Er-silicates will be presented.

4.1 Introduction

4.1.1 Si Photonics and Light Sources

In the developing field of Si photonics, great strides have been made in developing compact active photonic devices such as ultra-fast modulators [1, 2], optical buffers [3], and detectors [4, 5] that seemed impossible just a few years ago. By leveraging the astronomical infrastructure for Si processing, these devices hold the promise of achieving an electronic-photonic convergence on a Si platform that can overcome the "interconnect bottleneck."

J.S. Chang, K. Suh, and M.-S. Yang (✉)
Department of Physics, Korea Advanced Institute of Science and Technology (KAIST), Daejeon, Republic of Korea

J.H. Shin (✉)
Department of Physics; Graduate School of Nanoscience and Technology (WCU), Korea Advanced Institute of Science and Technology (KAIST), Daejeon, Republic of Korea
e-mail: jhs@kaist.ac.kr

D.J. Lockwood, L. Pavesi (Eds.): *Silicon Photonics II*.
Topics in Applied Physics **119**, 95–130 (2011)
DOI 10.1007/978-3-642-10506-7_4

However, a comparable advancement in developing a Si-based/compatible light source has so far been lacking, as no amount of advanced processing can overcome the fundamental limitation on light emission imposed by the indirect bandgap of Si. In a semiconductor, light emission typically involves recombination of an electron with a hole. As the momentum of a photon is negligible when compared with the crystal momentum of carriers in a semiconductor, electrons and holes need to have the same wave-vector, k, if their radiative recombination is to occur efficiently. This is the case for III–V compound semiconductors such as GaAs, whose valence band maximum and conduction band minimum both occur at the central Γ point of the first Brillouin zone. In case of Si, on the other hand, the conduction band minimum occurs along the [100] direction of the first Brillouin zone. Therefore, a third particle – such as a phonon – is required to supply the extra momentum, which greatly reduces the probability for radiative recombination of carriers in Si.

Thus, one obvious method of increasing the luminescence efficiency of Si is to suppress all other competing carrier recombination paths by starting with ultra-high purity Si wafers, and then thoroughly passivating all defects and surfaces. Using this approach, Green et al. have reported that Si light-emitting diodes (LEDs) can be made to luminesce as efficiently as conventional LEDs based on III–V compound semiconductors [6]. Another well-established method of increasing the luminescence efficiency of Si is to use Si nanostructures such as quantum dots and wires. Use of Si nanostructures modifies the light emission properties of Si in several important ways. First, the bandgap is increased due to the quantum confinement effect, leading to luminescence in the visible range [7]. Second, the radiative transition probability can be greatly increased [8]. Finally, the formation of individual nanocrystals (NCs) results in physical isolation of non-radiative defects, leading to an enhanced overall luminescence efficiency [9]. In fact, it had been reported that while the ensemble average of Si NCs luminescence efficiency can be quite low due to defective Si NCs that do not luminesce, the luminescence efficiency of individual luminescent Si NCs can be very close to 100% [10]. Since the first report on light emission from porous Si by Canham et al. [11], intense research into the subject of light emission from Si nanostructures has provided a wealth of impressive results, including the fabrication of visible LEDs and their integration with electronic circuits [12], photonic crystal LEDs [13], and optical gain under optical pumping [14, 15].

Yet despite the many impressive developments and the elegance of a monolithic all-Si electro-photonic convergence, the direct use of Si for light emission presents several difficulties. First, light emitted by Si is also absorbed by Si. Thus, the use of Si for a light source precludes its integration with developed Si electro-photonic devices, nearly all of which use the silicon-on-insulator (SOI) platform. Second, carrier recombination in Si remains indirect even for nanometer-sized clusters [16]. Thus, the high radiative efficiency is obtained under low-power or ultra-short pumping conditions only, as non-linear effects such as Auger-recombination quickly dominate under the high excitation conditions necessary for population inversion and lasing. This presents difficulties for the use of a Si-based light source for inter- and intra-chip communication applications that require high bandwidth and fast modulation speed.

High bandwidth and efficiency under high excitation conditions can be obtained by use of compound semiconductors dedicated for light generation. The traditional approach of heteroepitaxy of a compound semiconductor thin film on Si has the limitations imposed by the lattice mismatch. However, recent development in wafer-bonding technique may offer the possibility of wafer-scale integration of compound semiconductor photonic devices with Si electronic circuits regardless of their lattice mismatch [17]. Still, whether such a method can be used to reliably scale to CMOS production lines has yet to be determined. We note that there have been recent reports that strain-engineered Ge grown heterepitaxially on Si can be used for both the generation and detection of light at the technologically important wavelength range of 1.54 μm. [18]. At this moment, more research is needed before the feasibility of a Ge-based light source can be ascertained. But if successful, such Ge-based light source would have the advantage of easy wafer-scale fabrication with full CMOS compatibility.

4.1.2 Er as an Optical Dopant

An approach that is quite different from the ones discussed so far is the use of Er as an optical dopant for Si. The important point that sets Er-doping apart from other approaches is that even when Er is doped into a host, the light emission occurs via intra-4f-shell electronic transitions within the atom without involving carriers in the host matrix. These transitions are parity-forbidden, but occur due to the effects of the crystal field. This gives rise to stable luminescence properties and long luminescence lifetimes that allow easy population inversion. Furthermore, as these are core-level transitions, the overall luminescence properties are affected only weakly by external factors such as the host material and temperature. In particular, since the intra-4f transition from the first excited state to ground state emits light at the standard telecommunication wavelength 1.54 μm, use of Er enabled the development of Er-doped fiber amplifiers (EDFA) that have enabled the realization of all-optical long-distance optical telecom networks [19].

The most direct way of exciting Er^{3+} ions in any host material is through direct resonant optical pumping with a pump beam tuned to one of the higher optical absorption bands of Er^{3+}. On the other hand, in the case of a semi-conducting host material, it is possible to excite Er^{3+} ions indirectly via energy transfer from carriers to the 4f electrons [20, 21], as shown schematically in Fig. 4.1. In particular, Si could be used as an effective host material since its bandgap of 1.1 eV provides sufficient energy for excitation of 4f electrons of Er^{3+}, yet provides complete transparency to emitted 1.54 μm photons. Furthermore, using Si allows one to leverage the vast Si processing infrastructure and technological know-how as well, enabling a cost-effective large-scale production. These factors, together with the great success of EDFA in telecom applications, have led to investigation of Er-doping of Si as a viable means of developing an on-chip Si-compatible light source. Since the first work by Ennen et al. [22], a great explosion of research into Er doping of Si has led to demonstrations of not only photoluminescence (PL) but also electroluminescence (EL) at room temperature from Er-doped Si [23, 24].

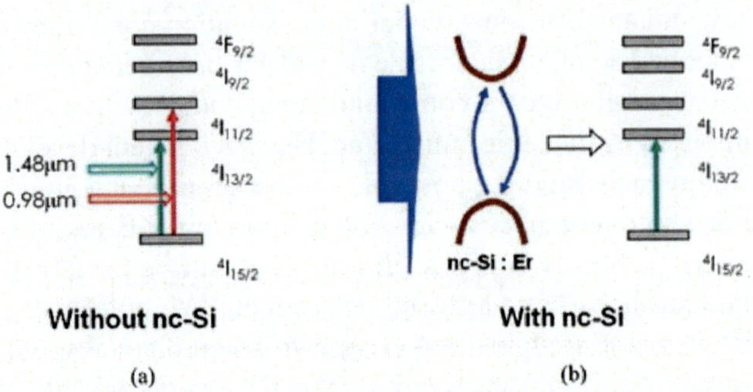

Fig. 4.1 Schematic description of (**a**) direct resonant optical excitation of Er without nc-Si sensitization; (**b**) Indirect carrier-mediated excitation of Er with nc-Si sensitization

Despite such success, Er-doping of *bulk* Si remains impractical because Er can react with Si to form non-luminescent configurations [25], and because excited Er^{3+} ions can back-transfer energy to the host electrons, resulting in a strong temperature quenching of Er^{3+} luminescence in Si [26]. These problems, however, can be solved by co-doping Er with activating ligands such as oxygen [25, 27] and using a host material with a wide bandgap [28]. Thus, silicon-rich silicon oxide (SRSO), which consists of nanocluster-Si (nc-Si) embedded inside an SiO_2 matrix, has emerged as an attractive host material that in effect combines the efficient Er^{3+} excitation properties of Si with the well-known and stable luminescent properties of Er^{3+} in SiO_2 [29–34]. In this case, nc-Si plays the role of classical sensitizers that efficiently absorb pump photons with a large absorption cross-section, then transferring it to nearby Er^{3+} ions. This process can be surprisingly efficient, with a transfer time of $\leq 1\,\mu s^{-1}$ and transfer efficiency of $\geq 55\%$ [35], and by now, efficient light emitting diodes [36] as well as signal enhancement under optical pumping have been demonstrated [37, 38].

Yet SRSO, in attempting to combine the best of Si and SiO_2, can also combine the worst of Si and SiO_2. Er-doped SRSO is usually formed by annealing Er-doped amorphous SiO_x both to precipitate excess Si into nc-Si and to optically activate Er [29–34]. Within this annealing process, Er atoms can both diffuse and react with excess Si, leading to a loss of optically active Er, as is observed for Er-doped bulk Si, and to breaking of the nc-Si/Er coupling necessary for nc-Si sensitization. [39–41]. There have been several reports that the number of Er atoms in SRSO that can be excited via nc-Si in SRSO can be very small – as low as only a few percent of the doped Er atoms – and that in many cases, less than half of the total Er can be excited at all, even with direct optical pumping [42], thereby precluding the possibility of optical inversion necessary both for lasing and optical amplification.

These issues are serious, but with the selection of proper materials and processing conditions, solutions may still be found. Indeed, there already exist several reports that a great improvement can be obtained by reducing the excess Si content and the annealing temperatures [37]. However, even then, the use of SRSO – or, in fact, any SiO_2-based host material – for a compact on-chip light source faces fundamental difficulties due to (1) the low solubility of Er in SiO_2, and (2) the low refractive index contrast achievable against possible cladding materials. First, because the light emission occurs via the atomic intra-4f transition, the amount of light that can be generated is limited strictly by the number of optically active Er^{3+} ions. Yet the concentration of optically active Er that can be doped into silica without clustering is limited to $\sim 10^{20}\,cm^{-3}$ [43]. Given that the emission cross-section of Er^{3+} is 10^{-21}–$10^{-20}\,cm^2$, this limits the maximum gain that can be obtained by any SiO_2-based material to a few dB/cm at most. Thus, obtaining a net gain from such a material will require a large device and/or highly advanced processing techniques to reduce any parasitic losses. Second, for a stable operation in integrated devices, all photonic devices need outer cladding layers. However, because the refractive index of SRSO is very close to that of SiO_2 with $\Delta n \leq 0.1$, it requires waveguides with a width $>1\,\mu m$, a bending radius $>100\,\mu m$, and an areal footprint $> 10^8\,\mu m^2$ for a 3 dB gain, as shown in Fig. 4.2 [44]. These values do not present any problem for stand-alone telecom devices, but are unacceptably large for electro-photonic integration. Indeed, it is instructional to realize that the only reported case of an on-chip light source based on Er-doped silica required a micro-disk resonator of 120-μm diameter with air-cladding [45].

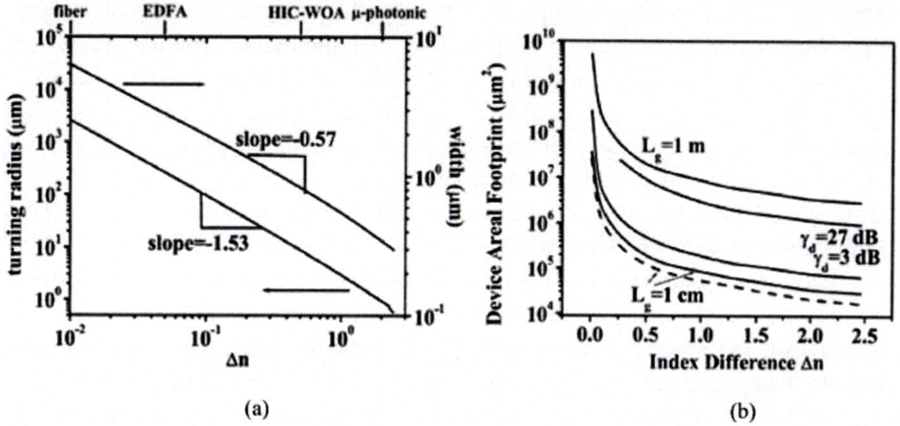

Fig. 4.2 (**a**) Plot of waveguide width and bend radius necessary for 0.042 dB/cm loss per turn; (**b**) plot of serpentine (*solid*) and coil (*dotted*) footprints versus Δn. Footprints are shown for constant device lengths and constant device gain. From [44]

4.1.3 Silicon-Rich Silicon Nitride and Er Silicates

In this chapter, silicon-rich silicon nitride (SRSN) and Er silicates will be introduced as novel materials that can solve the above problems, and enable development of a compact efficient light source based on Er.

Nitrides were chosen for several reasons. First, nitrides are used widely in the Si industry, and thus are fully CMOS compatible. Second, they have refractive indices that are much higher than that of oxide and can be controlled by varying the silicon content, thereby allowing compact optical devices on a chip without requiring cladding layers that are several microns thick. Third, nitrogen has been reported to be even more effective than oxygen in optically activating Er in Si [27], indicating that nitrides may offer an efficient environment of Er^{3+} luminescence. Finally, nitrides have a bandgap that is much smaller than that of oxide, which would allow easier injection of current for the possible development of electrically excited devices [46]. We find that nitrides do indeed offer all of the above advantages, and, in addition, are much more resistant to Er clustering, such that higher Er concentrations and annealing temperatures can be used. We note that there have been several recent reports on Er-doped SRSN that confirm the results presented here [47].

Er silicates present an interesting solution to the problem of low solubility of Er. By raising the Er concentration high enough, Er-rich crystalline rare-earth oxy-orthosilicates (R_2SiO_5, $R_2Si_2O_7$, etc), which are stable phases in the SiO_2–R_2O_3 system, can form [48]. They provide a combination of a very high concentration of optically active rare earth ions in a stable material, as the crystalline structure prevents clustering of Er on the atomic scale. In fact, the maximum possible Er^{3+} ion concentration is $2 \times 10^{22}\,cm^{-3}$ for Er_2SiO_5, nearly two orders of magnitude higher than that of silica. Furthermore, they have a high refractive index of about 1.825 [49], enabling compact optical devices with footprints that are only 1/70th of that of comparable silica-based devices. However, their compatibility with standard Si processes still needs to be resolved. We find that Er silicates can indeed offer a much higher Er concentration than is possible elsewhere. The maximum Er concentration that can be used is limited by cooperative upconversion. Still, the cooperative upconversion coefficient of Er silicates are much lower than that expected from silica-based thin films, demonstrating the promise of Er silicate thin films for high-gain compact light sources.

Finally, the applicability of Er-doped SRSN will be demonstrated by fabricating high-Q microdisks, which can form the basis for many important photonic components such as switches, filters, and light sources, and characterizing their optical performance.

4.2 Er-Doped Silicon-Rich Silicon Nitride

4.2.1 Experimental

SRSN:Er films of ~100 nm thickness were deposited on Si substrates by means of the ultra high vacuum (UHV) ion beam sputter method using a 750 eV N^+ beam from a Kaufmann-type ion gun. The base pressure of the system was 1×10^{-8} torr, and the stoichiometry of deposited films was controlled by varying the N_2 gas flow rate. The composition of the film was calibrated by in situ X-ray photoelectron spectroscopy (XPS) and Rutherford backscattering (RBS) analysis. The excess Si content ranged from 0 (pure Si_3N_4) to 17 at.%, while the Er content was fixed at 0.09 at.% for all films. After deposition, the films were furnace-annealed at 300–1,100 °C for 30 min in a flowing N_2 environment to precipitate nc-Si [50] and activate Er. A subsequent anneal at 650 °C in forming gas for 1 h was used to hydrogenate the film and passivate defects. For comparison, SRSO:Er and SiO_2:Er thin films of similar thickness, excess Si content, and Er content were also prepared using identical procedures.

The photoluminescence (PL) spectra were obtained using either the 477-nm line of an Ar laser or a 980-nm distributed feedback (DFB) laser diode. The 477-nm line excites Er^{3+} ions via nc-Si only, as it is non-resonant with optical transitions of Er^{3+}, while the 980-nm DFB laser excites Er^{3+} ions via direct optical absorption only, as it is resonantly absorbed by Er^{3+} ions but is hardly absorbed by nc-Si [51]. The low temperature PL spectra were measured using a closed-cycle helium cryostat, and the luminescence decay traces were measured using a digitizing oscilloscope. Finally, photoluminescence excitation (PLE) spectroscopy was performed using a Xe lamp (450 W) dispersed by a monochromator with 18-nn resolution.

4.2.2 Results and Discussion

Figure 4.3a shows the refractive indices of deposited SRSN:Er thin films, as measured by ellipsometry. We find that they all have refractive indices in excess of 2.0 in the visible range. By extrapolating the measured refractive index to the near-IR range using a single-pole Sellmeier equation, we estimate the refractive index at 1.54 μm to be in excess of 2.1 for all films considered (data now shown). However, the SRSN films remain highly transparent in the near-IR range. This is shown in Fig. 4.3b that shows the absorption coefficient of an SRSN film with excess Si content of 9 at.%. A clear absorption band edge near 650 nm can be observed.

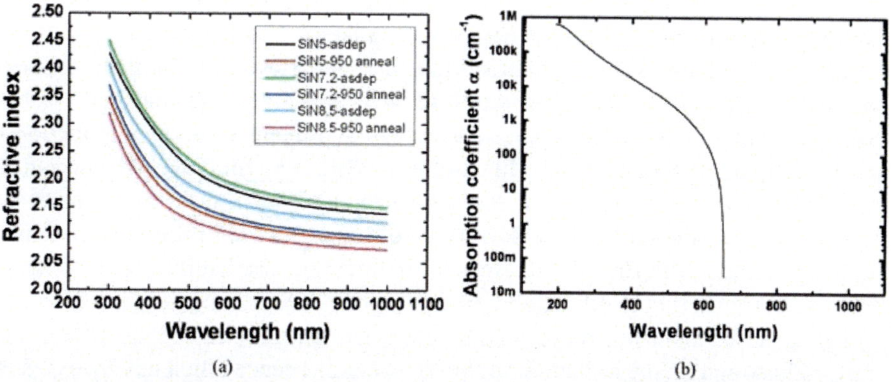

Fig. 4.3 (**a**) The refractive index of SRSN:Er thin films, as measured by ellipsometry. (**b**) The absorption coefficient of an SRSN:Er thin film with 8–9% excess Si. Note the clear absorption band edge near 650 nm

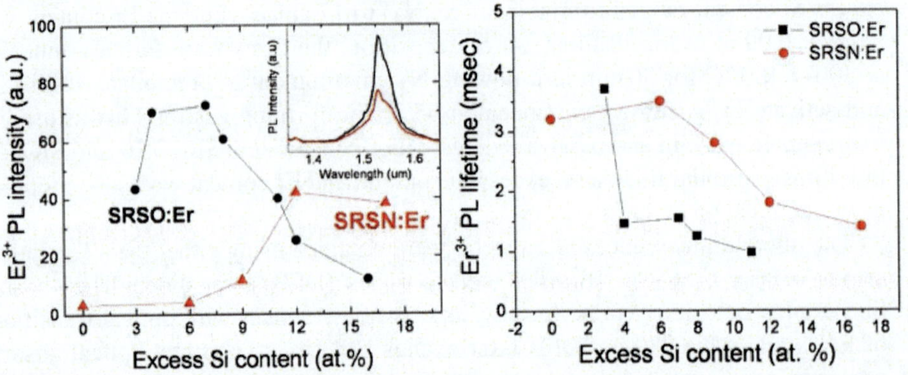

(a) (b)

Fig. 4.4 (**a**) The effect of excess Si content on nc-Si sensitized Er^{3+} luminescence from SRSO:Er and SRSN:Er thin films obtained with 200 mW of a 477-nm pump beam. All samples were annealed at 950 °C. The inset shows actual PL spectra from SRSO:Er and SRSN:Er films that displayed the maximum PL intensities. (**b**) The effect of excess Si content on the Er^{3+} luminescence lifetimes. Again, all samples were annealed at 950 °C

Figure 4.4a shows the effect of excess Si content on the Er^{3+} luminescence from SRSO:Er and SRSN:Er thin films, obtained with 200 mW of the 477 nm pump beam. All samples were annealed at 950 °C in this case. The inset shows actual PL spectra from SRSO:Er and SRSN:Er films that displayed the maximum PL intensities. Stoichiometric SiO_2 and Si_3N_4:Er show very little Er^{3+} PL, indicating that the presence of excess-Si is necessary for Er^{3+} sensitization for nitrides as well. In the case of SRSO:Er, the Er^{3+} PL intensity reaches a maximum near an excess Si content of 6 at.%, and decreases strongly as the excess Si content is increased further. In the case of SRSN:Er, a much higher excess Si content is needed for nc-Si sensitization to occur, and the maximum Er^{3+} PL intensity is reached near an excess Si content of 12 at.%. Note that the Er^{3+} PL intensity from SRSN:Er is lower than that from SRSO:Er, which seems to indicate that nc-Si sensitization is less effective in SRSN:Er. That is not the case, as will be shown later.

The effect of excess Si content on the Er^{3+} luminescence lifetimes is shown in Fig. 4.4b. Again, all samples were annealed at 950 °C. We find that for both SRSO:Er and SRSN:Er, increasing the excess Si content leads to a reduction of the Er^{3+} luminescence lifetimes. In case of the SRSO:Er film, the reduction in the Er^{3+} lifetime is quite drastic, as it decreases from nearly 3.7 to 1.5 ms as the excess Si content is increased by 1 at.% from 3 to 4 at.%. As the excess Si content is increased further, the Er^{3+} luminescence lifetime decreases further to 0.97 ms. In case of SRSN:Er, on the other hand, increasing the excess Si content from 0 all the way to 17 at.% results in a reduction in the luminescence lifetime from 3.2 to 1.4 ms only, indicating that the reduction of the Er^{3+} luminescence efficiency by excess Si is much less severe in SRSN, and can result in a luminescence efficiency that is even slightly higher than that in SRSO:Er.

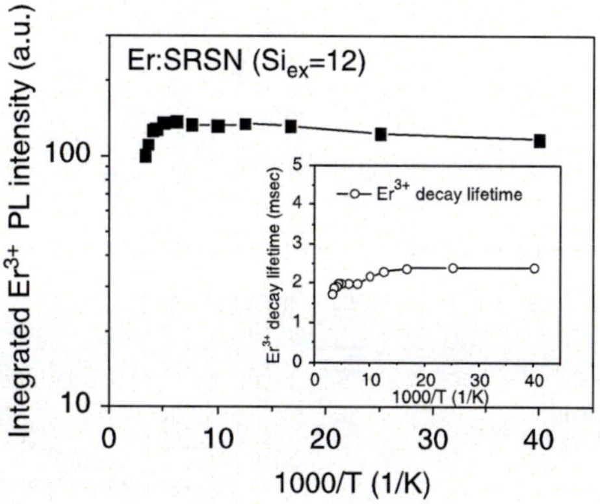

Fig. 4.5 The temperature dependence of the Er^{3+} PL intensity and lifetime for SRSN:Er. The Er^{3+} ions were excited using the 477-nm line of an Ar laser to ensure nc-Si sensitization

Figure 4.5 shows the effect of temperature on the Er^{3+} PL intensity and lifetimes from SRSN:Er. We find that neither the PL intensity nor the luminescence lifetimes from SRSN:Er show any significant temperature quenching, similar to what has been reported for SRSO:Er [52].

The results so far seem to indicate that while nc-Si sensitization does occur in SRSN:Er, it is not much different from, and does not possess a great advantage over, SRSO:Er in almost every way, especially when compared at the optimum condition of 6 at.% excess Si for SRSO:Er and 12 at.% excess Si for SRSN:Er. Such, however, is not the case when we compare the effect of increasing the annealing temperature beyond the optimum temperature of 950 °C. As shown in Fig. 4.6a, we observe a clear reduction of Er^{3+} PL intensity above the anneal temperature of 950 °C, reflecting the previously reported effect of Er clustering [12] and optical de-activation in the presence of nc-Si [20]. As shown in Fig. 4.6b, no such effect is observed from SRSN:Er, as the Er^{3+} PL intensity remains stable even after an anneal at 1,100 °C.

A clear difference that highlights the advantage of SRSN:Er, however, emerges in their Er^{3+} PLE spectra monitored at the Er^{3+} PL peak position of 1,535 nm, as is shown in Fig. 4.7. Also shown for comparison is the PLE spectrum from stoichiometric SiO_2:Er. The sharp kink near 490 nm is an optical artefact of the measurement system. In the case of SiO_2:Er, we clearly observe peaks at 380, 405, and 525 nm corresponding to ${}^4I_{15/2} \rightarrow {}^4G_{11/2}$, ${}^4I_{15/2} \rightarrow {}^2H_{9/2}$, and ${}^4I_{15/2} \rightarrow {}^2H_{11/2}$ direct optical transitions, respectively [53]. On the other hand, both SRSO:Er and SRSN:Er films show broad continuous PLE spectra, reflecting the broadband absorption spectrum of nc-Si. Note, however, that the nc-Si sensitized Er^{3+} PL intensity from the SRSN:Er film is much higher than that from the SRSO:Er film below the pump wavelength of 500 nm. Near the pump wavelength 350 nm, at which the nc-Si sensitized Er^{3+} PL is at the maximum for both SRSN:Er and SRSO:Er, the nc-Si sensitized Er^{3+} PL intensity from the SRSN:Er film is more than four times higher than that from the SRSO:Er film.

Part of the reason for the higher Er^{3+} PL intensity from SRSN:Er at shorter pump energies can be attributed to the larger bandgap of nitride-passivated nc-Si. The effect of surface passivation on the "bandgap" of nc-Si is well-documented

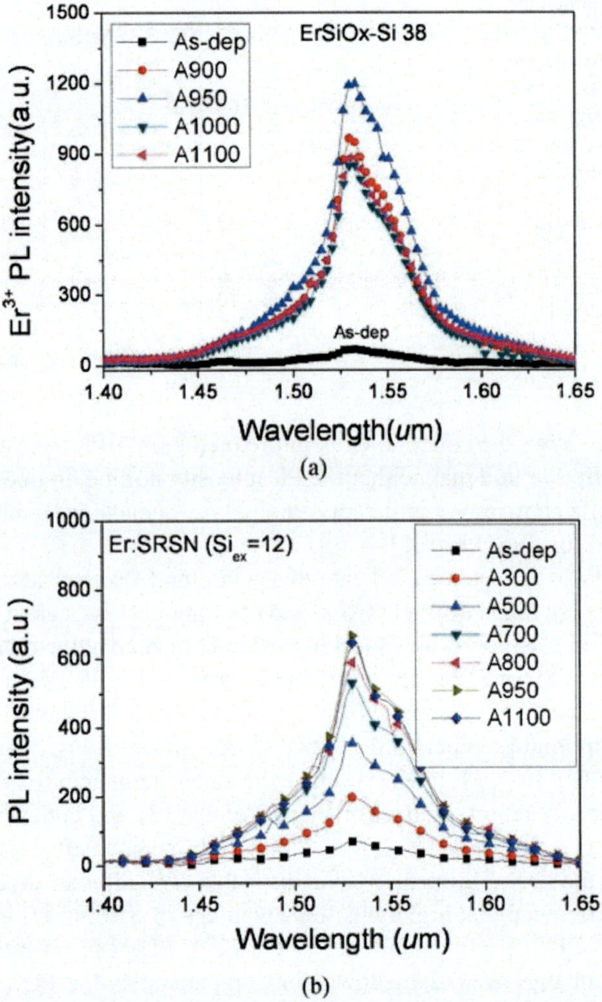

Fig. 4.6 (**a**) The effect of annealing temperature on nc-Si sensitized Er^{3+} emission from SRSO:Er. Note that the Er^{3+} PL intensity decreases for annealing temperatures above 950 °C; (**b**) the effect of annealing temperature on nc-Si sensitized Er^{3+} emission from SRSN:Er. The Er^{3+} PL intensity remains stable even after an anneal at 1,100 °C

[54], and it has been reported by several investigators that nitride passivated nc-Si shows visible luminescence at wavelengths that are much shorter than that from oxide-passivated nc-Si [55], and that host-sensitized excitation of a rare earth doped into nitride thin films requires pump wavelengths in the near UV range [56].

Still, the larger bandgap of nitride-passivated nc-Si does not explain the larger Er^{3+} PL intensity near the pump wavelength of 350 nm, when Er^{3+} PL intensities from both SRSO:Er and SRSN:Er are saturated. In fact, the higher Er^{3+} PL intensity from SRSN:Er is observed regardless of nc-Si sensitization. This is shown in Fig. 4.8, which shows the Er^{3+} PL spectra of the same films obtained under 980 nm

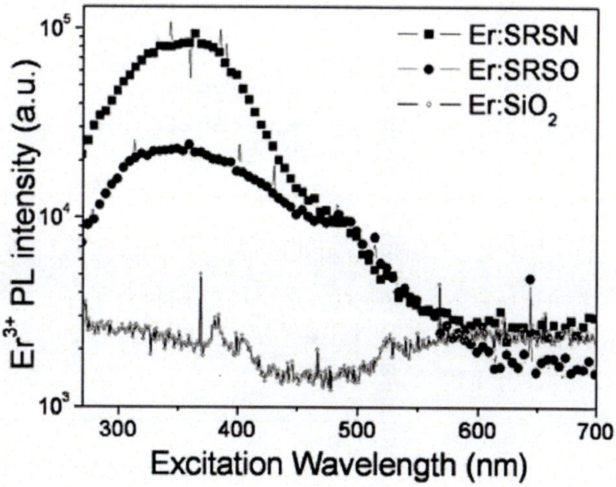

Fig. 4.7 The photoluminescence excitation (PLE) spectra of SRSN:Er, SRSO:Er, and SiO_2:Er, monitored at the Er^{3+} PL peak position of 1,535 nm. The sharp kink near 490 nm is an optical artefact of the measurement system. The sharp peaks at 380, 405, and 525 nm from SiO_2:Er correspond to the ${}^4I_{15/2}\rightarrow{}^4G_{11/2}$, ${}^4I_{15/2}\rightarrow{}^2H_{9/2}$, and ${}^4I_{15/2}\rightarrow{}^2H_{11/2}$ direct optical transitions, respectively

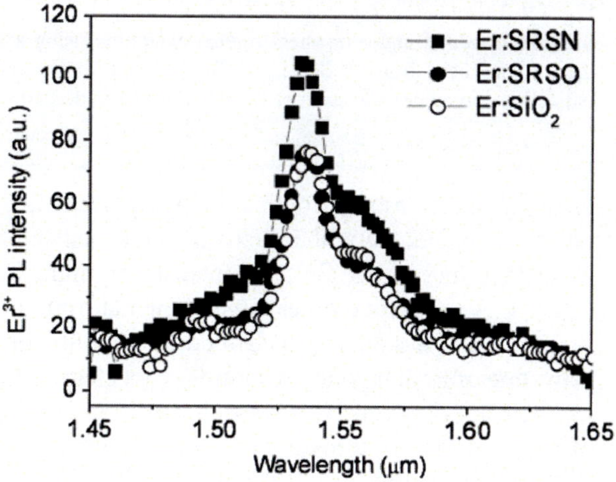

Fig. 4.8 Er^{3+} PL spectra from SRSN:Er, SRSO:Er, and SiO_2:Er obtained under resonant optical excitation using a 980-nm laser

excitation. As nc-Si sensitization is negligible at 980 nm [51], the Er^{3+} PL observed in this case is due to the direct optical excitation of Er^{3+} ions only. Furthermore, as the Er luminescence lifetimes from SRSO:Er and SRSN:Er are similar as well, the differences in PL intensities under direct optical pumping reflect the fraction of optically active Er^{3+} ions. We find that Er^{3+} PL intensity from SRSN:Er is higher than that from SRSO:Er even when excited directly. In fact, it is higher than even that from SiO_2:Er, which is often considered to be the standard for Er doping, and has an Er luminescence lifetime in the excess of 5 ms (data not shown).

A detailed quantitative comparison of PL intensities from different films can be difficult due to the possibility of refractive index-dependent optical artefacts [57]. However, the advantage of nitrides for Er-doping is real, as is further demonstrated in Fig. 4.9, which shows the anneal temperature dependence of the 980 nm pumped Er^{3+} PL intensity. We find that the Er^{3+} PL intensity from both SRSO:Er and SiO_2:Er decreases abruptly as the annealing temperature is increased beyond 950 °C, in agreement with previous reports of clustering and optical de-activation of

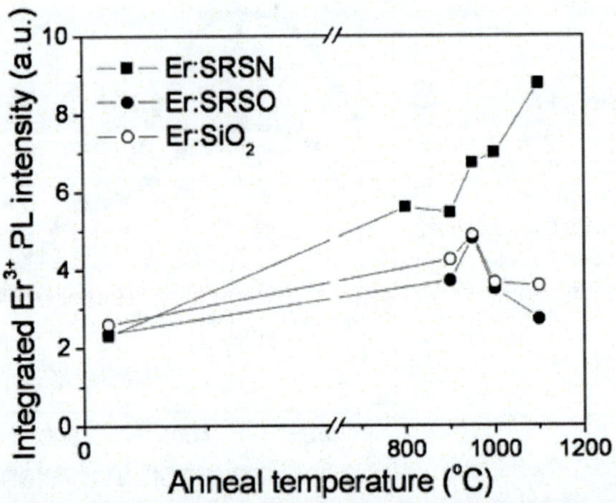

Fig. 4.9 The anneal-temperature dependence of the Er^{3+} PL intensity of SRSN:Er, SRSO:Er, and SiO_2:Er, obtained under resonant optical excitation using a 980-nm laser. Note the decrease in the Er^{3+} intensity from SRSO:Er and SiO_2:Er thin films at anneal temperatures above 1,000 °C, indicating optical de-activation of Er as distinct from de-coupling of Er from nc-Si

Er in silica at high temperatures [43]. In the case of SRSN:Er, however, the Er^{3+} PL intensity under resonant optical pumping keeps increasing with increasing anneal temperature, indicating that such optical de-activation of Er^{3+} is greatly suppressed in nitrides. Indeed, the Er^{3+} PL intensity from the SRSN:Er film after an anneal of 1,100 °C is nearly two times larger than that from the optimized SiO_2:Er film.

The exact reason for such thermal stability of SRSN:Er against optical de-activation of Er at high temperatures is not clear, but it is consistent with reports that metal diffusivity in nitrides is much lower than that in silica [58]. As atomic motion of both Er and Si is necessary for their reaction and subsequent optical de-activation of Er, suppression of Er diffusion in nitrides is consistent with the observed reduction of optical-deactivation of Er in SRSN even after high temperature annealing. For the purpose of Er doping, however, the implication of such suppression of optical de-activation of Er in nitrides is significant, as it opens up the possibility of a much greater degree of freedom in designing a nc-Si sensitized Er-doped Si thin film for use in optically active devices without needing to compromise between effective nc-Si sensitization and maintaining the optical activity of Er.

4.2.3 Conclusion

In conclusion, we have investigated the luminescence property and thermal stability of erbium-doped silicon-rich silicon nitride. We find nitrides offer both more effective nc-Si sensitization and much higher thermal stability for optically active Er than oxides, provided that higher pump energy is used to take into account the larger bandgap of nitride-passivated nc-Si. The results show that SRSN:Er is a promising alternative to SRSO:Er for a compact, low-cost Si-based light source.

4.3 Er Silicates

4.3.1 Experimental

Er silicates were formed by coating a dense array of Si nanowires (Si-NW) with a solution prepared by dissolving either pure $ErCl_3{\cdot}6H_2O$ or a mixture of $ErCl_3{\cdot}6H_2O$ and $YCl_3{\cdot}6H_2O$ in ethanol followed by rapid thermal annealing at high temperatures. Si-NWs were grown by the vapor-liquid-solid mechanism [59, 60] using an Au catalyst deposited on Si (111) substrates by sputtering. During growth, $SiCl_4$ was introduced into the growth furnace held at 900 °C by bubbling H_2 carrier gas through liquid $SiCl_4$ held at 0 °C at a flow rate of 1–10 sccm. Ar and H_2 gas was also introduced into the furnace at a flow rate of 100 sccm. The growth time was 20–30 min. After Si-NW growth, a solution prepared by dissolving either pure $ErCl_3{\cdot}6H_2O$ or a mixture of $ErCl_3{\cdot}6H_2O$ and $YCl_3{\cdot}6H_2O$ in ethanol was spin coated on the Si-NWs at 4,000 rpm for 1 min. Following spin-coating, a rapid thermal anneal at various temperatures and times were used to form the silicates. The use of Si-NWs is critical for uniform formation of silicates, as they provide the nanometer-sized diameter and large surface area that are better suited for chemical reaction and atomic transport between crystalline Si and Er solutions that are necessary for the formation of silicates [61]. Dilution of the Er with Y was done in order to control the Er concentration while keeping the crystalline structure constant, as Y forms silicate phases that have nearly the same lattice constants as Er silicates [62, 63]. Indeed, Czochralski-grown Er and Yb co-doped Y_2SiO_5 bulk crystals have been used successfully to fabricate solid state lasers [64]. Table 4.1 shows the weight and corresponding quantity in mol of $YCl_3{\cdot}6H_2O$ and $ErCl_3{\cdot}6H_2O$, both in solution and in the final silicates as determined by EDS.

The structure of the resulting silicate film was analyzed using fixed angle (θ) X-ray diffraction (XRD) and transmission electron microscopy (TEM). PL spectra were obtained using either the 488-nm line of an Ar laser or a 532-nm frequency-doubled Nd:YVO4 laser, a grating monochromator, and employing the standard lock-in technique. Note that in the case of silicates, excitation of Er occurs via direct

Table 4.1 The weights and corresponding quantities in mol of $YCl_3{\cdot}6H_2O$ and $ErCl_3{\cdot}6H_2O$ added to 2.0 g of ethanol. The actual concentrations of Er in the final silicate thin films were measured by energy-dispersive X-ray spectroscopy (EDS) to be 1.5, 2.5, 5, 8.8, 17, and 25 at.%, except for the sample with the nominal Er concentration of 0.26 at.% that was below the detection limit of EDS

Y (mg)	Er (mg)	*Y* (mol) $\times 10^{-4}$	Er (mol) $\times 10^{-4}$	Er at.%	Er at.% (EDS)
368.2	0	12.14	0	0	–
358.5	4.8	11.82	0.126	0.26	–
346.7	18.1	11.43	0.474	1.0	1.5
259.0	28.5	8.54	0.747	2.0	2.5
277.5	86.4	9.15	2.26	5.0	5.0
199.1	163.9	6.56	4.29	10	8.8
85.5	275.4	2.82	7.21	18	17
0	366.4	0	9.6	25	–

optical absorption only, as there is no excess Si to form nc-Si that could sensitize Er. We note, however, that Er ions in silicates can also be non-resonantly excited via carriers, if a Si layer can be made to intimately contact the silicate layers [65]. The pump beam size, as determined using a CMOS image sensor, was 0.15 mm^2 for the Ar laser and 0.14 mm^2 for the frequency-doubled Nd:YVO_4 laser. The PL decay traces were collected with a digitized oscilloscope. All spectra were corrected for the system response.

4.3.2 Results and Discussion

Figure 4.10 shows a scanning-electron microscope (SEM) image of as-grown Si-NWs at a tilt angle of (a) 45° and (b) 90°. We observe a dense array of straight Si-NWs of about 100 nm in diameter and ~10 μm in length with a Si-NW density of 1.4×10^9 cm^{-2}.

Figure 4.11a shows the SEM image of the sample that is spin-coated with pure Er solution and then annealed. We find that the Si-NWs have disappeared, and instead observe a "nano-bush" of roughly spherical grains. The formation of these grains is uniform across a large area, as is shown in Fig. 4.11b. In fact, the entire wafer turns uniformly white to the naked eye (data not shown).

The material phase of the "nano-bush" is identified by XRD analysis, as is shown in Fig. 4.12. From the position of the diffraction peaks, we identify the "nano-bush" to consist of nearly pure Er_2SiO_5 [62], which is a stable phase in the Er_2O_3–SiO_2 system [66] with a monoclinic symmetry of space group $P2_1/c$. More details about the shape of the silicate grains can be found in Fig. 4.13, which shows the TEM images of the fabricated silicate thin film. We find that the film consists of loose aggregates of Er_2SiO_5 nanocrystals that are about 50–200 nm in diameter. The nanocrystal grains are randomly oriented, but are single-crystalline, as is shown by the high-resolution TEM image shown in the inset.

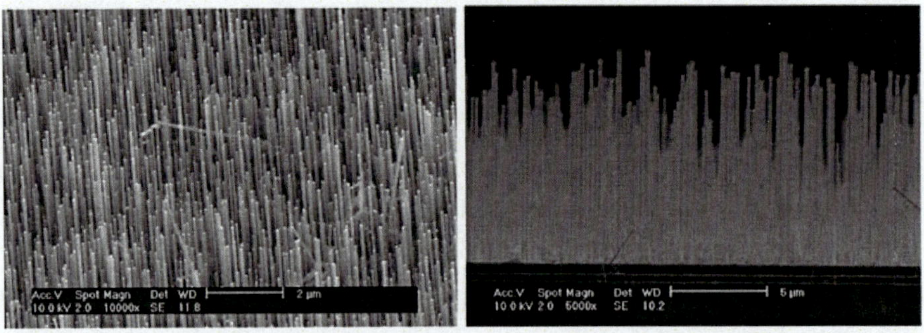

(a) (b)

Fig. 4.10 SEM images of Si-NWs grown via the VLS method for use in fabricating Er silicates (**a**) at a tilt angle of 45°, and (**b**) at a tilt angle of 90°

(a) (b)

Fig. 4.11 SEM images of Si-NWs that were spin-coated with pure Er solution and then annealed in order to fabricate Er silicates (**a**) at high resolution and (**b**) low resolution

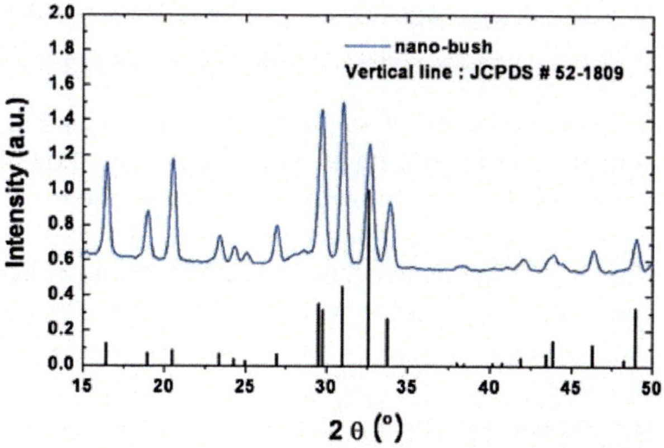

Fig. 4.12 The XRD spectrum of the fabricated silicate "nano-bush." Also shown are the expected peak positions and relative intensities for Er_2SiO_5. We obtain a near perfect match, indicating the formation of single-phase Er_2SiO_5

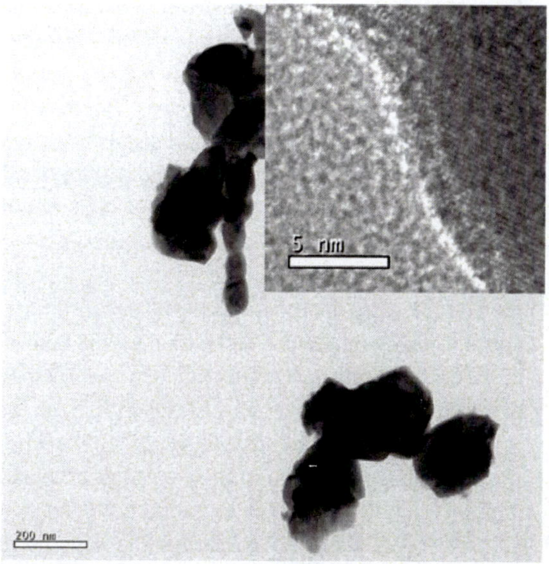

Fig. 4.13 The TEM images of fabricated Er_2SiO_5 nanograins. The scale bar represents 200 nm. The inset shows the high-resolution TEM image of an individual nanograin, showing well-developed lattice fringes indicative of high crystalline quality. The *scale bar* represents 5 nm

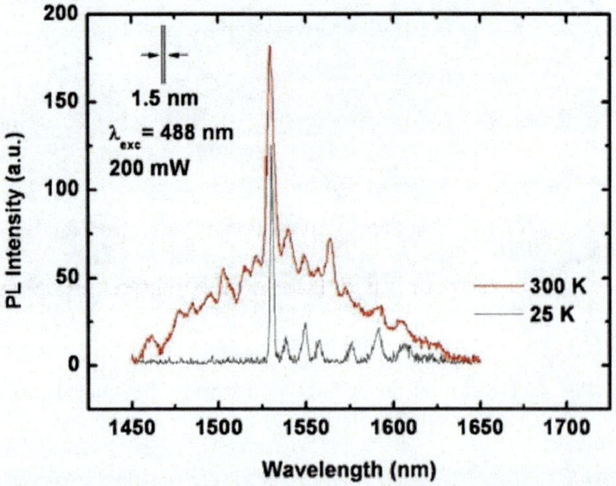

Fig. 4.14 PL spectra of the Er_2SiO_5 nanocrystals, measured both at room temperature and at 25 K using the 488 nm line of an Ar ion laser to optically excite the $^4I_{15/2} \rightarrow ^4F_{7/2}$ transition of Er^{3+} ions. At 25 K, the width of the central peak at 1,530 nm is limited by the system resolution, which is 1.5 nm

This formation of single-phase single-crystalline Er_2SiO_5 nanograins is confirmed from PL spectra as well. Figure 4.14 shows the PL spectra of the Er_2SiO_5 nanocrystals, measured both at room temperature and at 15 K using the 488 nm line of an Ar ion laser to optically excite the $^4I_{15/2} \rightarrow ^4F_{7/2}$ transition of Er^{3+} ions. We find that at low temperature, the PL spectrum is nearly atom-like with a series of sharp peaks, consistent with the single-crystalline nature of the fabricated silicate nanograins. The width of the central peak at 1,530 nm is limited by the system resolution, which is 1.5 nm. Even at room temperature, the central peak dominates the PL spectrum with a full width at half maximum (FWHM) of less than 4.5 nm, or 3 meV. It should be noted here that the spectra are not intentionally offset – the Er^{3+} PL intensity increases strongly as the temperature is raised. The peak intensity increases by 50%, while the integrated PL intensity increases more than sevenfold.

Such a sharp atom-like luminescence is not an inevitable byproduct of using a crystalline material. In many cases where a polycrystalline thin film is used, the presence of grain boundaries and strain often results in a very large inhomogeneous broadening not only from Er-doped polycrystalline Al_2O_3 [67], but also from pure Er_2O_3 as well [68]. In contrast, no such inhomogeneous broadening is observed from the Er_2SiO_5 film investigated here as it consists of a loose aggregate of high-quality single-crystalline nanograins. We also would like to note that that obtaining pure single-phase silicate material requires a careful control over the processing conditions. For instance, if we change the spin-coating speed to 1,000 rpm such that more of the solution remains on the Si nanowires, the resulting "nano-bush" consists of a mixture of Er_2SiO_5 and Er_2O_3 grains, as is shown in Fig. 4.15.

Given the Er concentration of 25 at.% in Er_2SiO_5, the maximum possible gain from Er_2SiO_5 would be in the excess of at least 200 cm^{-1}. In reality, however, such a high concentration leads to cooperative upconversion (CU) that limits the possible gain [69–71]. Figure 4.16 shows the schematic description of the CU process at the $^4I_{13/2}$ energy level. The CU is a dipole-dipole interaction between (a) two Er^{3+}

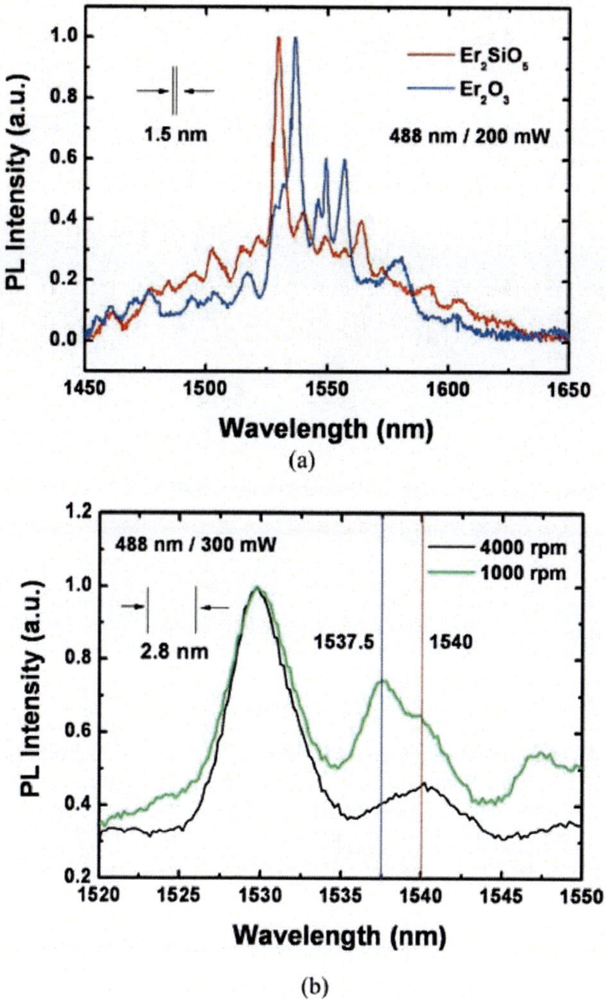

Fig. 4.15 Room temperature PL spectra of (**a**) pure Er_2SiO_5 and Er_2O_3 thin films; (**b**) Er silicate thin films formed by 4,000 and 1,000 rpm spin-coating. The one formed with 1,000 rpm spin coating shows peaks due to both Er_2SiO_5 and Er_2O_3 phases

ions in the first excited state, in which (b) an Er^{3+} ion in the first excited state ($^4I_{13/2}$) decays non-radiatively to the ground state ($^4I_{15/2}$) by exciting another Er^{3+} ion in the first excited state to the third excited state ($^4I_{9/2}$), which then (c) thermalizes rapidly to lower-lying states. Indeed, such CU can also occur between any two excited states of Er^{3+}. As CU annihilates one excited Er^{3+} ion without photon emission, it is an important loss mechanism that limits optical gain at a given pump power that needs to be investigated before the merits of Er silicates can be fully determined.

The effect of CU is investigated by observing the changes in the Er^{3+} luminescence properties as the Er^{3+} concentration is varied via Y dilution. Figure 4.17

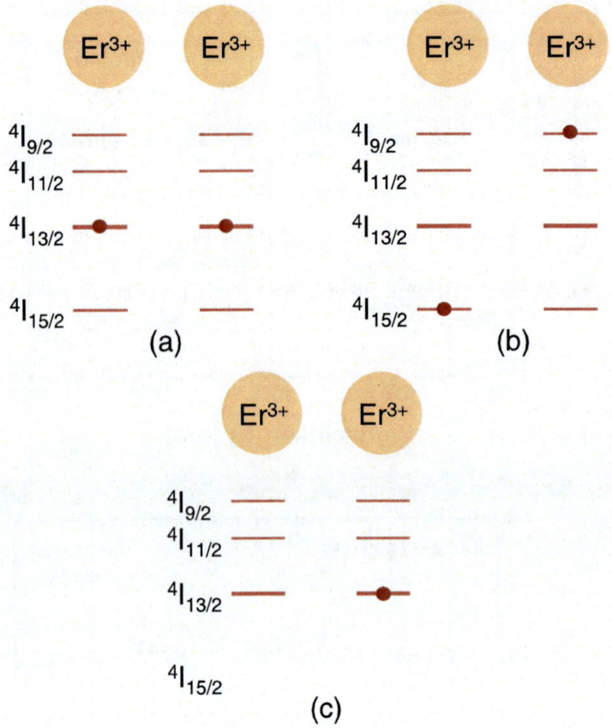

Fig. 4.16 The schematic description of the cooperative upconversion process at the Er^{3+} energy level $^4I_{13/2}$. The cooperative upconversion is a dipole-dipole interaction between (**a**) two Er^{3+} ions in the first excited state, in which (**b**) an Er^{3+} ion in the first excited state ($^4I1_{3/2}$) decays non-radiatively to the ground state ($^4I_{15/2}$) by exciting another Er^{3+} ion in the first excited state to the third excited state ($^4I_{9/2}$), which then (**c**) thermalizes rapidly to lower-lying states

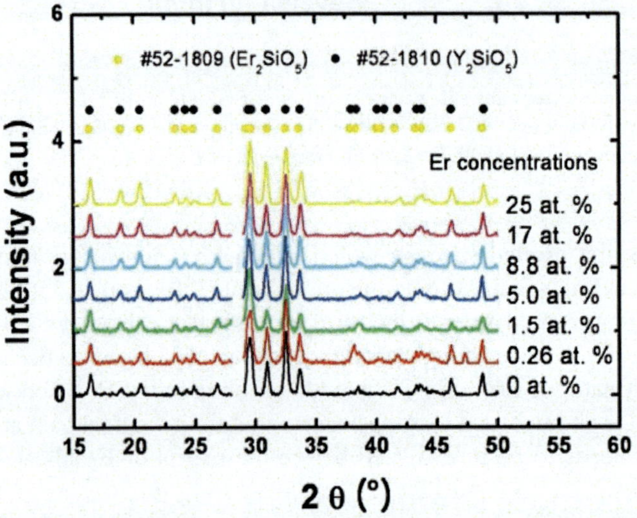

Fig. 4.17 The fixed θ X-ray diffraction (XRD) spectra from nanocrystals fabricated with various Er concentrations. Also shown are the expected XRD peak positions of Er_2SiO_5 and Y_2SiO_5

shows the X-ray diffraction (XRD) spectra from fabricated Er-Y silicate nanocrystals. The reported peak positions of Er_2SiO_5 and Y_2SiO_5 from the Joint Committee on Power Diffraction Standards (JCPDS) are also indicated. We observe identical spectra from all samples ranging from pure Y_2SiO_5 to pure Er_2SiO_5, with peaks that agree well with the reported peak positions from JCPDS [62].

Figure 4.18 shows the near-infrared PL spectra from the fabricated nanocrystals, normalized to the intensity at 1.53 μm. The shape of the PL peak near 1.53 μm due to the $^4I_{13/2} \rightarrow ^4I_{15/2}$ transition is the same for all Er concentrations, and is very similar to PL spectra from pure Er silicate thin films, as shown above. As the intra-4f transitions of Er^{3+} are parity forbidden but become allowed by the crystal field, the changes in the atomic environment of Er are reflected in the details of the Er PL spectrum shape. Thus, the fact that we observe the same sharp PL spectra for all Er concentrations, together with the XRD spectra shown in Fig. 4.17, clearly demonstrates that the crystal structure of the fabricated $Er_xY_{2-x}SiO_5$ nanocrystals is the same for all Er concentrations, right down to the atomic level.

Note that, on the other hand, the PL peak near 0.98 μm due to the $^4I_{11/2} \rightarrow ^4I_{15/2}$ transition is observable only for an Er concentration of 5.0 at.% or higher, even though the pump power was the same for all samples. A similar concentration dependence is shown in Fig. 4.19a, which shows the decay traces of the 1.53 μm Er^{3+} PL intensity for different Er^{3+} concentrations. At an Er^{3+} concentration of 0.26 at.%, we observe a single-exponential decay with a luminescent lifetime of 7.4 ms. As the Er concentration is increased further, the decay traces become non-exponential characterized by a fast initial decay. Figure 4.19b shows the decay traces of the 1.53 μm Er^{3+} PL intensity from the sample with 1.5 at.% Er at different pump powers. At 2 mW pump power, the decay trace is a single-exponential, with a luminescent lifetime of 8 ± 0.5 ms. As the pump power is increased, the decay

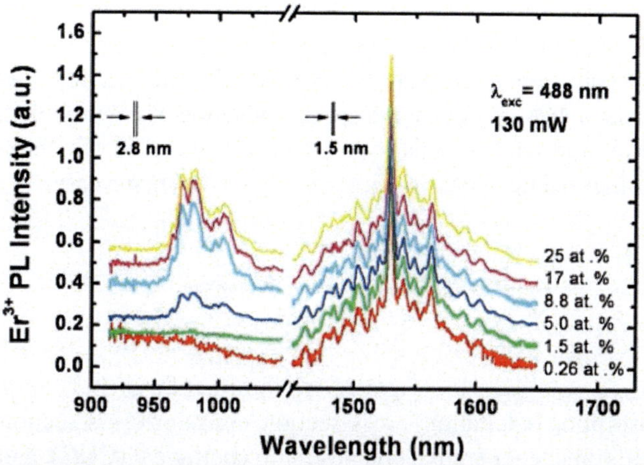

Fig. 4.18 The near-infrared PL spectra from the fabricated nanocrystals with various Er concentrations, normalized to the intensity at 1.53 μm. The curves are offset for clarity

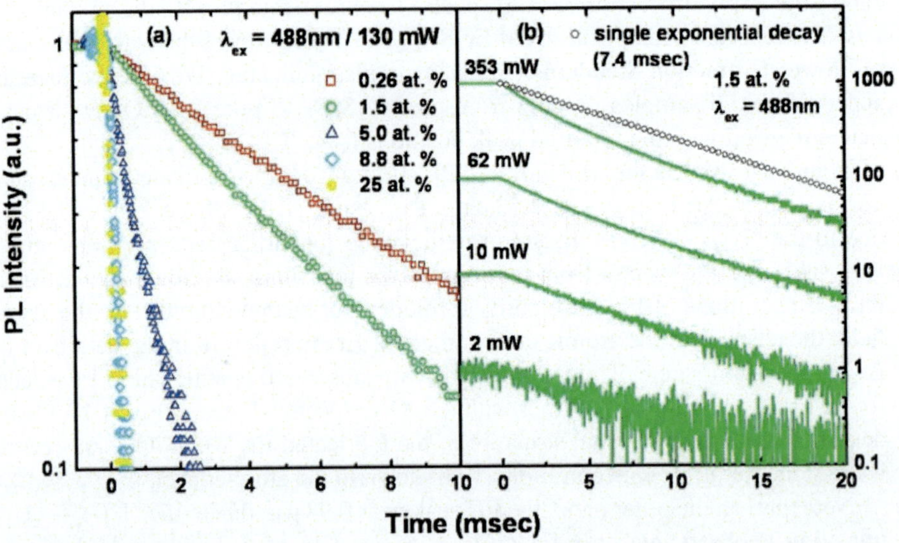

Fig. 4.19 (**a**) The time decay traces of the 1.53-μm Er^{3+} PL intensity for different Er^{3+} concentrations, after reaching steady state under a pump power of 130 mW. (**b**) The time decay traces of the 1.53-μm Er^{3+} PL intensity from the sample with 1.5 at.% Er at different pump powers

traces become non-exponential with shorter lifetimes. We note, however, that all decay traces, after the fast initial decay, approach a single exponential decay with a luminescent lifetime of 7.4 ms.

Such a concentration-dependent 0.98 μm Er^{3+} PL intensity, together with the concentration- and pump-power-dependent 1.53-μm Er^{3+} PL decay traces, bears the hallmark of CU. In investigating CU in $Er_xY_{2-x}SiO_5$, we first note that we do not observe a 0.98-μm PL peak at low Er concentrations even though we are exciting the 4f electrons to the higher-lying $^4F_{7/2}$ level by using the 488-nm pump beam. This indicates that the thermalization of the 4f electrons down to the $^4I_{13/2}$ level is very rapid, consistent with the high phonon energies reported for rare-earth oxyorthosilicates. [69, 72] Thus, we approximate the 4f electron distribution in low Er^{3+} concentration nanocrystals by a two-level model in which we consider the ground- and first-excited states of Er^{3+} only [70]. In such a case, we can write

$$\frac{dn(t)}{dt} = \sigma_{abs}\phi(1-n) - \sigma_{em}\phi n - \frac{n}{\tau} - CNn^2 \quad (4.1)$$

where n, σ_{abs}, σ_{em}, ϕ, τ, C, and N are the fraction of Er^{3+} in the first excited state, the absorption (excitation) cross-section, emission cross-section, pump photon flux, decay lifetime, cooperative upconversion coefficient (CUC), and concentration of Er, respectively. Note that since we are using the 488 or 532 nm pump, the emission cross-section at 1.54 μm is assumed to be 0. The analytic solution of Eq. (4.1) is

$$n(t) = \frac{1}{\tau}\left[\left(\frac{1}{\tau n(0)} + CN\right)\exp(t/\tau) - CN\right]^{-1} \tag{4.2}$$

$$n(0) = \frac{\sigma_{abs}\phi + \sigma_{em}\phi + 1/\tau}{2CN}\left\{\left[1 + \frac{4CN\sigma_{abs}\phi}{(\sigma_{abs}\phi + \sigma_{em}\phi + 1/\tau)^2}\right]^{1/2} - 1\right\} \tag{4.3}$$

In Eqs. (4.2) and (4.3), $n(0)$ represents the steady-state value of n, and $n(t)$ represents the time evolution n at time t after the pump beam has been turned off at $t = 0$. As $n(0)$ is proportional to the 1.53 μm Er^{3+} PL intensity, C can be found by fitting Eqs. (4.2) and (4.3) to the luminescence decay curve and the pump power dependence of the 1.53 μm Er^{3+} luminescence intensity, provided that σ and τ are known.

For τ, we use 7.4 ms, which is the lifetime observed under conditions where CU is negligible. To obtain σ, we have measured the rise time of the 1.53 μm Er^{3+} PL from the sample with 0.26 at.% Er at different pump powers, under the condition that no CU effects appear. The results are shown in Fig. 4.20. The slope of the curve then gives σ, which was found to be (a) $(4.5 \pm 0.2) \times 10^{-20}$ cm^2 and (b) $(2.3 \pm 0.1) \times 10^{-20}$ cm^2 at 488 and 532 nm, respectively. This value is much higher than the value of $\sim(2\text{–}3) \times 10^{-21}$ cm^2 typically reported [9] for the $^4F_{7/2}$ level of Er^{3+}, which we attribute to the multiple scattering of the pump beam by nanocrystal aggregates. Because of the multiple scattering, the effective excitation cross-section can vary from sample to sample, even though the true atomic excitation cross-section should be the same for all samples. To investigate the variation, we measured the Er^{3+} PL properties from more than 30 samples with 2.5 at.% Er and found the variation in the PL intensity to be about ±50% and the variation of the Er^{3+} luminescence decay time to be ±10% (data not shown). Based on these data, we calculate the excitation cross-section of Er^{3+} in $Er_xY_{2-x}SiO_5$ to be $(4.5 \pm 2.3) \times 10^{-20}$ cm^2 at 488 nm and $(2.3 \pm 1.2) \times 10^{-20}$ cm^2 at 532 nm.

Figure 4.21 shows (a) the luminescence decay curve at the pump power of 650 mW and (b) the pump power dependence of the 1.53 μm Er^{3+} luminescence

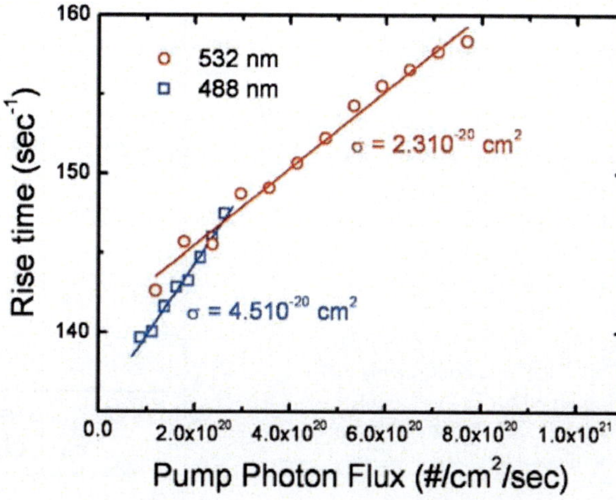

Fig. 4.20 The rise time of 1.53 μm Er^{3+} PL from the sample with 0.26 at.% Er at different pump powers at the pump wavelengths of 488 and 532 nm, under the condition that no CU effects appear. The slope then gives the effective excitation cross-section indicated

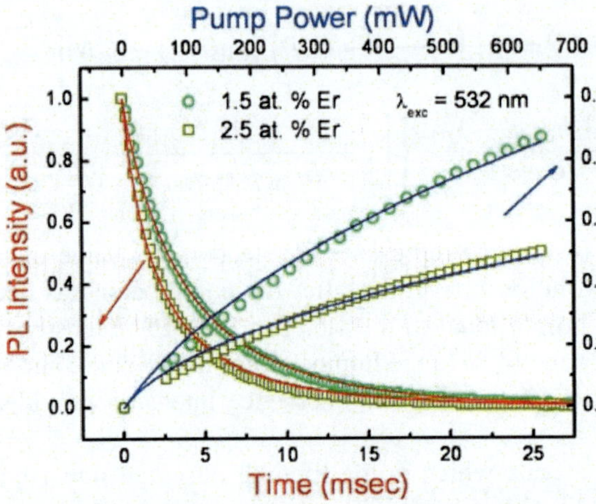

Fig. 4.21 The luminescence time decay curve at a pump power of 650 mW and the pump power dependence of the 1.53-μm Er^{3+} luminescence intensity for 1.5 and 2.5 at.% Er at the pump wavelength of 532 nm. The fitting results of $C = 1.7 \times 10^{-18}$ cm^3/s with $n(0) = 0.086$ for 1.5 at.% Er and $C = 5.1 \times 10^{-18}$ cm^3/s with $n(0) = 0.049$ for 2.5 at.% Er (when $\sigma = 2.3 \times 10^{-20}$ cm^2 is used) are shown by the *solid curves*

intensity for 1.5 and 2.5 at.% Er. By using a single set of parameters to fit both curves to Eqs. (4.2) and (4.3) with a σ value in the range of $(2.3\pm1.2)\times10^{-20}$ cm^2, we obtain values of $C = (2.2\pm1.1)\times10^{-18}$ cm^3/s with $n(0) = 0.08\pm0.04$ for 1.5 at.% Er and $C = (5.4\pm2.7)\times10^{-18}$ cm^3/s with $n(0) = 0.05\pm0.03$ for 2.5 at.% Er. The fitting results of $C = 1.7\times10^{-18}$ cm^3/s with $n(0) = 0.08$ for 1.5 at.% Er and $C = 5.1 \times 10^{-18}$ cm^3/s with $n(0) = 0.05$ for 2.5 at.% Er (when $\sigma = 2.3 \times 10^{-20}$ cm^2 is used) are shown by solid curves in Fig. 4.21

These values are very low, especially given the high Er concentrations used. Figure 4.22 shows the previously reported values of CUC [73], together with the

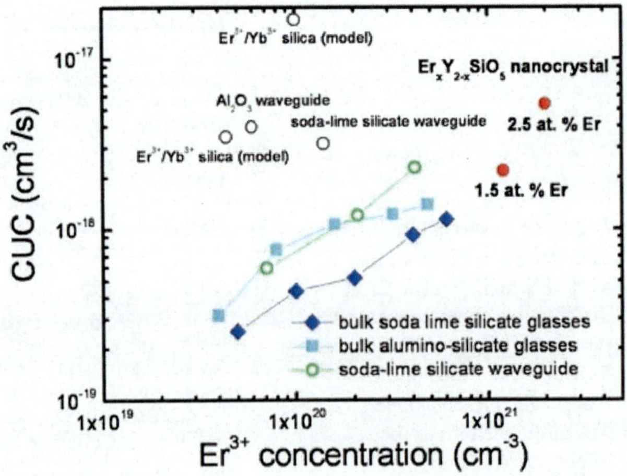

Fig. 4.22 Reported values of the cooperative upconversion coefficient at the $^4I_{13/2}$ energy level of Er in various hosts from [73]. For comparison, the cooperative upconversion coefficients obtained for $Er_xY_{2-x}SiO_5$ with 1.5 and 2.5 at.% Er are indicated by *filled circles*

value of CUC obtained here. In case of Er-doped silica, C was reported to be as high as $1.7\times10^{-17}\,cm^3/s$ at an Er concentration of $1\times10^{20}\,cm^{-3}$. [74] A value of $4\times10^{-18}\,cm^3/s$ was reported for Er-doped Al_2O_3, but at an Er concentration of only $6\times10^{19}\,cm^{-3}$. [75] A lower value of $3\times10^{-18}\,cm^3/s$ had been reported for Ge/Al doped fused-silica optical fibers, but only at a low Er concentration of $\sim2\times10^{19}\,cm^{-3}$. [76] A low value of $1.2\times10^{-18}\,cm^3/s$ at an Er concentration of $6\times10^{20}\,cm^{-3}$ has been reported, but in the case of bulk soda-lime silicate glasses prepared from batches melted at 1,400 °C for as long as 50 h. [73]. In other words, the CUCs obtained for $Er_xY_{2-x}SiO_5$ nanocrystals are among the lowest ever reported, at an Er concentration that is among the highest ever reported.

Such a low value of C is critical for efficient optical amplification at the high Er concentrations necessary for compact optical amplifiers, and demonstrates the viability of using $Er_xY_{2-x}SiO_5$ nanocrystals. The exact reason for such a low value of C is not yet clear. We postulate, however, that the high crystalline quality of the fabricated $Er_xY_{2-x}SiO_5$ nanocrystals is at least partially responsible, as the crystal structure of oxyorthosilicates has only two sites for rare earth ions that are separated by either a bridging oxygen or a SiO_4 tetrahedron [77], and thereby precludes the possibility of the formation of Er clusters that are known to lead to high values of C. [71] In addition, the near-perfect lattice matching between Er_2SiO_5 and Y_2SiO_5 allows dilution of Er within the matrix without the formation of Er_2SiO_5 domains.

For amplifier applications, it is more likely that either a 0.98 μm or 1.48 μm pump beam will be used. We can use Eq. (4.3) to obtain the inversion curve under the 1.48 μm pump, as shown in Fig. 4.23. The inversion level is defined as the fraction of the population at the $^4I_{13/2}$ state, N_2/N. In case of the 1,480 nm pump, stimulated emission as well as absorption of pump light should be considered. The values of absorption and emission cross-section of Er in $Er_xY_{2-x}SiO_5$ at 1.48 μm are obtained based on the peak absorption cross-section of Er^{3+} in $Er_xY_{2-x}SiO_5$ ($\sim 1.1\times10^{-20}\,cm^2$) and the McCumber theory. The simulation results indicate that population inversion can be achieved at a pump intensity of 0.6 mW/μm^2, and

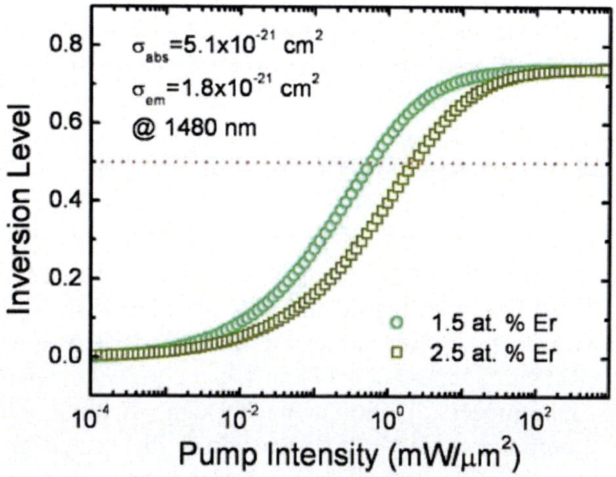

Fig. 4.23 The simulated inversion level as a function of 1,480 nm pump intensity of $Er_xY_{2-x}SiO_5$ with 1.5 and 2.5 at.% Er in considering only cooperative upconversion at the $^4I_{13/2}$ state. Here, we consider stimulated emission as well as absorption of pump light, which limits the maximum inversion level to ~0.7

23 dB/cm gain, corresponding to 70% inversion, can be achieved at a pump intensity of 6.5 mW/μm^2 for 1.5 at.% Er. Even higher gain could be possible by using a higher Er concentration. In 2.5 at.% Er, population inversion can be achieved at a pump intensity of 2.2 mW/μm^2, and 38 dB/cm gain, corresponding to 70% inversion, can be achieved at a pump intensity of 25 mW/μm^2. Note, however, that we only consider the cooperative upconversion for the simulation result. The inclusion of other effects such as excited state absorption and higher order cooperative upconversion can lead to less than optimum results. Furthermore, that the silicates need yet to be fabricated in a dense continuous thin film before they can be used for photonic devices, and it still needs to be determined whether such a dense thin film will have the advantageous properties of nanograins.

4.4 Applications of Er-Doped SRSN: High-Q Microdisks

4.4.1 Experimental

A 330 nm thick Er-doped a-SiN_x film (8–9 at.% excess Si) with 0.2 at% (2×10^{20} cm^{-3}) Er was deposited on Si substrate at room temperature by means of the reactive ion beam sputtering deposition method using a combination of 600 eV Ar^+ and N^+ beams. Subsequently, a 280-nm-thick SiO_2 layer was deposited on the top to be used as a hard mask. After deposition, the films were annealed at 950 °C for 30 min in a flowing Ar environment to both precipitate nc-Si and activate Er. Finally, the films were hydrogenated by a 30 min anneal in a flowing forming gas environment (10% H_2/90% N_2) at 650 °C to passivate defects. The finished samples all showed strong Er^{3+} luminescence with a luminescence lifetime of 0.69 ms when pumped off-resonantly at 477 nm that did not coincide with any optical transitions of Er^{3+} ions, indicating both successful optical activation and nc-Si sensitization of doped Er ions.

Disk patterns 25.2 μm in diameter were formed using an ELS-7000 electron beam lithography system. Afterwards, disks were formed by dry-etching using a combination of CHF_3/O_2 gases in an inductively coupled plasma (ICP). Finally, an HNO_3/HF wet etch was used to undercut the Si layer and form a pedestal.

4.4.2 Results and Discussion

Figure 4.24 shows the SEM image of the fabricated microdisk. The pedestal is 11 μm high, with a center contact patch diameter of 5.2 μm. The surface roughness of the film before and after the microdisk fabrication is shown in Fig. 4.25. Prior to microdisk fabrication, the film is very smooth, with a root mean square (RMS) roughness of 0.14 nm with a correlation length of 22.1 nm. After the fabrication process, the film is much rougher, with an RMS roughness of 1.15 nm with a correlation length of 25 nm. Using the analytical expression derived by Payne et al. [78] that relates surface roughness to scattering loss, we estimate that the

Fig. 4.24 SEM image of a fabricated SRSN:Er microdisk. The pedestal is 11 μm high, with a center contact patch diameter of 5.2 μm

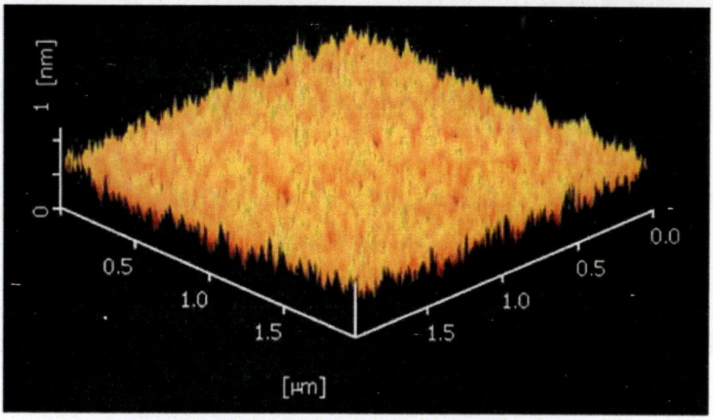

(a)

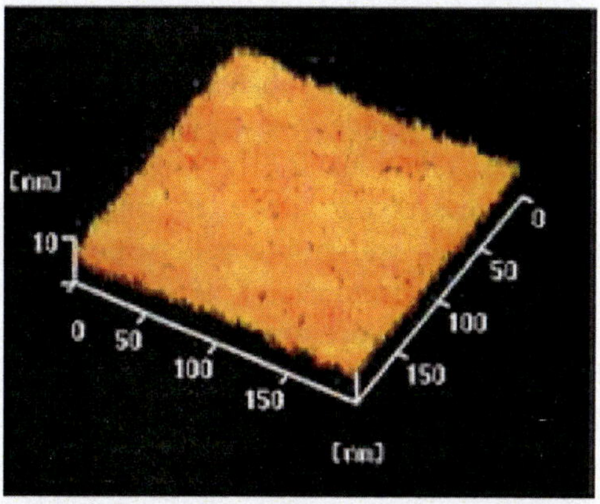

(b)

Fig. 4.25 AFM image of the SRSN:Er film (**a**) prior to and (**b**) after microdisk fabrication. The fabrication process increases the rms roughness from 0.14 to 1.15 nm

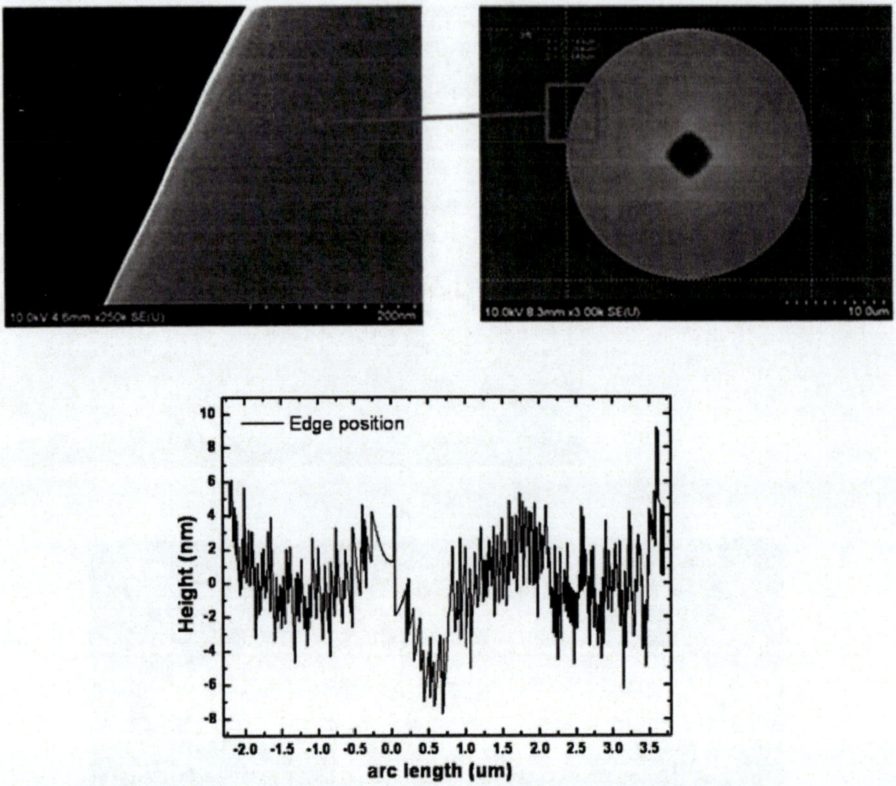

Fig. 4.26 *Top*: High-resolution SEM image of the microdisk edge. *Bottom*: The position of the disk edge, as measured from the SEM image. Positive height denotes position beyond the mean radius, and negative height denotes position below the mean radius

surface roughness contributes about 1 dB/cm to the loss factor. Another factor that can contribute to scattering loss is the edge roughness. The sidewall roughness was measured by extracting the contour from a high-resolution SEM image of the disk edge, as is shown in Fig. 4.26. Using the analytical expressions in [79] and [80] that relate edge roughness of a waveguide to scattering loss, we estimate that the edge roughness will contribute an additional loss factor of about 1 dB/cm.

The $|E|^2$-field distributions of the first-order transverse electric-like (TE) and transverse magnetic-like (TM) whispering gallery modes (WGM) of the fabricated microdisk, with azimuthal numbers of 79 and 60, respectively, simulated by the three-dimensional finite-difference time-domain (FDTD) method are shown in Fig. 4.27. We find that the TE-like mode is confined very strongly within the disk with a mode overlap of 0.94 ± 0.01. Such tight confinement leads to a negligible radiation loss such that the theoretical Q-factor greatly exceeds 10^6. Note, however, the large mode overlap also leads to a large absorption by Er. Assuming a typical Er absorption cross-section of 0.8×10^{-20} cm^2 [81] we expect the Er absorption to contribute about 6.5 dB/cm to the cavity loss at the wavelength of at 1.54 μm. The TM-like mode, on the other hand, extends far out of the disk, resulting in a mode overlap of 0.50 ± 0.02, and a radiation limited theoretical Q-factor of only 80,000.

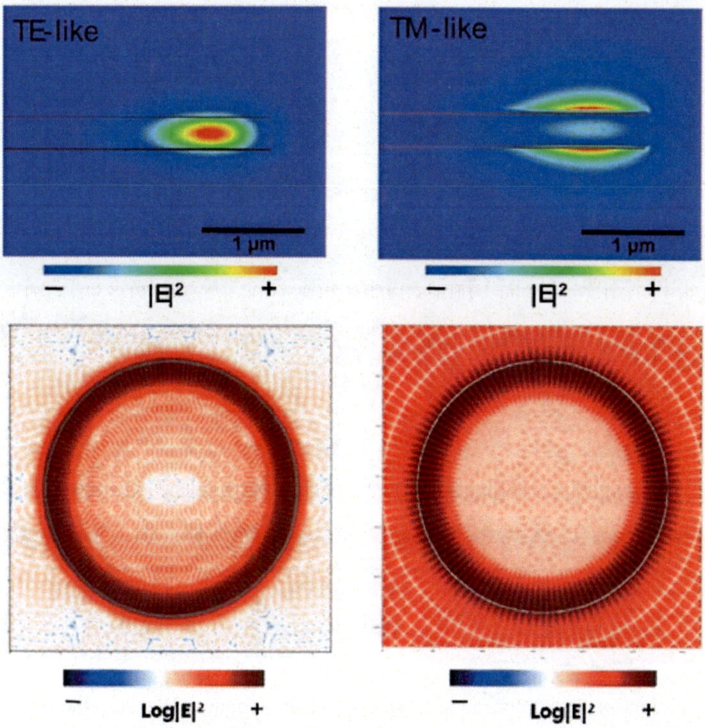

Fig. 4.27 FDTD-simulated $|E|^2$-field distributions of the first-order transverse electric-like (TE) and transverse magnetic-like (TM) WGM modes of the fabricated microdisk, with an azimuthal number of 79 and 60, respectively. Note that the plan-view image is given in log scale to accentuate the large radiation loss of the TM-like mode

The actual Q-factors of the fabricated microdisks were measured using a U-bent tapered fiber [82] mounted on a three-axis translator with sub-nm resolution. The diameter of the tapered section was measured to be 1.36 μm. Due to the strong electrostatic force between the fiber and the disk, all measurements were performed with the fiber directly in contact with the disk edge. Light from a tunable laser (1,475–1,545 nm) with a linewidth <300 kHz and controlled polarization was then coupled into the tapered fiber. Resonance modes of the microdisk were investigated by measuring the change in the transmitted beam intensity as the signal wavelength was swept. The details of the measurement setup are shown in [83].

Figure 4.28 shows a CCD camera image of the disk in contact with the tapered fiber. The U-bent fiber is essentially straight at this scale, and is resting on the top edge of the disk. Also shown are the transmission spectra for TE- and TM-polarized input signals. We obtain clear dips in the transmitted signal intensity. In particular, we obtain major dips occurring in pairs near 1,490, 1,505, and 1,520 nm, corresponding to a free spectral range of 14.4 nm. This corresponds to a modal index and effective resonator diameter of 1.98 and 25.3 μm, respectively, in excellent agreement with the data presented above. Furthermore, the position of these transmission dips agree very well with the expected positions of the first- and second-order radial TE-like modes, calculated by FDTD simulations (indicated

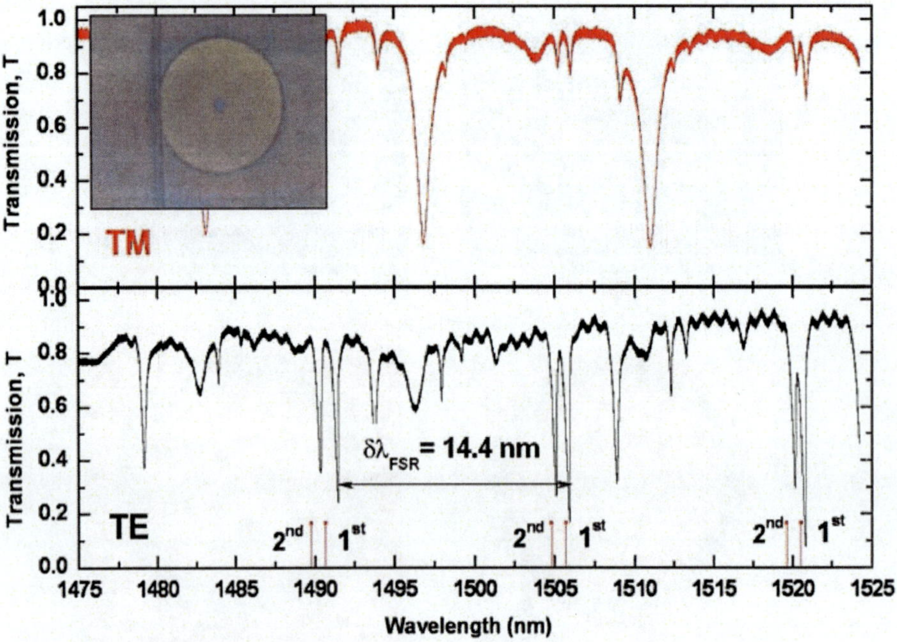

Fig. 4.28 The transmission spectra for TM- and TE-polarized input light, together with the expected position for first-order radial modes for TE-like modes. Shown in the *inset* is the optical microscope image of the SRSN:Er microdisk in contact with the coupling fiber

by short vertical lines). Thus, we identify them to be the first- and second-order whispering gallery modes. The remaining transmission dips are ascribed to be higher order radial modes. The measured Q-factors of the first-order TE-like modes, obtained by dividing the resonance wavelength by the transmission dip width, are $\sim 5{,}000$ and $\sim 8{,}900$ for the disks with 0.2 and 0.02 at.% Er, respectively. For the higher order TE-like modes, the Q-factor can be as high as 13,000. For the TM-like mode, however, the Q-factors are much lower, being around 1,000 only.

These Q-factors are comparable to those reported previously from SiN microring resonators [84], but much lower than the intrinsic Q-factors estimated by FDTD. Surface and edge scattering and absorption by Er cannot be the reason for such low values, as it can induce a loss of at most 10 dB/cm, corresponding to a Q-factor of 46,900 for the TE-like mode. On the other hand, the microdisk cavity is coupled very strongly to the fiber due to the direct contact of the fiber with the disk. Furthermore, the fact that the weakly coupled higher-order modes consistently show a much higher Q-factor suggests that the coupling fiber itself is a significant source of the loss. Therefore, we have extracted the fiber-induced loss using the coupled mode theory [85, 86]. As shown in Fig. 4.29, the theory considers the loss due to the coupling with the fiber in fitting the Lorentzian dip in the transmission spectrum.

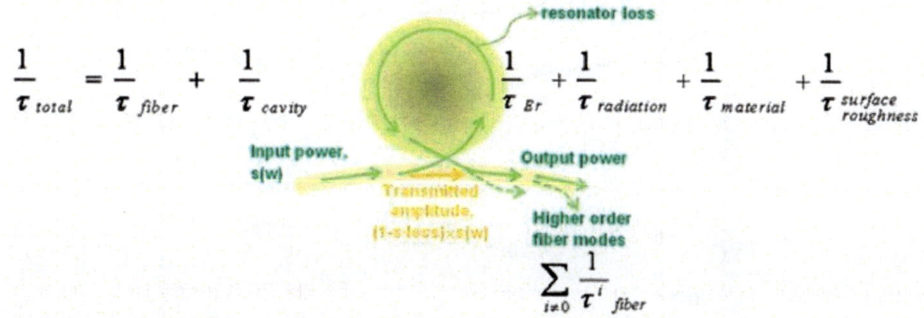

Fig. 4.29 Terms considered in coupled mode theory for extracting the cavity Q-factor from measured transmission spectra

Since the coupling fiber is very thin, we neglect the higher order mode of the fiber, in which case we can describe the Lorentzian dip by

$$\begin{aligned} \text{FWHM} &= 1/\tau_{\text{cavity}} + 1/\tau^{0}_{\text{fiber}} \\ \text{Lorentzian dip} &= (1 - s \cdot \text{loss}) \times \left\{ 1 - \frac{\left(1/\tau_{\text{cavity}} - 1/\tau^{0}_{\text{fiber}}\right)^2}{\left(1/\tau_{\text{cavity}} + 1/\tau^{0}_{\text{fiber}}\right)^2} \right\} \end{aligned} \tag{4.4}$$

where $1/\tau_{\text{cavity}}$ represents the entire cavity loss (including scattering, absorption, and radiation), $1/\tau_{\text{fiber}}$ represents the fiber coupling loss, and s·loss represents the loss of the input power at the fiber-cavity contact.

Figure 4.30 shows a detailed fit to a transmission dip due to a first-order radial mode of the microdisk, and the results of the fits. We find that TE-like modes coupling with the fiber dominate the overall loss. The intrinsic cavity Q-factor for the first-order radial modes can be as large as 26,580. Since this contains the Er absorption loss, the intrinsic Q factor of the resonator structure itself is much higher, and is estimated to be 4.5×10^4. This corresponds to a loss of about 9 dB/cm, which is significantly larger than the calculated values. For the TM-like modes, on the other hand, the Q-factor is dominated by the intrinsic cavity Q-factor, which can be as low as 2,100, corresponding to a very large cavity loss of >180 dB/cm that is several orders of magnitude larger than the calculated values.

In order to investigate the origin of the large cavity losses, we have performed FDTD simulations that included the coupling fiber. The results of the simulation, together with the structure used in the simulation, are shown in Fig. 4.31. As shown in Fig. 4.31b, we find that with the fiber in contact with the disk the TE-like mode excited in the disk decays exponentially with a decay time of 0.02 ns, corresponding to a Q-factor of only 26,520 even though no scattering or absorption losses were included, indicating the coupling with the fiber is a major source of optical loss for the microcavity. In addition, as Fig. 4.31c shows, the shape of the H_z field is no

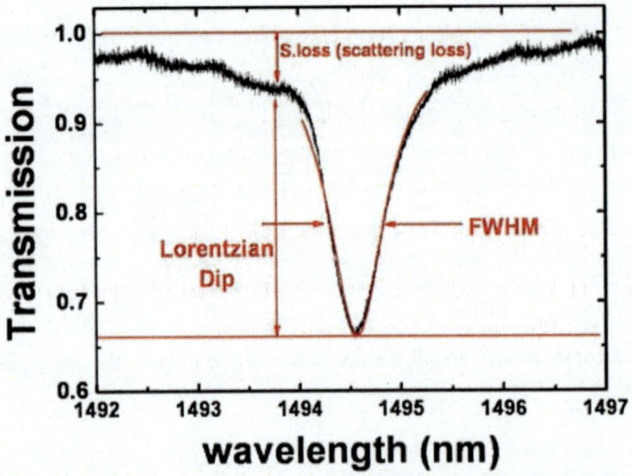

	Er 0.2 at.%			
TE-like	$Q^0{}_{fiber}$	Q_{cavity}	$Q_{cavity\ without\ Er}$	$Q_{measured}$
1st order @ 1492 nm	6990 (57 dB/cm)	26,580 (14 dB/cm)	45,000	5700 (max)
Higher order @ 1498 nm	1.4×10^5 (2.7 dB/cm)	10,017 (38 dB/cm)	12,900	9400 (max)
TM-like				
1st order @ 1483 nm	2104	5745	6100	1540

Fig. 4.30 A typical fit to the transmission spectrum using the coupled mode theory, and results of the fits for TE-like and TM-like modes

longer circular even though only the first-order radial mode was excited, indicating that the fiber induces a cavity loss of the first-order TE-like mode via coupling with higher-order modes as well. On the other hand, as Figs. 4.31d and e show, the values of $|H_z|^2$ and $|H|^2$ are identical, indicating that the overall mode remains TE-like.

With the fiber in contact with the disk, the TM-like mode decays even more rapidly with a decay time of 3.75 ps, corresponding to a Q-factor of 5,560, as shown in Fig. 4.31f. This is partly due to the large mode overlap of the TM mode with the fiber that results in a stronger coupling with the fiber, as indicated by the large light intensity in the fiber shown in Fig. 4.31g. Another important reason for such low Q-factor is shown in Figs. 4.31h and i. We find that the value of $|H_z|^2$, which corresponds to the TE-like component, is within 99.5% of the total $|H|^2$, indicating that the overall steady-state mode is TE-like even though the disk was excited via a TM-polarized source within the disk. This indicates that the contact with the fiber

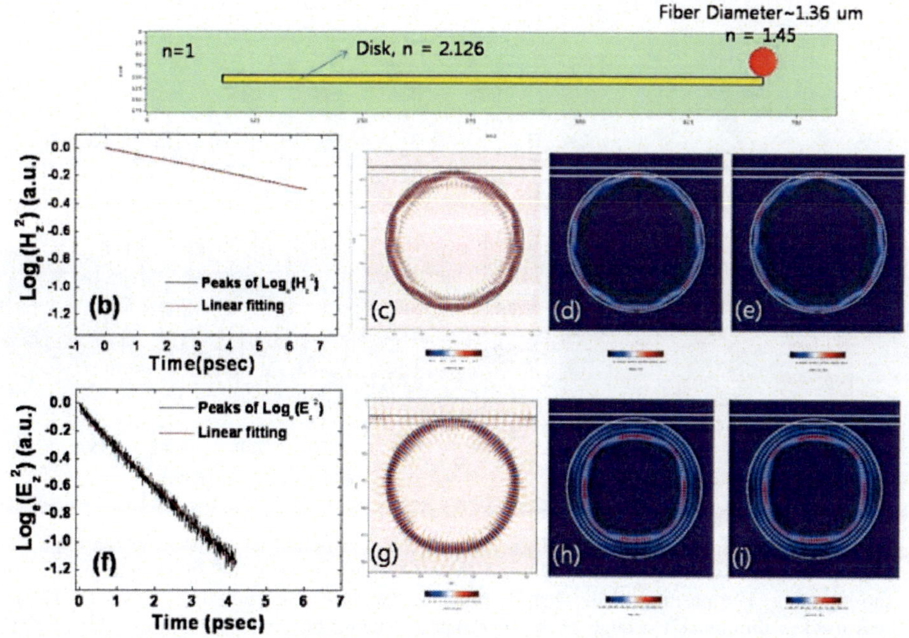

Fig. 4.31 (**a**) The structure used in calculations that include the coupling fiber in contact with the disk. A Gaussian-shaped source located inside the disk was used to excite the cavity; (**b**) calculated decay of the excited TE-like mode; (**c**) H_z field; (**d**) $|H_z|^2$ distribution, with a maximum value of 1 a.u.; and (**e**) $|H|^2$ distribution, with a maximum value of 1 a.u. of the excited TE-like mode. (**f**) Calculated decay of the excited TM-like mode; (**g**) E_z field; (**h**) $|H_z|^2$ distribution, with a maximum value of 0.995 a.u.; and (**i**) $|H|^2$ distribution, with a maximum value of 1 a.u. of the excited TM-like mode

breaks the symmetry to such a degree that the disk effectively ceases to support the TM-like modes, resulting in a large cavity loss (as distinct to coupling loss) for the TM-like modes.

Such mode coupling, and the resulting cavity loss, depends sensitively on the details of the fiber-disk coupling [84]. Thus, the values obtained above cannot be compared directly with experimental values. Still, the simulated values agree quite well with the experimentally obtained values, indicating that the contact with the coupling fiber, rather than scattering or absorption in the microdisk, plays a major role in determining the Q-factor of the fundamental WGM of the microdisk, by inducing not only the coupling loss, but also a large cavity loss due to symmetry-breaking induced coupling of the fundamental mode with other cavity modes. This is consistent with recent results obtained by several investigators who have reported that with optimization SiN-based waveguides can have propagation losses that are less than 1 dB/cm and a bending loss of less than 1 dB for a 1-μm radius bend [5, 87]. Thus, the results indicate that with an optimized processing condition and better coupling scheme (e.g., with a well-defined waveguide instead of a fiber in contact with the disk), it should be possible to increase the disk Q-factor to a point where net optical gain is possible.

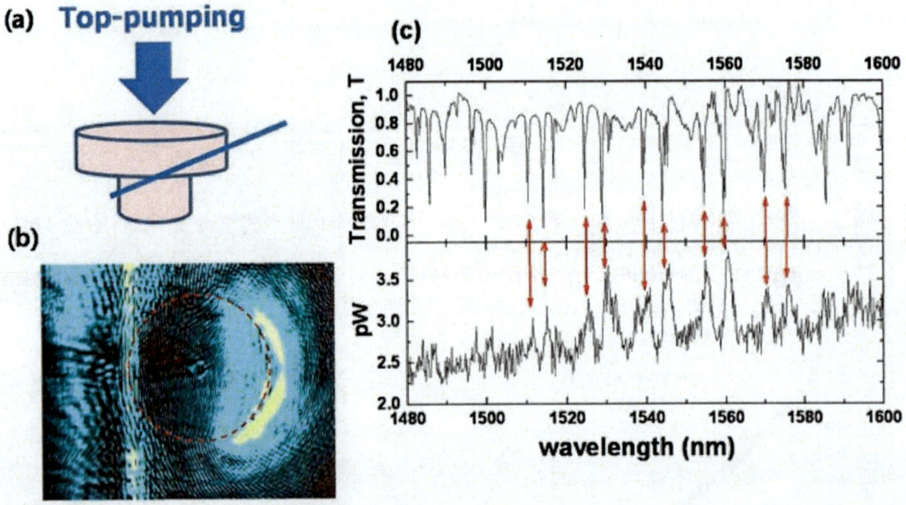

Fig. 4.32 (**a**) A schematic description and (**b**) a CCD image of an SRSN:Er microdisk under top pumping. The position of the disk is indicated by the *dotted circle*. A 473-nm DPSS laser was used to illuminate the edge of the disk, and a tapered fiber was used to measure both the transmission and emission spectra shown on the right. Note the exact correspondence between the dips in transmission and the peaks in emission. The width of the emission peaks is limited by the detector resolution, which was set at 2 nm

Still, the fabricated microdisks posses the critical advantage of SRSN:Er microdisk resonators – the ability to be pumped off-resonantly from the top without the need for a laser tuned precisely to one of the resonant modes and to be coupled via a fiber or waveguide. This is demonstrated in Fig. 4.32, which shows the Er^{3+} emission spectrum collected via a tapered fiber and observed by an optical spectrum analyzer (OSA) at a resolution bandwidth (RBW) of 2 nm. When illuminated from the top by a 473 nm laser, we observe peaks precisely at wavelengths that correspond to high-Q fundamental whispering gallery modes in the transmission spectrum obtained by the same setup. This demonstrates the core advantage of a nc-Si sensitized Er microdisk over conventional Er-doped microdisks, as excitation of whispering gallery modes does not require a pump laser tuned precisely to one of the whispering gallery modes of the disk together with a coupling fiber/waveguide whose coupling needs to be optimized for both excitation and emission wavelengths. Furthermore, the pumping scheme is quite robust, as changes in temperature or coupling will not affect the pump efficiency.

It should be noted, however, that such nc-Si sensitization suffers from built-in inefficiencies such as (a) a low degree of absorption within the thin microdisk (only 17% of the pump beam is absorbed); (b) a low excitation efficiency of the WGM (pump light not absorbed within the mode volume does not lead to excitation of the WGM); and (c) an intrinsically low quantum efficiency (a 2.62 eV photon is used to obtain an 0.8 eV photon). Furthermore, increasing the absorption while maintaining uniform excitation is difficult. Also, shaping the top-pumping beam into a well-aligned annulus is non-trivial [88]. In other words, while nc-Si sensitization plus top-pumping offer advantages in terms of cost and robust pumping, it also introduces factors that can substantially lower the overall efficiency of the final device even under the best conditions, and this should be taken into account when developing this technology into a practical application.

4.5 Conclusion

In conclusion, we have introduced Er-doped silicon-rich silicon nitride (SRSN:Er) and Er silicates as novel materials that can provide high and stable Er concentrations and a high refractive index for compact on-chip light sources. SRSN:Er allows sensitization of Er by nc-Si without optical de-activation of Er, and Er silicates can provide gain in excess of 30 dB/cm, while at the same time providing a much suppressed cooperative upconversion. The actual realization of light sources using these materials still needs more research, but as the work on SRSN:Er demonstrates, many promising on-chip photonic devices can already be realized with them.

Acknowledgments This work was supported in part by grant No. (R11-2003-022) through Optics and Photonics Elite Research Academy (OPERA) and by grant No. (R01-2007-000-21036-0) of the Korea Science and Engineering Foundation (KOSEF), and by KINC. The authors also acknowledge Dr. K.-J. Kim, G. Y. Sung, and Y. H. Lee for valuable discussion, knowledge, and help.

References

1. Q. Xu, B. Schmidt, S. Pradhan, M. Lipson, Nature **435**, 325 (2005) 95
2. H.S. Rong, R. Jones, A.S. Liu, O. Cohen, D. Hak, A. Fang, M. Paniccia, Nature **433**, 725 (2005) 95
3. F. Xia, L. Sekaric, Y. Vlasov, Nat. Photonics **1**, 65 (2007) 95
4. J. Fujikata, K. Nose, J. Ushida, K. Nishi, M. Kinoshita, T.Shimizu, T. Ueno, D. Okamoto, A. Gomyo, M. Mizuno, T. Tsuchizawa, T. Watanabe1, K. Yamada, S. Itabashi, K. Ohashi. Appl. Phys. Express **1**, 022001 (2008) 95
5. D. Ahn, C.-Y. Hong, J. Liu, W. Giziewicz, M. Beals, L.C. Kimerling, J. Michel, J. Chen, F.X. Kärtner, Opt. Express **15**, 3916 (2007) 95, 125
6. M.A. Green, J. Zhao, A. Wang, P.J. Reece, M. Gal, Nature **412**, 805–808 (2001) 96
7. B. Delley, E.F. Steigmeier, Appl. Phys. Lett. **67** 2370 (1995) 96
8. M.S. Hybertsen, Phys. Rev. Lett. **72** 1514 (1994) 96
9. S.-H. Choi, J.N. Kim, H.Y. Kim, Y.-K. Hong, J.-Y. Koo, J.H. Seok, J.Y. Kim, Appl. Phys. Lett. **80**, 2520 (2002) 96, 115
10. S. Miura, Y. Nakamura, M. Fujii, M. Inui, S. Hayashi, Phys. Rev. B. **73**, 245333 (2006) 96
11. L.T. Canham, Appl. Phys. Lett. **57**, 1046 (1990) 96
12. K.D. Hirschman, L. Tsybeskov, S.P. Duttagupta, P.M. Fauchet, Nature **384**, 338 (1996) 96, 103
13. C. Presti, A. Irrera, G. Franzò, I. Crupi, F. Priolo, F. Iacona, Appl. Phys. Lett. **88**, 033501 (2006) 96
14. L. Pavesi, L. Dalnegro, C. Mazzoleni, G. Franzo, F. Priolo, Nature **408**, 440 (2000) 96
15. H. chen, J.H. Shin, P.M. Fauchet, J.-Y. Sung, J-H. Shin, G.Y. Sung, Appl. Phys. Lett. **91**, 173121 (2007) 96
16. P.D.J. Calcott, K.J. Nash, L.T. Canham, M.J. Kane, J. Lumin. **57**, 257 (1993) 96

17. A. Fang, H. Park, O. Cohen, R. Jones, M.J. Paniccia, J.E. Bowers, Opt. Express, **14**, 9203 (2006) 97
18. J. Liu, X. Sun, D. Pan, X. Wang, L.C. Kimerling, T.L. Koch, J. Michel, Opt. Express **15**, 11272 (2007) 97
19. National Research Council, *Condensed-Matter and Materials Physics: Basic Research for Tomorrow's Technology* (National Academy Press, Washington, 1999) 97
20. S. Schmitt-Rink, C.M.Varma, A.F.J. Levi, Phys. Rev. Lett. **66**, 2782 (1991) 97, 103
21. I.N. Yassievich, L.C. Kimerling, Semicond. Sci. Technol. **8**, 718 (1991) 97
22. H. Ennen, J. Schneider, G. Pomrenke, A. Axmann, Appl. Phys. Lett. **43**, 943 (1983) 97
23. G. Franzo, F. Priolo, S. Coffa, A. Polman, A. Carnera, Appl. Phys. Lett. **64**, 2235 (1994) 97
24. B. Zheng, J. Michel, F.Y.G. Ren, L.C. Kimerling, D.C. Jacobson, J.M. Poate, Appl. Phys. Lett. **64**, 2842 (1994) 97
25. A. Polman, G.N. van den Hoven, J.S. Custer, J.H. Shin, R. Serna, P.F.A. Alkemade. J. Appl. Phys. **77**, 1256 (1995) 98
26. F. Priolo, G. Franzò, S. Coffa, A. Polman, S. Libertino, R. Barklie, D. Carey. J. Appl. Phys. **78**, 3874 (1995) 98
27. J. Michel, J.L. Benton, R.F. Ferrante, D.C. Jacobson, D.J. Eaglesham, E.A. Fitzgerald, Y.-H. Xie, J.M. Poate, L.C. Kimerling. J. Appl. Phys. **70**, 2627 (1991) 98, 100
28. P.N. Favennec, H.L' Haridon, M. Salvi, D. Moutonnet, Y. Le Guillou, Luminescence of erbium implanted in various semiconductors: IV, III–V, and II–VI materials. Electron. Lett. **25**, 718 (1989) 98
29. A.J. Kenyon, P.F. Trwoga, M. Federighi, C.W. Pitt, J. Phys.: Condens. Matter **6**, L319 (1994) 98
30. T. Kimura, A. Yokoi, H. Horiguchi, R. Saito, T. Ikoma, A. Sato, Appl. Phys. Lett. **65** 983 (1994) 98
31. M. Fujii, M. Yoshida, Y. Kanzawa, S. Hayashi, K. Yamamoto, Appl. Phys. Lett. **71**, 4525 (1997) 98
32. J.H. Shin, M.-J. Kim, S.-Y. Seo, C. Lee, Appl. Phys. Lett. **72**, 1092 (1998) 98
33. M. Stepikhova, W. Jantsch, G. Kocher, L. Palmetshofer, M. Schoisswohl, H.J. von Bardeleben, Appl. Phys. Lett. **71**, 2975 (1997) 98
34. G. Franzò, V. Vinciguerra, F. Priolo, Appl. Phys. A **69**, 3 (1999) 98
35. P.G. Kik, M.L. Brongersma, A. Polman, Appl. Phys. Lett. **76**, 2325 (2000) 98
36. M.E. Castagna, S. Coffa, M. Monaco, A. Muscara, L. Caristia, S. Lorenti, A. Messina, Mat. Sci. Eng. B **105**, 83 2003 98
37. N. Daldosso, D. Navarro-Urrios, M. Melchiorri, C. Cargia, P. Pellegrino, B. Garrido, C. Sada, G. Battaglin, F. Gourbilleau, R. Rizk, L. Pavesi, IEEE J. Sel. Top. Quantum Electron. **12**, 1607 (2006) 98, 99
38. H.S. Han, S.-Y. Seo, J.H. Shin, N. Park, Appl. Phys. Lett. **81**, 3720 (2002) 98
39. B. Garrido, C. Garcia, P. Pellegrino, D. Navarro-Urrios, N. Daldosso, L. Pavesi, F. Gourbilleau, R. Rizk, Appl. Phys. Lett. **89**, 163103 (2006) 98
40. P. Pellegrino, B. Garrido, J. Arbiol, C. Garcia, Y. Lebour, J.R. Morante, Appl. Phys. Lett. **88**, 21915 (2006) 98
41. J.S. Chang, J.-H. Jhe, M.-S. Yang, J.H. Shin, K.J. Kim, D.W. Moon, Appl. Phys. Lett. **89**, 181909 (2006) 98
42. S. Minissale, T. Gregrokiewicz, M. Forcales, R.G. Elliman, Appl. Phys. Lett., **89**, 171908 (2006) 98
43. A. Polman, A Lidgard, D.C. Jacobson, P.C. Becker, R.C. Kistler, G.E. Blonder, J.M. Poate, Appl. Phys. Lett. **57**, 2859 (1990) 99, 106
44. S. Saini, J. Michel, L.C. Kimerling, J. Lightwave Technol. **21**, 2368 (2003) 99
45. T.J. Kippenberg, J. Kalkman, A. Polman, K.J. Vahala, Phys. Rev. A **74**, 051802(R) (2006) 99
46. K.-S. Cho, N.M Park, T.Y Kim, K.H. Kim, G.Y. Sung, J.H. Shin, Appl. Phys. Lett. **86**, 071909 (2005) 100
47. L. Dal Negro, R. Li, J. warga, S.N. Basu, Appl. Phys. Lett. **92**, 181105 (2008) 100
48. F.F. Lange, S.C. Singhal, R.C. Kuznicki, J. Am. Ceram. Soc. **60**, 249 (1977) 100

49. N.A. Toporov, F.Ya. Galakhov, S.F. Konovalova, Russian Chem. Bull. **10**, 1271 (1961) 100
50. M.-S. Yang, K.-S. Cho, J.-H. Jhe, S.-Y. Seo, J.H. Shin, K.J. Kim, D.W. Moon, Appl. Phys. Lett. **85**, 3408 (2004) 101
51. J.H. Shin, S.Y. Seo, S. Kim, S.G. Bishop, Appl. Phys. Lett. **76**, 1999 (2000) 101, 105
52. S.-Y. Seo, J.H. Shin, App. Phys. Lett. **75**,4070 (1999) 103
53. T. Makimura, K. Kondo, H. Uematsu, C. Li, K. Murakami Appl. Phys. Lett. **83**, 5422 (2003) 103
54. A. Puzder, A.J. Williamson, J.C. Grossman, G. Galli, J. Chem. Phys. **117**, 6721 (2002) 104
55. T.-W. Kim, C.-H. Cho, B.-H. Kim, S.-J. Park, Appl. Phys. Lett. **88**, 123102 (2006) 104
56. H. Jeong, S.-Y. Seo, J.H. Shin, Appl. Phys. Lett. **88**, 161910 (2006) 104
57. R. Ferre, B Garrido, P. Pellegrino, M. Peralvarez, C. Garcia, J.A. Moreno, J Carreras, J.R. Morante, J. Appl. Phys., **98**, 084319 (2005) 105
58. G. Raghavan, C. Chiang, P.B. Anders, S. Tzeng, R. Villasol, G. Bai, M. Bohr, D.B. Fraser, Thin Film Solid **262**, 168 (1995) 106
59. R. S. Wagner, W.C. Ellis, Appl. Phys. Lett. **4**, 89 (1964) 107
60. A.M. Morales, C.M. Lieber, Science **279**, 208 (1998) 107
61. K. Suh, J.H. Shin, S.-J. Seo, B.-S. Bae, Appl. Phys. Lett. **89**, 223102 (2006) 107
62. JCPDS, International Center for Diffraction Data (www.icdd.com), ICDD Card No. 00-052-1809 (Er_2SiO_5), No. 00-052-1810 (Y_2SiO_5) 107, 108, 113
63. C. Li, C. Wyon, R. Moncorgé, IEEE J. Quantum Electron. **28**, 1209 (1992) 107
64. C. Li, R. Moncorgé, J.C. Souriau, C. Borel, C. Wyon, Opt. Commun. **107**, 61 (1994) 107
65. H.-J. Choi, J.H. Shin, K. Suh, H.-K. Seong, H.-C. Han, J.-C. Lee, Nano Lett. **5**, 2432 (2005) 108
66. F.F. Lange, S.C. Singhal, R.C. Kuznicki, J. Am. Ceram. Soc. **60**, 249 (1977) 108
67. G.N. van den Hoven, E. Snoeks, A. Polman, J.W.M. van Uffelen, Y.S. Oei, M.K. Smit, Appl. Phys. Lett. **62**, 3065 (1993) 110
68. S. Saini, K. Chen, X. Duan, J. Michel, L.C. Kimerling, M. Lipson, J.Electron. Mater. **33**, 809 (2004) 110
69. P.C. Becker, N.A. Olsson, J.R. Simpson, *Erbium-Doped FiberAmplifiers: Fundamentals and Technology* (Academic, San Diego, 1999) 110, 114
70. E. Snoeks, G.N. van den Hoven, A. Polman, B. Hendriksen, M.B.J. Diemeer, F. Priolo, J. Opt. Soc. Am. B **12**, 1468 (1995) 110, 114
71. P.G. Kik, A. Polman, J. Appl. Phys. **93**, 5008 (2003) 110, 117
72. L. Fornasiero, K. Petermann, E. Heumann, G. Huber, Opt. Mater. **10**, 9 (1998) 114
73. M.P. Hehlen, N.J. Cockroft, T.R. Gosnell, A.J. Bruce, G. Nykolak, J. Shmulovich, Opt. Lett. **22**, 772 (1997) 116, 117
74. M. Federighi, F. Di Pasquale, IEEE Photon. Tech. Lett. **7**, 303 (1995) 117
75. G.N. van den Hoven, A. Polman, C. van Dam, J.W.M. van Uffelen, M.K. Smit, J. Appl. Phys. **68**, 1886 (1996) 117
76. J. Nilsson, P. Blixt, B. Jaskorzynska, J. Babonas, J. Lightwave Technol. **13**, 341 (1995) 117
77. J. Felsche, Struct. Bond. **13**, 99 (1973) 117
78. F.P. Payne, J.P.R. Lacey, Opt. Quantum Electron. **26**, 977 (1994) 118
79. M. Borselli, T.J. Johnson, O. Painter, Opt. Express **13**, 1515 (2005) 120
80. R.D. Kekatpure, M.L. Brongersma, Phys. Rev. A, **78**, 023829 (2008) 120
81. W.J. Miniscalco, J. Lightwave Technol. **9**, 234 (1991) 120
82. I.-K. Hwang, S.-K. Kim, J.-K. Yang, S.-H. Kim, S.H. Lee, Y.-H. Lee, Appl. Phys. Lett. **87**, 131107 (2005) 121
83. J.S. Chang, M.-K. Kim, Y.-H. Lee, J.H. Shin, G.Y. Sung, Proc. SPIE **6897**, 68970O (2008) 121
84. S. Zheng, H. Chen, A.W. Poon, IEEE J. Sel. Top. Quantum Electron. **12**, 1380 (2006) 122, 125
85. H.A. Haus, *Waves and Fields in Optoelectronics* (Prentice-Hall, New Jersey, 1984) 122
86. M.-K. Kim, J.-K. Yang, Y.-H. Lee, I.-K. Hwang, J. Korean Phys. Soc. **50**, 1027–1031 (2007) 122
87. N. Daldosso, M. Melchiorri, F. Riboli, M. Girardini, G. Pucker, M. Crivellari, P. Bellutti, A. Lui, L. Pavesi, J. Lightwave Technol. **22**, 1734–1740 (2004) 125
88. H. Lee, J.H. Shin, N. Park, Opt. Express **13**, 9881 (2005) 127

Index

Chapter 5
Germanium as a Material to Enable Silicon Photonics

R. Ichikawa, S. Takita, Y. Ishikawa, and K. Wada

Abstract Germanium has been an enabler of the information age. Ge on Si nucleates Si photonics as well as high-speed CMOS electronics. Recently, Ge has played a significant role in integrating materials such as III–Vs on Si. The structure of GaAs on a thick Ge layer on Si has been studied for many years to expand its device application menu such as lasers, high-performance transistors, and tandem solar cells on Si. However, an ultra-thin Ge buffer layer (referred to as (Ge) hereafter) technology described in this chapter has created new fields for applications. One of the emerging fields is the structure and properties of AlGaAs/GaAs/(Ge)/Si/Ge, which has been impossible to create previously using the thick Ge buffer on Si technology. Here, we demonstrate an application as a new green power generation platform, i.e., high-efficiency cost-effective tandem solar cells using Si as a cell as well as the mechanical substrate. The (Ge) thickness has not been fully optimized yet, but is in the range 10–20 nm. Our design for a tandem solar cell shows that its theoretical efficiency reaches 43%. The key attributes are the crystalline quality and surface roughness of ultrathin (Ge). We have experimentally optimized the (Ge) buffer thickness to achieve both requirements and prototyped Ge solar cells on Si. The Ge solar cells have successfully reproduced their ideal external quantum efficiency. This is the proof of concept of the success of the Ge challenge as the material enabler to integrate Si and GaAs.

5.1 Introduction

Germanium has nucleated solid-state electronics as a material for transistors [1], and has now empowered current electronics as a stressor of Si complementary metal oxide semiconductor (CMOS) circuits [2]. It is considered that Ge will be a material

R. Ichikawa, S. Takita, and Y. Ishikawa (✉)
Department of Materials Engineering, University of Tokyo, Hongo, Bunkyo, Tokyo 113-8656, Japan

K. Wada (✉)
Department of Materials Engineering, University of Tokyo, Hongo, Bunkyo, Tokyo 113-8656, Japan
e-mail: kwada@material.t.u-tokyo.ac.jp

D.J. Lockwood, L. Pavesi (Eds.): *Silicon Photonics II.*
Topics in Applied Physics **119**, 131–141 (2011)
DOI 10.1007/978-3-642-10506-7_5

of p-MOS in the scheme referred to as the extension of “Moore’s Law” [3]. Ge has enabled CMOS-compatible photodetectors on a Si chip [4] and, recently, modulators [5], and even an optical gain media in the optical communication wavelength range [6]. Thus, Ge has been regarded as a material enabler to converge electronics and photonics on the Si CMOS platform, i.e., Si photonics. In the current stage of Si photonics development, such an ultrathin layer of Ge, hereafter referred to as (Ge), can perform as an excellent buffer layer between Ge and Si, i.e., highly lattice mismatched material systems. The natural consequence of the (Ge) technology is to combine heterogeneous materials such as GaAs on Si, since Ge perfectly lattice-matches with GaAs-based devices that can be light emitters, high-speed transistors, and solar cells. In this chapter, we will explore another challenge of (Ge) as a material integration platform of Si and GaAs, which can be referred to as a “green power generation platform.” We design a four-junction AlGaAs(1)/AlGaAs(2)/(Ge)/Si/Ge tandem solar cell, that could not be done previously without the ultrathin (Ge) technology. We present the theoretical design of the tandem cells, the experimental results of the (Ge) buffer layer optimization, and a prototype Ge/Si solar cell as a proof of concept.

5.2 Material Design

Ge is an element that lies between Ga and As in the periodic table, and thus there are many similarities in the physical and chemical properties of Ge and GaAs. Typical examples are the lattice constant, linear expansion coefficient, and thermal conductivity, as given in Table 5.1 [7]. These similarities favor hetero-epitaxial growth of GaAs on Ge or vice versa. Despite the fact that Ge is a non-polar material while GaAs is a polar material, anti-phase domain boundaries (APDs) of GaAs on Ge have naturally been suppressed by introducing a miss-orientation of the substrate surface orientation. High-quality GaAs epilayers on Ge bulk substrates and its hetero-structures have been achieved [8].

On the other hand, Ge is limited in natural abundance and in technological maturity, compared with Si where 30-cm-diameter dislocation-free wafers are commercially available. Thus, the first material integration was attempted for GaAs on Si. As calculated from Table 5.1, the lattice mismatch between GaAs and Si is as large as 4%, which naturally lead to a high density of dislocations in GaAs on Si [9]. These defects have made the material integration of Si and GaAs unsuccessful. On the other hand, it has recently been reported that dislocation-free Ge mesas can be grown on Si [1]. In this method a (Ge) buffer layer was first grown on Si at a

Table 5.1 Material parameters of Ge, GaAs, and Si at 300 K

Materials	Lattice constants (nm)	Linear expansion coefficient (10^{-6}/°C)	Thermal conductivity (W/cm/°C)	Bandgap (eV)
Ge	0.5658	5.9	0.58	0.661
GaAs	0.565325	5.73	0.55	1.424
Si	0.5431	2.6	1.3	1.12

low temperature (around 300–400 °C) and a Ge epilayer was then grown on it at a medium temperature (around 600 °C), hereafter referred to as two-step growth. The (Ge) buffer layer is highly dislocated but the subsequently grown Ge epilayers contain a reduced density of threading dislocations, i.e., $\sim 10^9\ \mathrm{cm}^{-3}$. Such a high density of threading dislocations was reduced during a high temperature post-growth annealing. The mechanism should be a strain-driven dislocation glide and reaction: Ge has a linear expansion coefficient twice that of Si, as indicated in Table 5.1, inducing thermal mismatch during the post-growth annealing. Threading dislocations in the Ge epilayer become mobile in terms of stress generated in the Ge layer annealed at high and low temperatures cyclically. In other words, heating the Ge epilayers on Si substrates to a higher temperature near the melting point introduces compressive strain within the Ge epilayer. The strain mobilizes threading dislocations, gliding the dislocations to relax in the epilayer. Cooling the structure down to a lower temperature around 700 °C where threading dislocations are still mobile in Ge introduces tensile strain in the Ge epilayer that also glides back dislocations to relax in the epilayer. These motions of threading dislocations eventually reduce the dislocation density in terms of dislocation annihilation in pairs with opposite Burgers vectors. When the Ge epilayers are smaller than the gliding distance for dislocations, dislocation-free Ge epilayers on Si should be obtained [1]. In addition, Ge never grows on silica masks in terms of the reduction reaction of silica by Ge, resulting in Ge on Si patterns with no islands on silica patterns. Currently, square Ge mesas of $10 \times 10\,\mu\mathrm{m}^2$ have been reported dislocation-free even on Si. This suggests that the dislocation glide distance would be about $5\,\mu\mathrm{m}$. Larger mesas can be dislocation-free, when the larger stress is accumulated in the mesas in terms of rapid thermal processing.

These experiments are strongly suggestive that dislocation-free GaAs on (Ge) on Si should be possible provided that the Ge on Si is dislocation-free. AlGaAs is a lattice matching system of GaAs, and indeed dislocation-free on GaAs. Thus, AlGaAs/GaAs hetero-structures should be possible on Si using the (Ge) buffer and mesa technology. Since threading dislocations in GaAs and AlGaAs systems act as efficient recombination centers to degrade device performance, the dislocation-free Ge on Si would be a new material integration platform to integrate Si with GaAs. The material integration would enable electronic and photonic convergence on the Si CMOS platform. Si has a twice larger thermal conductivity (see Table 5.1), acting as an excellent heat sink as well for the high injection devices if realized on Si. Actually, such material integration has been performed to demonstrate lasers and light emitting diodes on Si [10]. It required thick (Ge) buffer layers to grow GaAs on Si and is difficult to apply to tandem solar cell stacks. We will come back this point later on.

5.3 Device Design: Tandem Cells

Highly efficient, series-connected tandem solar cells using Ge and III–V compounds such as GaAs/Ge, InGaP/InGaAs/Ge, and InGaP/GaAs/Ge have attracted much attention due to their lattice matching characteristics [11]. However, it was quite challenging to achieve four-junction tandem solar cells with the materials system of interest: Ge and GaAs. It is because of the current mismatch due to the large difference in the bandgap energy (E_g) between GaAs (1.424 eV) and Ge (0.661 eV), as shown in Table 5.1. Recently, GaInNAs has attained technological interest because of the following reasons: first, the bandgap can be controlled to be ~1 eV at the middle of GaAs and Ge enabling current matching, and second, the lattice constant can be matched to Ge and GaAs. Indeed, the four-junction GaInP/GaAs/GaInNAs/Ge tandem solar cell has been prototyped [12]. However, there is still the challenging issue associated with Ge substrates, i.e., low thermal conductivity (see Table 5.1). Its technical immaturity is always the issue. Si is an ideal material to breakthrough these issues and to meet the bandgap requirement of ~1 eV, as indicated in Table 5.1, provided that the lattice mismatch issue is solved.

We have designed a four-junction tandem solar cell using Si highlighting a middle bandgap material and a substrate with technical maturity. In other words, the four tandem cell structure can basically be AlGaAs/GaAs/(Ge)/Si/Ge and GaInP/GaAs/(Ge)/Si/Ge. We have chosen the former cell structure to design the structure and the layer thicknesses together with the alloy content of the AlGaAs layer on a quantitative basis. This tandem solar cell was designed, focusing on the current matching by calculating the photocurrent generation at each thickness, using the following equation [13].

$$J_{sc} = \int \phi(\lambda) \times (1 - \exp[-\alpha(\lambda) \times L])d\lambda, \tag{5.1}$$

where J_{sc} denotes the short-circuit current density (mA/cm^2), $\phi(\lambda)$ is the number of incident photons of AM1.5G (/nm) [14], $\alpha(\lambda)$ is the optical absorption coefficient (/cm), and L is the thicknesses of the cells (cm). For simplicity, we assumed the following:

(a) The ideal current collection conditions of all the layers, i.e., external quantum efficiency (EQE) = 1.
(b) Light passes through each cell once with internal reflections ignored.

The absorption coefficients of GaAs and AlGaAs employed are taken from [15] and [16]. It turned out that the GaAs layer should be an Al-lean AlGaAs layer for perfect current matching.

Figure 5.1 shows the structure of a tandem solar cell $Al_xGa_{1-x}As/Al_yGa_{1-y}As/Si/Ge$ with a Ge buffer layer on Si, which meets the current matching requirement. As in Fig. 5.2, the photocurrent accumulated from 300 to 1,550 nm is 54.05 mA, assuming the AM1.5G current spectrum (100 mW/cm^2, 1 sun). The dotted and dashed lines are referred to as the current division by four cells and the Ge buffer layer, respectively. The bottom cell is Ge and its bandgap ($E_{g,\mathrm{Ge\ direct}}$) is 0.80 eV. The reason for using the direct gap here is our assumption that the Ge layer is thin, because of epitaxial growth. Ge generates 54.05 mA when it absorbs light shorter than 1,550 nm (0.8 eV) in wavelength. However, the weak Ge absorption near the

Fig. 5.1 Schematic of the $Al_xGa_{1-x}As/Al_yGa_{1-y}As/Si/Ge$ tandem solar cell

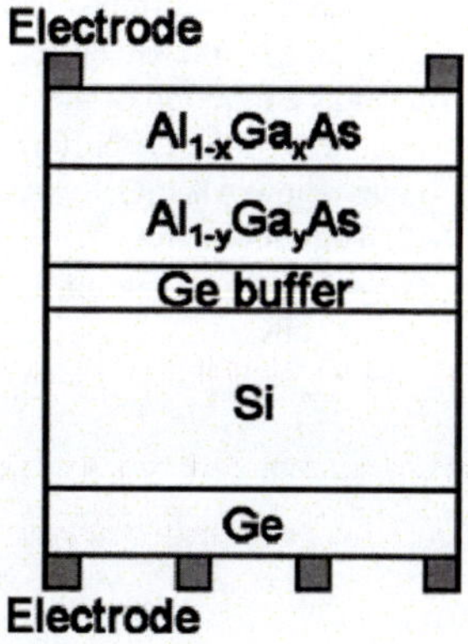

Fig. 5.2 The AM1.5G current spectrum division in the tandem cell of Fig. 5.1

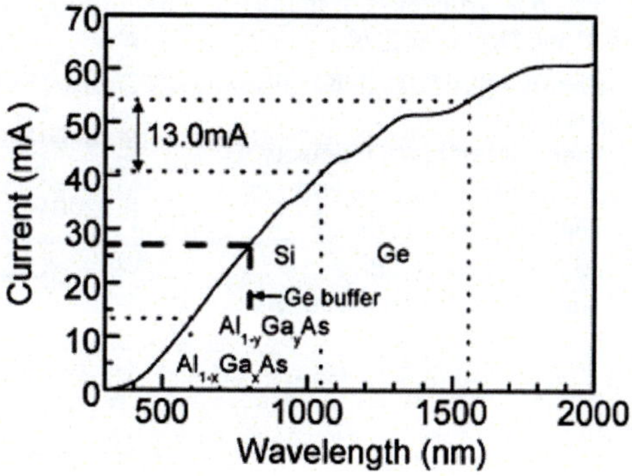

band edge generates a slightly lower current. For example, it is 53.3 mA/cm^2 instead when the Ge epilayer is 3 μm thick. Therefore, J_{sc} of each cell is about 13 mA/cm^2, as in Fig. 5.2. Here, we ignored the absorption by the (Ge) layer, which does not alter the final thickness determination, as described next.

The (Ge) buffer accumulates most defects in the layer, especially in the vicinity of the Si and (Ge) interface, allowing a nearly perfect subsequent $Al_yGa_{1-y}As$ layer to grow on it. However, the Ge buffer layer should act as an absorber causing light loss. This is because photo-excited charges generated in Ge buffer layer may easily recombine through defects in the layer and do not contribute to the short-circuit current. This loss of photocurrent reduces the total J_{sc} in terms of the lower photocurrents in the lower-positioned Si and Ge cells. The photocurrent calculated to be generated in the (Ge) buffer layer is shown in Fig. 5.3. Here, we assumed that the top and second-to-top cells are $Al_{0.30}Ga_{0.70}As$ (223 nm thick) and $Al_{0.05}Ga_{0.95}As$ (878 nm thick), respectively. Although the photocurrent generated in the (Ge) buffer layer increases with the (Ge) buffer layer thickness, the photocurrent loss is below 2 mA/cm^2 when the (Ge) buffer thickness is a typical one, i.e., 10–30 nm in the growth of Ge on Si. Thus, the (Ge) layer can be neglected in the calculation of the photocurrents for current matching. Nonetheless, we have rigorously counted its loss effect in the design of our tandem structures. Figure 5.4 shows the thickness of the Si and Ge cells as a function of the thickness of the (Ge) buffer layer.

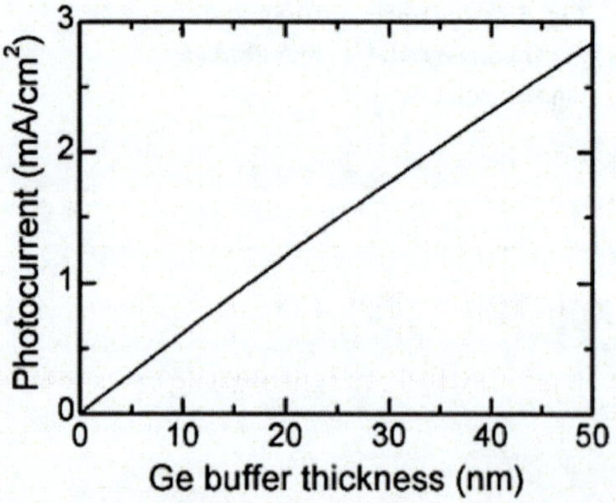

Fig. 5.3 Photocurrent generated by the Ge buffer layer as a function of the thickness of the Ge buffer layer below $Al_{0.30}Ga_{0.70}As$ (223 nm thick) and $Al_{0.05}Ga_{0.95}As$ (878 nm thick) cells

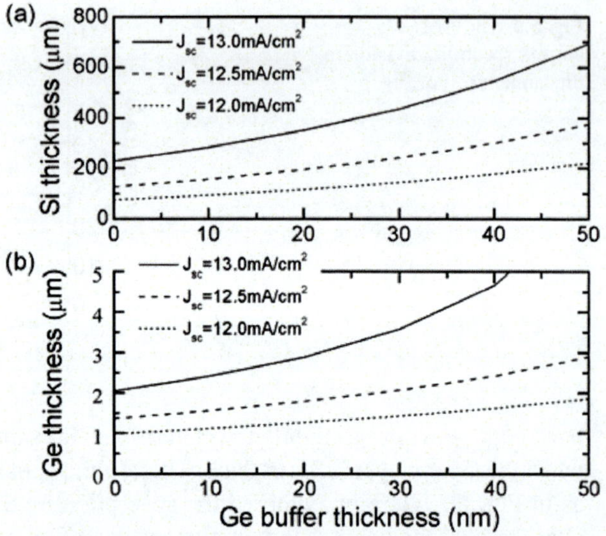

Fig. 5.4 Thickness of the (**a**) Si and (**b**) Ge cell as a function of the thickness of the Ge buffer layer with short-circuit current density J_{sc} in the case that the top and the second cells are $Al_{0.30}Ga_{0.70}As$ and $Al_{0.05}Ga_{0.95}As$, respectively

The variable parameter is the short-circuit current density when the top and the second cells are $Al_{0.30}Ga_{0.70}As$ and $Al_{0.05}Ga_{0.95}As$, respectively. It is clear that the Si and Ge cells should thicken with the (Ge) buffer layer. One of the material set properties is that the Si layer should be 101–307 μm thick when the (Ge) buffer layer is ~15 nm thick, generating J_{sc} = 12–13 mA/cm^2 in the Si cell. Since Si solar cells are currently around 200 μm thick, the Si wafer can be used as both a substrate and one of tandem cells as previously noted. Furthermore, the Ge layer should be 1.168–2.580 μm thick when the (Ge) buffer layer is again ~15 nm thick. This material set property is also realistic according to our experience. The Ge cell generates J_{sc} = 12–13 mA/cm^2 when meeting the current matching condition.

The final issue is to design the composition and thickness of the upper AlGaAs cells on the (Ge) buffer to get a J_{sc} of 12–13 mA/cm^2. Figure 5.5 shows the thickness of these AlGaAs cells as a function of the short-circuit current density of 12–13 mA/cm^2. The variable parameter is the Al composition x and y. Figure 5.5a is for the top cell, and Fig. 5.5b is for the second cell when the top cell is

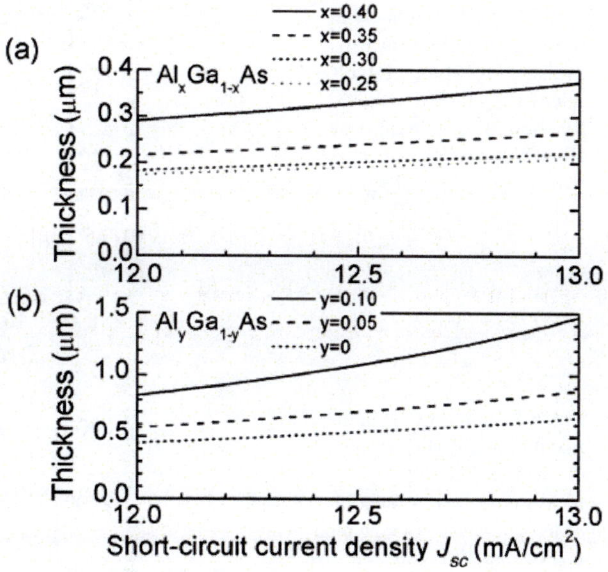

Fig. 5.5 Thickness of (**a**) $Al_xGa_{1-x}As$ and (**b**)$Al_yGa_{1-y}As$ cells with Al composition x and y as a function of short-circuit current density J_{sc}. **b** is for the case that the top cell is $Al_{0.30}Ga_{0.70}As$

$Al_{0.30}Ga_{0.70}As$. To achieve a high open-circuit voltage, the Al composition should be high. It is found from Fig. 5.5a that the top cell's thickness should increase with Al composition because the Al rich AlGaAs causes a blue-shift of the optical absorption edge. However, the high Al composition ($x > 0.3$) AlGaAs frequently results in Al oxidation, which could generate unwanted recombination centers and lead to current mismatch between the cells. To compromise on the above mentioned issues, we have chosen $Al_{0.30}Ga_{0.70}As$ as the top cell. For a high short-circuit current density, the $Al_yGa_{1-y}As$ cell must be thick, but it is better to keep the thickness under about 1 μm when considering a practical epitaxial growth time. These considerations determine that the best composition for the second cell is $Al_{0.05}Ga_{0.95}As$.

5.4 Materials and Device Characteristics: (Ge) Buffer on Si and Ge Solar Cell on Si

The quality of the (Ge) buffer layer is the key to realize this tandem cell, because $Al_{1-y}Ga_yAs$ growth on Ge and $Al_{1-x}Ga_xAs$ growth on $Al_{1-y}Ga_yAs$ are thus possible. To obtain a high-quality AlGaAs layer, the surface roughness of Ge buffer layer should be low. We implemented the Ge growth experiment on a non-vicinal p^- Si (100) wafer at 370 °C without annealing using ultra-high vacuum chemical vapor deposition (UHV – CVD). The (Ge) buffer layer thicknesses were 15, 30, and 60 nm. Figure 5.6 shows the root-mean-square (RMS) roughness of the Ge surface measured by atomic force microscopy (AFM) as a function of the Ge thickness. It is clear that the RMS roughness increases with the (Ge) buffer thickness. Figure 5.7 shows the full width at a half maximum (FWHM) of the (Ge) buffer peak in X-ray diffraction (XRD) measurement as a function of the Ge thickness. The FWHM

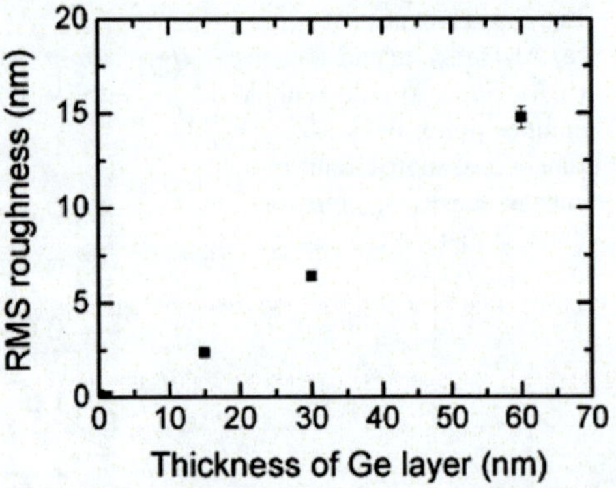

Fig. 5.6 RMS roughness of the as-grown Ge surface measured by atomic force microscopy (AFM) as a function of the thickness of Ge, which was grown at 370 °C on a p^- Si (100) wafer by UHV–CVD

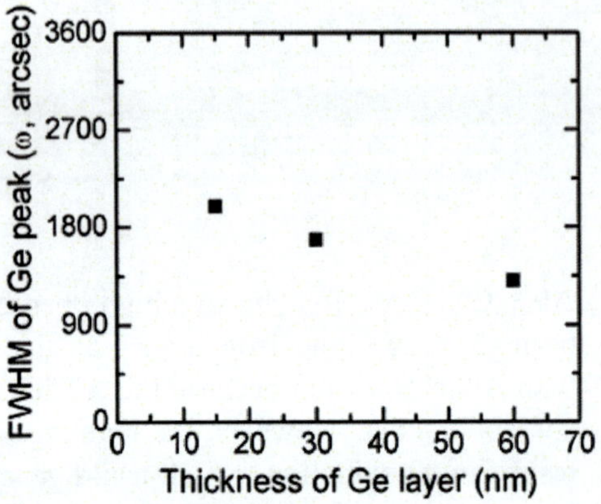

Fig. 5.7 Full width at half maximum (FWHM) of the Ge peak in XRD measurements as a function of the thickness of Ge, which is grown at 370 °C on a p^- Si (100) wafer by UHV–CVD

provides information on the variation of the Ge lattice, which reflects its crystalline quality. The FWHM decreases with the (Ge) buffer thickness, indicating that better crystalline quality is obtained with the thicker (Ge) buffer.

To decide the optimum thickness, we have defined the figure of merit (FOM) as a qualitative measure:

$$\text{FOM} = \frac{1}{\text{RMS} \cdot \text{FWHM} \cdot \text{absorption}}. \tag{5.2}$$

Here, the absorption is a measure of the photocurrent generated by the (Ge) buffer layer. The RMS roughness, crystalline quality (FWHM), and light loss are shown in Figs. 5.6, 5.7, and 5.3, respectively. Figure 5.8 shows the FOM as a function of the thickness of the (Ge) buffer layer. From consideration of the FOM, a thin (Ge) buffer layer 15 nm thick is optimal, as shown in this experiment.

To check the abovementioned buffer layer technology, we have fabricated the Ge solar cells on a Si substrate and measured the efficiency of Ge solar cells on Si. The solar cell structure from the top down is the following: 50-nm Si cap/1-μm Ge with 15 nm (Ge) buffer layer/p^+ Si substrate. The growth temperatures of the (Ge) buffer and upper Ge epilayer were 370 and 600 °C, respectively. After epitaxial growth, phosphorous ions were implanted at 70 keV at a dose of $6 \times 10^{14}\,\text{cm}^{-2}$ into the

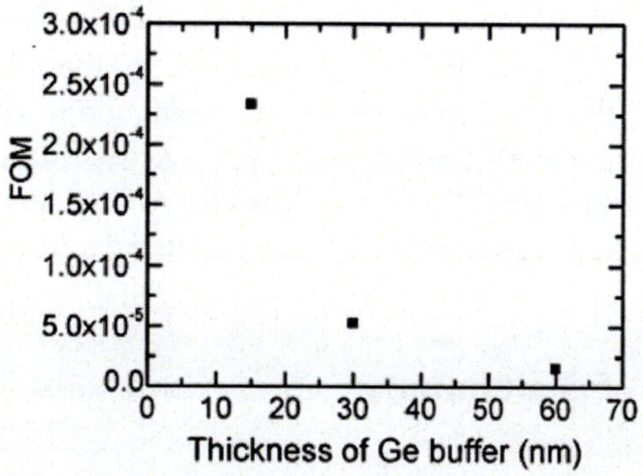

Fig. 5.8 Figure of merit (FOM = 1/[RMS roughness × crystalline quality{FWHM} × light loss]) as a function of the thickness of the Ge, which was grown at 370 °C on a p^- Si (100) wafer by UHV–CVD

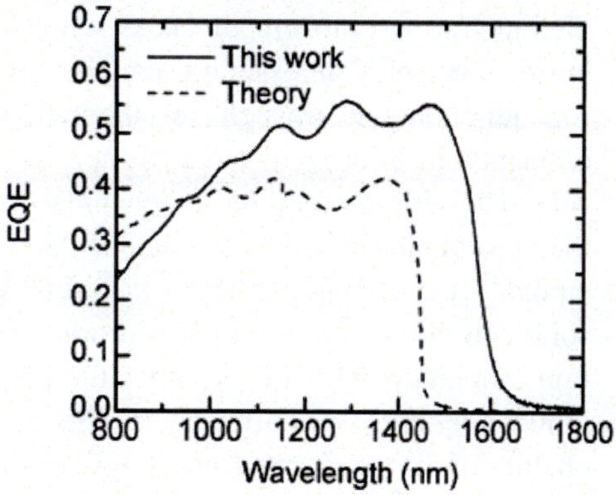

Fig. 5.9 External quantum efficiency (EQE) of our photodiode and the theoretically predicted response as a function of wavelength

Si cap layer. Figure 5.9 shows the external quantum efficiency (EQE) of the Ge solar cell on Si. The solid and dashed lines are the EQE of the solar cell and of a theoretical calculation assuming 1 μm Ge as an active layer, respectively. It is clear that the experimental data have been reproduced by the theory. The red-shift of the experiment compared with the theoretical calculation is the bandgap shrinkage in terms of the tensile strain induced by thermal mismatch between the Si and Ge expansion coefficients [17].

Table 5.2 summarizes the optimum thickness of each cell at $J_{sc} = 13\,mA/cm^2$ considering the experiment noted above. The top $Al_{0.30}Ga_{0.70}As$ cell is 223 nm thick, which fits with a short growth time and allows the formation of a thin p^+ and n^+ layer at the top and bottom subsurface layer. The second $Al_{0.05}Ga_{0.95}As$ cell is suitable for a reasonable growth time. The thickness of the Si and Ge cells are also appropriate as a wafer and for epitaxial growth, respectively.

Table 5.2 The optimum thickness of each cell for the tandem solar cell at $J_{sc} = 13\,mA/cm^2$

Layer	Thickness (μm)
$Al_{0.3}Ga_{0.7}As$	0.223
$Al_{0.05}Ga_{0.95}As$	0.878
(Ge) buffer	0.015
Si	307
Ge	2.58

We have estimated that the total power efficiency of this tandem cell is 43.07% at $J_{sc} = 13\,\text{mA/cm}^2$. For simplicity, the fill factors (FF) have been derived assuming that the ideality factor of the pn diodes would be 1 and $V_{oc} = E_g - 0.4$. Here $E_g = 1.798\,\text{eV}$ for $Al_{0.30}Ga_{0.70}As$, $E_g = 1.486\,\text{eV}$ for $Al_{0.05}Ga_{0.95}As$, $E_g = 1.12\,\text{eV}$ for Si, and $E_g = 0.8\,\text{eV}$ for Ge. A very high-power efficiency should be obtained by the present tandem solar cell using a Si wafer.

5.5 Summary

We have described the use of Ge as a material enabler for Si photonics, especially for green power generation technology based on Si. The role of Ge as a materials integration platform of GaAs with Si and the newly developed ultrathin (Ge) buffer layer play an essential part in this solar cell application. We have designed four-junction tandem cells of the type $Al_xGa_{1-x}As/Al_yGa_{1-y}As$/(Ge)/Si/Ge, as described in Table 5.2. Here, Si acts as a solar cell with $E_g = 1.12\,\text{eV}$ and as a substrate with the required mechanical strength. A theoretical calculation indicates the cell efficiency to be 43% under AM1.5G and 1-sun conditions. This should be equivalent to the tandem cell of a GaInNAs cell on Ge instead of the Si cell. The Ge solar cell fabricated on Si has reproduced the theoretically predicted external quantum efficiency, which overcomes the potential weak point of the tandem structure induced by the (Ge) buffer layer. This gives us clear evidence that Ge should be an enabler of materials integration of GaAs with Si.

The materials integration based on Ge will stimulate GaAs-based photonic and electronic devices as well. As shown in Table 5.1, the thermal conductivity of Si is twice as high as that of Ge and GaAs, which should be another benefit from the use of Si. In other words, the heat penalty is always the factor limiting the performance and reliability of high injection devices such as lasers and high electron mobility transistors (HEMTs).

Acknowledgments The basic concept was created at the Massachusetts Institute of Technology (MIT) and submitted as the MIT case number 10445 while one of the authors (KW) was at the MIT Microphotonics Center. We are grateful to our collaborators, D.T. Danielson, T. Montalbo, J. Graham, J. Michel, and L.C. Kimerling at MIT, and Y. Okita and J. Osaka at the University of Tokyo.

References

1. W.H. Braten, Entry of 15 Dec. 1947, laboratory note, case number 38139-7, Bell Laboratories archives 131, 132, 133
2. J. Welser, J.L. Hoyt, J.F. Gibbons, IEEE Electron. Dev. Lett. **15**, 100, (1994) 131
3. G. Moore, Proc. SPIE **2440**, 2 (1995) 132
4. H.-C. Luan, D.R. Lim, K.K. Lee, J.G. Sandland, K. Wada, L.C. Kimerling, Appl. Phys. Lett. **75**, 2909 (1999) 132
5. J.-F. Liu, M. Beals, A. Pomerenke, S. Bernardis, R. Sun, J. Cheng, L.C. Kimerling, J. Michel, Nat. Photonics. **2**, 433 (2008) 132

6. J. Liu, X. Sun, D. Pan, X. Wang, L.C. Kimerling, T.L. Koch, J. Michel, Opt. Express **15**, 11272 (2007) 132
7. http://www.ioffe.rssi.ru/SVA/NSM/Semicond/ 132
8. S.M. Sze, *Physics of Semiconductor Devices*, 2nd edn. (Wiley, New York, 1981) 132
9. M. Yamaguchi, Y. Ohmachi, T. Oh'hara, Y. Kadota, M. Imaizumi, S. Matsuda, Prog. Photovolt.: Res. Appl. **9**, 191 (2001) 132
10. M.E. Groenert, C.W. Leitz, A.J. Pitera, V. Yang, H. Lee, R.J. Ram, E. Fitzgerald, J. Appl. Phys. **93**, 362 (2003) 133
11. T. Takamoto, T. Agui, E. Ikeda, H. Kurita, *Conference Record of the 28th IEEE Photovoltaic Specialists Conference* (Anchorage, AK, USA, 2000), pp. 976–981 133
12. D.J. Friedman, S.R. Kurtz, Prog. Photovolt.: Res. Appl. **10**, 331 (2002) 134
13. S.R. Kurtz, A.A. Allerman, E.D. Jones, J.M. Gee, J.J. Banas, B.E. Hammons, Appl. Phys. Lett. **74**, 729 (1999) 134
14. http://rredc.nrel.gov/solar/spectra/am1.5/ ASTM Standard Tables for Reference Solar Spectral Irradiance at Air Mass 1.5: Direct Normal and Hemispherical for a 37° Tilted Surface, Standard G159-99, American Society for Testing and Materials, West Conshohocken, 1999 134
15. D.E. Aspnes, S.M. Kelso, R.A. Logan, R. Bhat, J. Appl. Phys. **60**, 754 (1986) 134
16. D.E. Aspnes, *Properties of Silicon* (INSPEC, IEE, London, 1988), p. 59 134
17. Y. Ishikawa, K. Wada, D.D. Cannon, J.F. Liu, H.C. Luan, L.C. Kimerling. Appl. Phys. Lett. **82**, 2044 (2003) 139

Index

Chapter 6
Silicon Photonics: The System on Chip Perspective

Alberto Scandurra

Abstract This chapter describes possible applications of silicon photonics to the System on Chip (SoC) domain. Systems on Chip (SoCs) are complex systems containing billions of transistors integrated in a unique silicon-chip, implementing even complex functionalities by means of a variety of modules communicating with the system memories and/or between them through a proper communication system. The higher and higher integration density is becoming such that many issues arise when a SoC has to be integrated, and electrical limits of interconnect wires are a limiting factor for performance. According to this scenario, a new technology is required for the on-chip interconnect, in order to overcome current physical and performance issues; one possible solution is the optical interconnect, exploiting the many benefits of light to transport information across the chip. From an industrial point of view it is fundamental that such a new technology be fully CMOS compatible, in order to be able to continue to use current SoC design methodologies as well as present manufacturing equipment for the whole electronic part of the chip. Many semiconductor industries are today investigating such a novel field and a number of projects are currently running in order to demonstrate the feasibility of such a revolutionary on-chip communication system relying on both CMOS technology and photonics.

6.1 Introduction

Systems on Chip (SoCs) are complex systems containing billions of transistors integrated in a unique silicon-chip, implementing even complex functionalities by means of a variety of modules communicating with the system memories and/or between them through a proper communication system. The higher and higher integration density is becoming such that many issues arise when a System on Chip (SoC) has to be integrated, and electrical limits of interconnect wires are a limiting

A. Scandurra (✉)
Technical Staff and Advanced Programs Responsible, On-Chip Communication Systems Group, STMicroelectronics, Catania, Italy
e-mail: alberto.scandurra@st.com

D.J. Lockwood, L. Pavesi (Eds.): *Silicon Photonics II*.
Topics in Applied Physics **119**, 143–168 (2011)
DOI 10.1007/978-3-642-10506-7_6

factor for performance. According to this scenario, a new technology is required for the on-chip interconnect, in order to overcome current physical and performance issues; one possible solution is the optical interconnect, exploiting the many benefits of light to transport information across the chip. From an industrial point of view it is fundamental that such a new technology be fully CMOS compatible, in order to be able to continue to use current SoC design methodologies as well as present manufacturing equipment for the whole electronic part of the chip. Many semiconductor industries are today investigating such a novel field and a number of projects are currently running in order to demonstrate the feasibility of such a revolutionary on-chip communication system relying on both CMOS technology and photonics.

To cover all the topics introduced above, this chapter is organized as follows:

- Section 6.2, *The System on Chip paradigm*: describes the SoC as the modern approach for designing and integrating complex systems;
- Section 6.3, *On-chip communication*: deals with the SoC communication infrastructure, illustrating the concepts of the on-chip bus and Network on Chip;
- Section 6.4, *SoC integration issues:* describes physical and performance issues usually met during the SoC integration phase, and the origin of such issues, mainly at the physical level;
- Section 6.5, *On-chip optical interconnect*: describes the on-chip optical interconnect as a possible solution to the SoC integration issues, detailing the PICMOS project and the WADIMOS project as state-of-the-art examples of the work being carried out in this novel field;
- Section 6.6, *Conclusion*: contains a summary of what has been described in the overall chapter;
- *References*: contains the list of books, papers, and Internet websites where some information for writing this chapter has been taken from.

6.2 The System on Chip Paradigm

The SoC is the essential solution for delivering competitive performance and cost in today's challenging electronics market. Consumers using PCs, PDAs, cell-phones, games, toys, and all the other products you can think of, demand more features, instant communications, and massive data storage in ever smaller and more affordable products. The unstoppable drive in silicon fabrication has delivered technology to meet this demand – chips with 50 million gates using 130 nm processes are no more than the size of your thumbnail. These SoCs present one of the biggest challenges engineers ever faced; how to manage and integrate enormously complex designs that combine the richest imaginable mix of microprocessors, memories, busses, architectures, communications standards, protocol processors, interfaces, and other intellectual property components where system level considerations of synchronization, testability, conformance, and verification are crucial. Integrated circuit (IC) design has become a multi-million-gate challenge for which only the best design teams are good enough.

The techniques used in designing multi-million-gate SoCs employ the world's most advanced electronic design automation (EDA), with a level of sophistication that requires highly trained and experienced engineers. Key issues to be managed in the design process include achieving timing closure that accounts for wire delays in the metal interconnects inside the chip, and designs for tests so that the chips can be manufactured economically. Early prediction of the right architecture, design-flow, and best use of EDA solutions is required to achieve the first silicon success that can decrease the time-to-market from years to months.

The building-blocks of a SoC can be distinguished in initiators, or processing elements (PEs) and targets, or storage elements (SEs); *initiators* are all the blocks able to generate traffic, i.e., write data into a SE and read data from a SE; *targets* are blocks able to manage the traffic generated by the initiators.

Among the initiators of the system the following classes can be identified:

- processors;
- real time initiators;
- DMAs (Direct Memory Access).

Processors, like ST20, ST40, ST50, and LX from STMicroelectronics, have strict requirements in terms of latency and bandwidth. Anyways, their bandwidth must be in some way limited to allow the other initiators to be serviced.

Real time initiators, like audio/video blocks, are more latency tolerant than the processors, but have strict needs in terms of bandwidth.

DMAs do not have any particular requirements in terms of latency or bandwidth, and normally they can work using the remaining bandwidth, that is the part of the bandwidth not consumed by the processors and the real time initiators.

Among the targets the following classes can be identified:

- external fast memories;
- internal slow memories;
- peripherals.

External fast memories comprise high performance memories such as SDRAM (synchronous dynamic random access memory) and DDR (dual data rate) SDRAM, used mainly for real time applications like video, and today working at around 400 MHz. Their speed is limited by physical constraints imposed by pads.

Slow memories are usually low performance memories like SRAM and Flash, used for storage of huge amounts of data, whose access is managed by caches, working at around 200 MHz. Their speed is limited by application reasons.

Peripherals are slow memories such as I^2C and Smartcard, used where no high performances are required, and working at around 50/100 MHz.

Normally the CPUs run at the highest speed and the memory system represents the SoC bottleneck in terms of performance.

The different clock frequencies required to operate the different subsystems are generated by the clockgen, i.e., the clock generator; this approach consisting of having within a chip different islands running at different frequencies is called GALS (globally asynchronous locally synchronous) and is widely used today.

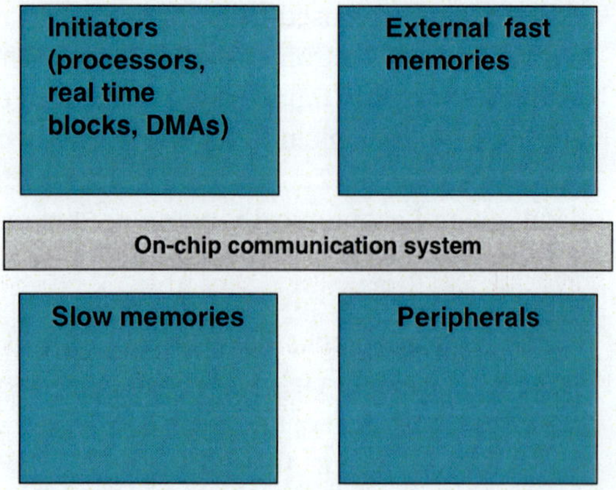

Fig. 6.1 SoC typical organization

The different subsystems working at different frequencies are linked together by the on-chip interconnect (Fig. 6.1), such as the STBus/STNoC [1] in the case of STMicroelectronics products.

Typically the on-chip interconnect optimizes the CPU path; that is the interconnect normally works at the same frequency as the CPU.

Proper frequency converters have to be placed between the interconnect and the other subsystems to perform the required frequency conversion in order to make the different subsystems communicate with each other.

6.3 On-Chip Communication

From a communication point of view, a typical SoC can be seen as a number of processing elements and storage elements connected by proper communication architectures.

As shown in Fig. 6.2, a central and key element in a complex SoC is the global on-chip communication architecture (OCCA), the infrastructure that interconnects these devices and provides the communication mechanisms necessary for distributed computation among different processing elements. The throughput and

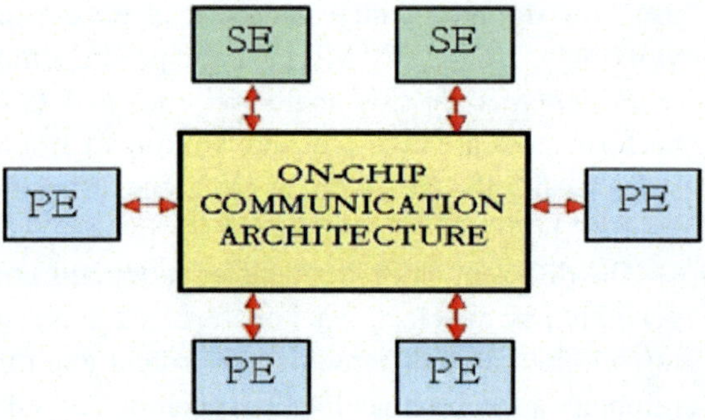

Fig. 6.2 Generic SoC architecture

latency of the communication infrastructure, and also the relevant power consumption, often limit the overall SoC performance.

Till now the prominent type of OCCA has been the on-chip bus, such the STBus from STMicroelectronics, the AMBA bus from ARM, CoreConnect from IBM or SiliconBackplane from Sonics, which represent the traditional shared-communication medium; such an OCCA, not scalable at all, has been able to fulfill SoC requirements because the performance bottleneck has always been the memory system.

With the growing requirements of more modern SoCs and CMOS technology scaling, the performance bottleneck is moving from memories to interconnect, as detailed in Sect. 6.4.

To overcome this limit, a new generation architecture, called Network on Chip (NoC), has been deeply studied and proposed; it is an attempt to translate the networking and parallel computing domain experience into the SoC world, relying on a packet-switched micro-network backbone based on a well-defined protocol stack. Innovative NoC architectures include VSTNoC from STMicroelectronics, Æthereal (Philips Research Lab), and Xpipe (University of Bologna).

6.3.1 On-Chip Bus

On-chip busses are communication systems composed of an intelligent logic, responsible for arbitration among the possible traffic flows injected by the different SoC initiators (PEs able to generate traffic), and a set of physical channels through which the traffic flows are routed from initiators to targets (PEs able to receive and process traffic) and vice versa.

The peculiarities of a bus, which are also the main drawbacks, are

- limited available bandwidth, given by the product of the bus size by the bus operating frequency; getting a higher available bandwidth would result in widening the bus size, thereby amplifying physical issues such as wire congestion, or increasing the operation frequency, having an impact on the power consumption, and in turn limited by physical issues such as capacitive load and capacitive coupling;
- lack of bandwidth scalability, since connecting more IPs to the bus means to divide the total available bandwidth among all the IPs, thereby allocating to each of them a lower bandwidth;
- limited system scalability, since connecting more IPs to the bus results in an increase of the capacitive load, responsible for an operating frequency drop;
- limited quality of service, since there is no possibility to treat in a different way different traffic classes, such as low latency CPUs, high bandwidth demanding video/audio processors, and DMAs;
- high area occupancy, due to the high number of wires required to transport all the protocol information, i.e., data and control signals (STBus interfaces, for example, are characterized by hundreds of wires);
- high power consumption, determined by the switching activity potentially affecting all the wires of the bus.

6.3.2 Network on Chip

As already mentioned, to overcome the disadvantages of on-chip busses, both the classic shared-bus (like AMBA AHB) and the more advanced communication systems supporting crossbar structures (such as the STBus) used in current SoCs, highlighted by the new requirement and needs of modern applications in conjunction with the most recent technology features, a novel on-chip communication architecture called NoC has been proposed.

It is important to highlight that the NoC concept is not a mere adaptation to the SoC context of the parallel computing or the wide area network domains; many issues are in fact still open in this new field, and the highly complex design space requires a deep exploration. Key open points are, for instance, the choice of the network topology, the message format, the end-to-end services, the routing strategies, the flow control and the queuing management. Moreover which kind of Quality of Service (QoS) to provide is another open issue, as well the most suitable software view for allowing the applications to exploit the NoC infrastructure peculiarities.

Exploiting the lesson learned by the telecommunication community, the global on-chip communication is decomposed into layers similar to the ISO-OSI Reference Model (see Fig. 6.3). The protocol stack enables different services and allows QoS, providing to the programmer an abstraction of the communication framework. Layers interact through well-defined interfaces and they hide the low-level physical DSM (Deep SubMicron) issues.

The Physical layer refers to all that concerns the electronic details of wires, the circuits and techniques to drive information (drivers, repeaters, layout); data link ensures a reliable transfer despite the physical unreliability and deals with medium access (sharing/contention). At the Network level there are issues related to the topology and the consequent routing scheme, while the Transport layer manages the end-to-end services and the packet segmentation/re-assembly. The other levels, up to the Application layer, can be viewed as a sort of merged adaptation layer that implements (in HW or through part of an OS) services and exposes the NoC infrastructure according to a proper programming model, e.g., the message passing (MP) paradigm.

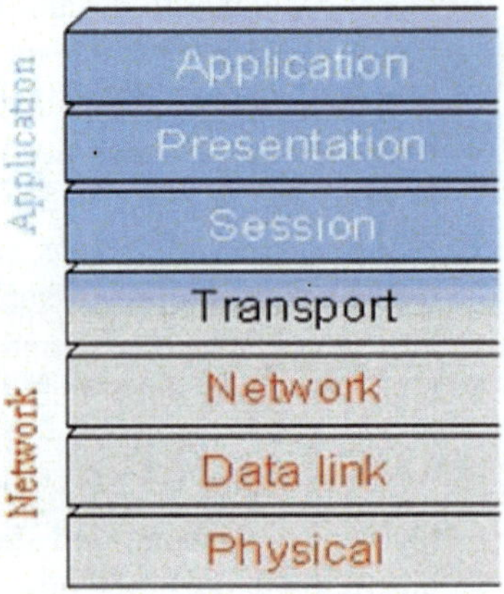

Fig. 6.3 ISO-OSI protocol stack

Despite the similarity discussed above, it is clear that the micro-network in the single chip domain differs from the wide-area network; NoC unique features are the spatial locality of connected modules, the reduced non-determinism of the on-chip traffic, the energy and latency constraints, the possibility of application specific stack services, and the need for low cost solutions.

An open issue in the NoC literature is the trade-off between the QoS provided by the network and the relevant implementation cost; on the other hand, a best effort service allows for a better average utilization but it cannot support a QoS. Users demand predictability of the application, so to mix both approaches could be a smart solution. Some papers provided an exhaustive overview of a QoS approach. QoS must be supported at all layers, and basic services are a fixed bandwidth, a maximum latency, and the correctness (no errors) and the completion (no packet loss) of the transmission.

The NoC communication is packet-based and the generally accepted forwarding scheme is a wormhole because it allows for a deeper pipeline and a reduced buffering cost. Packets are divided into basic units called flits; the queues in each node have the flit granularity and the physical node-to-node links are managed by a flow control that works on a flit per flit basis.

Another key point is the network topology, which has to be regular and simple. Literature foresees hybrid solutions with the local cluster based on shared buses and NoC for global communication. Some NoC state-of-the-art projects are based on the simple ring, two-dimensional mesh, fat tree [2], and Octagon [3] topologies, as shown in Fig. 6.4.

As far as the routing policy is concerned, it is possible to choose between deterministic, adaptive, source, arithmetic, or table-driven schemes; the deadlock handling is topology dependent. Input queues are suitable for a low cost implementation, but they show limited performance with respect to output buffering. In terms of control flow many solutions select a simple request/grant scheme, others a more efficient credit-based one. Links can be noise channels, so the literature begins to present work on error detection code or error correction code applied to on-chip interconnections, with distributed or end-to-end error recovery strategies.

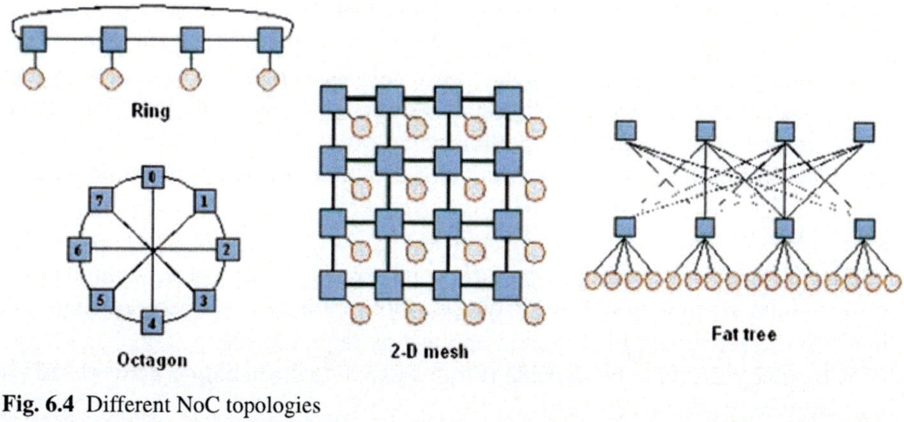

Fig. 6.4 Different NoC topologies

Besides routers, a significant amount of area is consumed by the so-called network interface (NI) that is the "access" to the NoC, translating the connected IP transactions to packets that are exchanged in the network. The NI hides network-dependent aspects to the PE, covering the Transport layer (connection handling, de-assembling of messages, higher level services).

To summarize, here is a list of the main benefits of the NoC approach:

- modularity, thanks to standard basic components, the NI, and the Router;
- abstraction as an inherent property of the layered approach, fitting also the demands of QoS;
- flexibility/scalability of the network as a benefit of a packet-based communication;
- regular and well-controlled structure to cope with DSM issues;
- re-use of the communication infrastructure viewed as a platform.

6.3.2.1 Topology

Two different approaches can be followed for the specification of the topology of a NoC:

- topology dependent
- topology independent

The former approach specifies the network architecture and its building blocks assuming a well-defined topology. The latter one aims at providing flexibility to the SoC architect in choosing the topology for the interconnect, depending on the application. This means that it is possible to build any kind of topology by plugging together the NoC building-blocks in the proper way. Even though it is more versatile because of the higher configurability allowed, this second approach has the following drawbacks:

- a very wide design and verification space, which would require a very big effort to ensure to the NoC user a high-quality product;
- the complexity of the network layer design (including issues such as deadlock) would be exposed to the SoC architect, thus requiring novel specific competencies and high effort in defining an effective (in terms of performance) and deadlock-free architecture;
- the need of high parametric building blocks, with few cost optimization possibilities.

Moreover the NoC built on top of a specific topology still presents a high degree of flexibility (routing, flow control, queues, QoS) in order to properly configure the interconnect for supporting different application requirements.

A topology is said to be scalable if it is possible to create larger networks of any size, by simply adding new nodes.

6.3.2.2 Routing Algorithms

Routing algorithms are responsible for the selection of a path from a source node to a destination node in a particular topology of a network.

A good routing algorithm balances the load across the various network channels even in the presence of non-uniform and heavy traffic patterns. A well-designed routing algorithm also keeps path lengths as short as possible, reducing then the overall latency of a message.

Another important aspect of a routing algorithm is its ability to work in the presence of faults in the network. If a particular algorithm is hardwired into the routers and a link or node fails, the entire network fails. If the algorithm can instead be reprogrammed or adapted to bypass the failure, the system can continue to operate with only a slight loss in performance.

Routing algorithms are classified depending on how they select between the possible paths from a source node to a destination node. Three main categories are specified:

- deterministic, where always the same path is chosen between a source and a destination node, even if multiple paths exist;
- oblivious, where the path is chosen without taking into account the present state of the network; oblivious routing algorithms include deterministic routing algorithms as a subset;
- adaptive, where the current state of the network is used to select the path.

6.3.2.3 Deadlock

A deadlock is the condition occurring in an interconnection network when a set of packets are unable to make any progress because they are waiting for one another to release network resources, such as buffers or channels.

Deadlock is a catastrophic event for the network. After a few resources are kept busy by deadlocked packets, other packets get blocked on these resources, thus paralyzing the network operation. To prevent such a problem, two solutions can be put in place:

- deadlock avoidance, a method to guarantee that the network cannot become deadlocked;
- deadlock recovery, a method consisting of detecting and correcting deadlock.

If deadlock is caused by dependencies external to the network, it is called high level deadlock or protocol deadlock (hereafter we call low-level deadlock the one related to the dependencies of the topology plus the relevant routing algorithm). For instance a simple request/response protocol could lead to deadlock conditions when dependencies occur in target devices between the incoming requests and the outgoing responses.

A network must always be free of deadlock, livelock, and starvation. A livelock refers to packets circulating the network without making any progress toward their destination. Starvation refers to packets indefinitely waiting at a network buffer (due to an unfair queuing policy). Both livelock and starvation reflect problems of fairness in network routing or scheduling policies.

As far as deadlock is concerned, in the case of deterministic routing, deadlock is avoided by eliminating cycles in the resource dependency graph; this is a directed graph, which depends on the topology and the routing, where the vertices are the resources and the edges represent the relationships due to the routing function. In the case of wormhole packet switching, these resources are the virtual channels; so we talk about a virtual channel dependency graph. A virtual channel (VC) provides logical links over the same shared physical channels, by establishing a number of independently allocated flit buffers in the corresponding transmitter/receiver nodes. When the physical link is not multiplexed among different VCs, the resource dependency graph could be simply called a channel dependency graph.

The protocol (or high level) deadlock refers to a deadlock condition due to resource dependencies external to the network. For instance, when a request-response protocol, such as STBus from STMicroelectronics or AMBA AXI from ARM, is adopted as end-to-end in the network, a node connected as target introduces dependencies between incoming requests and outgoing responses: the node does not perform as a sink for incoming packets, due to the finite size of the buffers and the dependencies between requests and responses.

In shared memory architectures, complex cache-coherent protocols could lead to a deeper level of dependencies. The effect of these protocol dependencies can be eliminated by using disjoint networks to handle requests and replies; the following two approaches are possible:

- two physical networks, i.e., separated physical data buses for requests and responses;
- two virtual networks, i.e., separated virtual channels for requests and responses.

6.3.2.4 Quality of Service

The set of services requested by the IPs connected to the network (called network clients) and the mechanisms used to provide these services are commonly referred to as QoS.

Generally, it is useful to classify the traffic across the network into a number of classes, in order to efficiently allocate network resources to packets. Different classes of packets usually have different requirements in terms of importance, tolerance to latency, bandwidth, and packet loss.

Two main traffic categories are specified:

- guaranteed service;
- best effort.

Traffic classes belonging to the former category are guaranteed a certain level of performance as long as the injected traffic respect a well-defined set of constraints.

Traffic classes belonging to the latter category do not get any strong guarantee from the network that instead will simply make its best effort to deliver the packets to their destinations. Best effort packets may then have arbitrary delay, or even be dropped.

The key quality of service concern in implementing best effort services is providing fairness among all the best effort flows. Two alternative solutions exist in terms of fairness:

- Latency-based fairness, aiming at providing equal delays to flows competing for the same resource;
- Throughput-based fairness, aiming at providing equal bandwidth to flows competing for the same resource.

While latency-based fairness can be achieved implementing a fair arbitration scheme (such as round-robin or less recently used (LRU,)), throughput-based fairness can be achieved in hardware by separating each flow requesting a resource into a separate queue, and then serving the queues in round-robin fashion. The implementation of such a separation can be expensive; in fact while physical channels (links) do not have to be replicated because of their dynamic allocation, virtual channels and buffers, requiring FIFOs, have to be replicated for each different class of traffic. So it is very important to choose the proper number of classes needing true isolation, keeping in mind that in many situations it may be possible to combine classes without a significant degradation of quality of service but gaining a reduction in hardware complexity.

6.3.2.5 Error Recovery

A high-performance, reliable, and energy-efficient NoC architecture requires a good utilization of error-avoidance and error-tolerance techniques, at most levels of its layered organization.

Modern technologies used to implement the present day systems, in order to reduce power consumption, adopt lower values of the power supply voltage, thus reducing the margins of noise immunity for the signals transmitted over the communication network of the system.

This leads to a noisy interconnect, behaving as an unreliable transport medium, and introducing errors in the transmitted signals. So the communication process needs to be fault tolerant to ensure correct information transfer. This can be achieved through the use of channel coding; channel coding schemes introduce a controlled amount of redundancy in the transmitted data, increasing its noise immunity.

Linear block codes are commonly used for channel encoding. Using an (n,k) linear block code, a data block k bits long is mapped onto an n bit code word, which is transmitted over the channel. The receiver examines the received signal and declares an error if it is not a valid code word.

Once an error has been detected, it can be handled in one of two different ways:

- Forward error correction (FEC), where the properties of the code are used to correct the error;
- Retransmission, also called automatic repeat request (ARQ), where the receiver asks the sender to retransmit the code word affected by the error.

FEC schemes require a more complex decoder, while ARQ schemes require the existence of a reverse channel from the receiver to the transmitter, in order to ask for the retransmission.

6.4 SoC Integration Issues

Using the newest CMOS technologies, where the DSM effects are so evident, the physical design of a SoC is facing more and more two types of issues:

- *Physical issues*, related to the difficulties encountered mainly during the placement of the hard macros and the standard cells, and during the routing of clock nets and communication system wires;
- *Performance issues*, related mainly to the bandwidth requirements of the different IPs, that in order to be fulfilled, would require SoCs to run at very high speeds.

Figure 6.5 is an illustration of the physical issues; it shows the floorplan of an sample CMOS chip for consumer application and the interconnect implemented with the VSTNoC solution.

In this figure the blue squares are the network Interfaces of the various IPs of the chip (both initiators and targets), the blue rectangles are the nodes, responsible for arbitration and propagation of information, and the blue lines are the physical channels connecting the different NoC devices. All these elements must be physically located in the grey area, representing the physical space available for the interconnect. Because of the shape (quite irregular and with thin regions) and the area size, it is evident that the placement of the interconnect standard cells can be difficult, and the routing of the wires that can be also very long can suffer from congestion.

As far as performance is concerned, two main factors influence the overall operating frequency of a SoC, i.e., device switching times and the interconnect offered bandwidth. Current technologies can achieve unprecedented transistor transition frequencies due to short transistor lengths. However, the same is not true for interconnects. Indeed, continually shrinking feature sizes, higher clock frequencies, and growth in complexity are all negative factors as far as switching charges on metallic interconnect are concerned. This situation is shifting the IC design bottleneck from computing capacity to communication.

Feature sizes on integrated circuits and also, therefore, circuit speed have followed Moore's law for over 3 decades and the CMOS integration capability is still increasing. In this respect, according to the international technology roadmap for

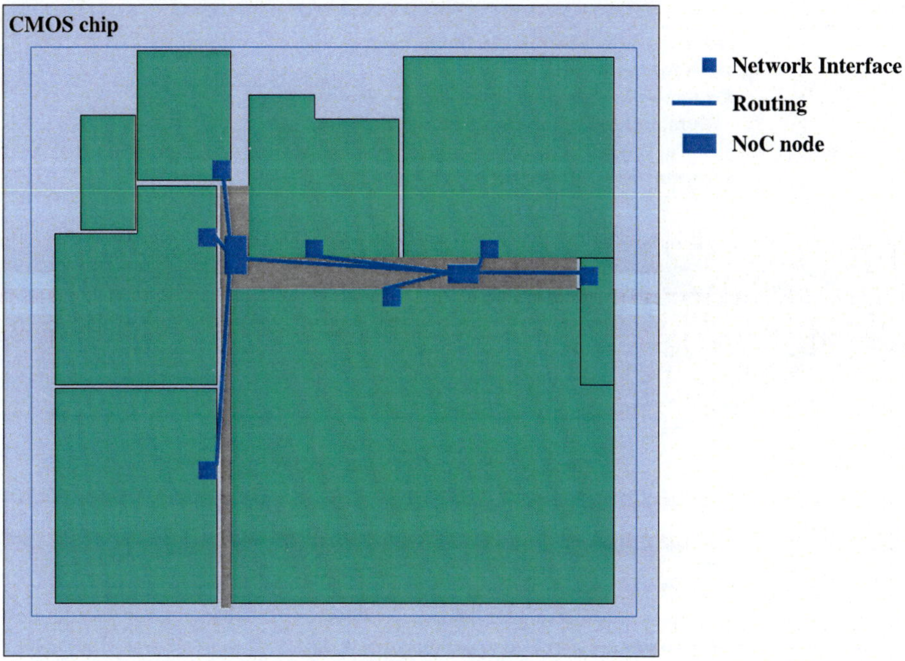

Fig. 6.5 CMOS chip with VSTNoC physical design

semiconductors (ITRS [4]), the RC time constants aSoCsiated with metallic interconnects will not be able to decrease sufficiently for the high-bandwidth applications destined to appear in the next few years (see Fig. 6.6). Internal data rates of processors fabricated in deep submicron CMOS technology have exceeded gigahertz rates. While processing proceeds at GHz internally, off-chip wires have held inter-chip clock rates to hundreds of MHz.

The rate of inter-chip communication is now the limiting factor in high performance systems. The function of an interconnect or wiring system is to distribute clock and other signals to and among the various circuits/systems on a chip. The fundamental development requirement for interconnects is to meet the high-speed transmission needs of chips despite further scaling of feature sizes. This scaling down, however, has been shown to increase severely the signal runtime delays in the global interconnect layers. Indeed, while the reduction in transistor gate lengths increases the circuit speed, the signal delay time for global wires continues to increase with technology scaling, primarily due to the increasing resistance of the wires and their increasing lengths. Current trends to decrease the runtime delays, the power consumption, and the crosstalk, focus on lowering the RC-product of the wires, by using metals with lower resistivity (i.e., copper instead of aluminum) and by the use of insulators with lower dielectric constant. Examples of the latter include nanoporous SiOC-like or organic (SiLK type) materials, which have dielectric constants below 2.0 or air gap approaches, which reach values close to 1.8–1.7. Integration of these materials results in an increased complexity, however, and they

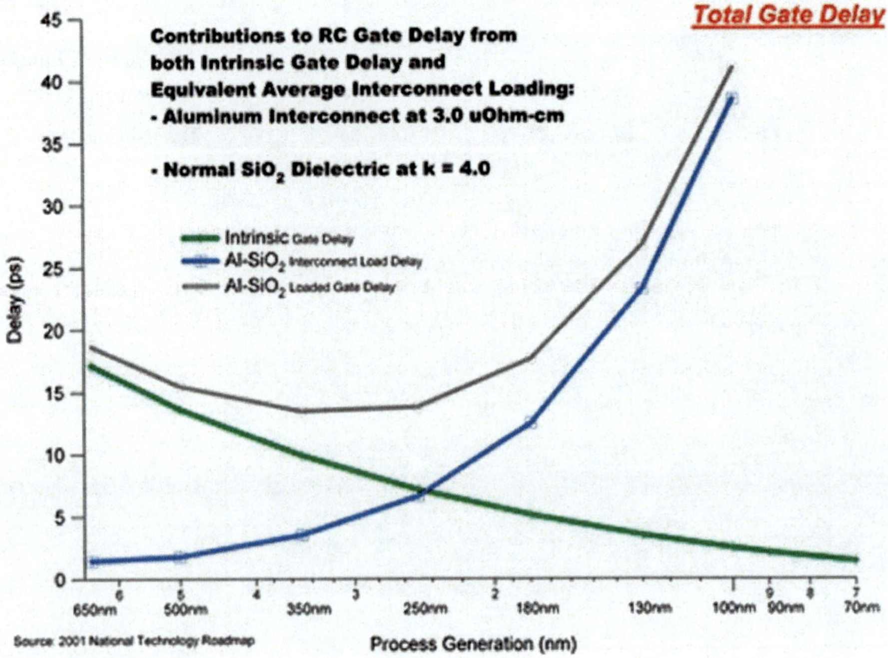

Fig. 6.6 Average interconnect delay as a function of process

have an inherent mechanical weakness. Moreover, introducing ultra low dielectric constant materials reaches its fundamental physical limit when one considers that the film permittivity cannot be less than 1.

6.4.1 Electrical Interconnect Classification

From a technological point of view, interconnects can be classified in the following categories (see Fig. 6.7):

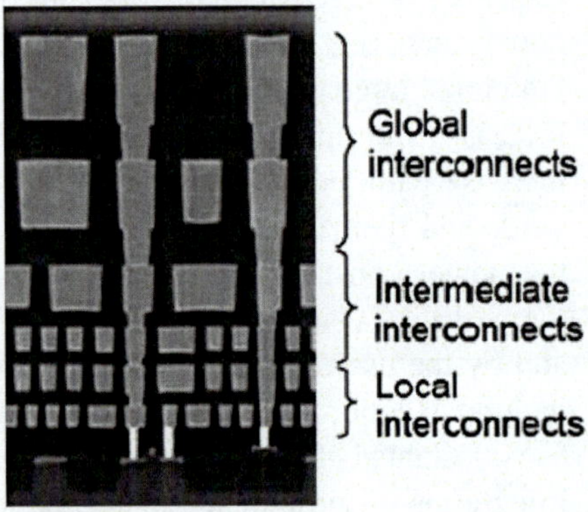

Fig. 6.7 Interconnect classification from a technological point of view

- *local interconnect*, used for short-distance communication, typically between logic units, and comprising the majority of on-chip wires; they have the smallest pitch and a delay of less than one clock cycle;
- *global interconnect*, providing communication between large functional blocks (IPs); they are fewer than local interconnects, but are no less important. Improving the performance of a small number of critical global links can significantly enhance the total system performance. Global interconnects have the largest pitch and a delay longer than one or two clock cycles typically.
- *intermediate interconnect*, having dimensions that are between those of local and global interconnects.

A key difference between local and global interconnect is that the length of the former scales with the technology node, while for the latter the length is approximately constant.

From a functional point of view, the two main important and performance-demanding applications of interconnects in SoC are signaling (i.e., the communication of different logic units) and clock distribution. In this context they can be classified as

- *point-to-point links*, used for critical data-intensive links, such as CPU-memory buses in processor architectures;
- *broadcast links*, representing physical channels where the number of receivers (and therefore repeaters) is high and switching activity is also high
- *network links*, targeted at system buses and reconfigurable networks, aiming at serving complete system architectures, whose typical communication is around several tens of Gb/s.

6.4.2 Electrical Interconnect Metrics

An ideal interconnect should be able to transmit any signal with no delay, no degradation (either inherent or induced by external causes), over any distance without consuming any power, requiring zero physical footprint and without disturbing the surrounding environment.

According to this, a number of metrics have been defined in order to characterize the performance and the quality of real interconnects, such as

- propagation delay
- bandwidth density
- power-delay product
- bit error rate

6.4.2.1 Propagation Delay

The propagation delay is the time required by a signal to cross a wire. Pure interconnect delay depends on the link length and the speed of propagation of the wavefront

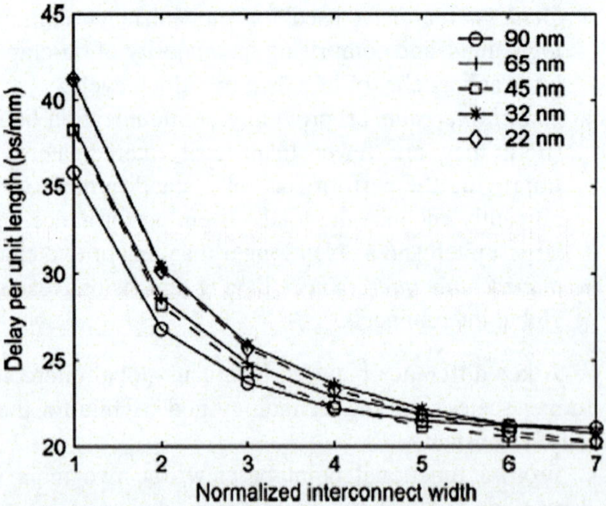

Fig. 6.8 Interconnect delay as function of interconnect width

(time of flight). Electrical regeneration introduces additional delay through buffers and transistor switching times. Additionally, delay can be induced by crosstalk.

It can be reduced by increasing the interconnect width at the expense of a smaller bandwidth density.

Technology scaling has insignificant effect on the delay of an interconnect with an optimal number of repeaters. The minimum achievable interconnect delay remains effectively fixed at approximately 20 ps/mm when technology scales from 90 (year 2004) to 22 nm (year 2016), as shown in Fig. 6.8.

6.4.2.2 Bandwidth Density

Bandwidth density is a metric that characterizes information throughput through a unit cross-section of an interconnect. Generally, it is defined by the pitch of the electrical wires.

6.4.2.3 Power-Delay Product

Signal transmission always requires power. In the simplest case, it is required to change the charge value on the equivalent capacitor of a metallic wire. In more realistic cases, power will also be required in emitter and receiver circuitry, and in regeneration circuits.

A distinction can also be made between static and dynamic power consumption by introducing a factor a representing the switching activity of the interconnect link $(0 < a < 1)$.

The power and delay product (PDP) is routinely used in the technology design process to evaluate circuit performance.

6.4.2.4 Bit Error Rate

The bit error rate (BER) may be defined as the rate of error occurrences and is the main criterion in evaluating the performance of digital transmission systems.

For an on-chip communication system a BER of 10^{-15} is acceptable; electrical interconnects typically achieve BER figures better than 10^{-45}. That is why the BER is not commonly considered in integrated circuit design circles. However, future operation frequencies are likely to change this, since the combination of necessarily faster rise and fall times, lower supply voltages, and higher crosstalk increases the probability of wrongly interpreting the signal that was sent.

Errors come from signal degradation. Real signals are characterized by their actual frequency content and by their voltage or current value limits. The frequency content will define the necessary channel bandwidth, according to Shannon – Hartley theorem. Analogue signals are highly sensitive to degradation and the preferred mode of signal transmission over interconnect is digital.

Signal degradation can be classed as time-based, inherent, and externally induced.

- Time-based: non-zero rise-time, overshoot, undershoot, and ringing time-based degradation can be incorporated into the delay term for digital signals. While the whole of these degradations can be assimilated into a quasi-deterministic behavior that does not exceed the noise margins of the digital circuits, a transformation in temporal space is possible (to contribute to the regeneration delay term). This assumption is however destined to disappear with nanometric technologies because of a more probabilistic behavior and especially of weaker noise margins.
- Inherent: attenuation (dB/cm), skin effect, and reflections (dB).
- Externally induced: crosstalk (dB/cm) and sensitivity to ambient noise.

The allowable tolerance on signal degradation and delay for a given bandwidth and power budget forces a limit to the transmission distance. The maximum interconnect segment length can in fact be calculated, a segment being defined as a portion of interconnect not requiring regeneration at a receiver point spatially distant from its emission point.

Signal regeneration in turn leads to a further problem, i.e., the energy used to propagate the signal in the transmission medium can escape into the surrounding environment and perturb the operation of elements close to the transmission path.

6.5 On-Chip Optical Interconnect

Because of the electrical interconnect limitations highlighted in previous section, several researchers have come to the conclusion that "the global interconnect performance needed for future generations of ICs cannot be achieved even with the most optimistic values of metal resistivity and dielectric constants." Currently, low values of resistivity are 2.2 μOhm × cm offered by copper and 1.6 μOhm × cm offered by silver, while low values of dielectric constant, lower than 2 (between

1.7 and 1.8) are offered by SiOC-type nanoporous materials or SiLK-type organic materials. However, evolutionary solutions will not suffice to meet the performance roadmap and therefore radical new approaches are needed.

Several different such possibilities are now envisaged, the most prominent of which are the use of RF or microwave interconnects, optical interconnects, three-dimensional interconnects, and cooled conductors.

As industrial partner of two funded European projects (PICMOS and WADIMOS), STMicroelectronics is currently investigating the optical interconnect solution because of the potential intrinsic advantages of such a technology, i.e.,

- high bandwidth available thanks to the light frequency (THz);
- immunity to noise;
- no crosstalk between light beams;
- removal of physical issues affecting CMOS chip fabrication thanks to a separated photonic interconnect chip.

The communication system architecture and the materials used for realizing the optical interconnect play a key role in such a novel approach.

A smart choice of the interconnect architecture helps to bypass some physical problems (such as beams crossing, the need for multiple wavelengths emitters, and even the need for extremely high frequencies such as the light frequency), and on the opposite side some intrinsic physical issues drive the choice of the communication system architecture. The key factors are illustrated in the following:

- *Bandwidth*: in current SoCs the bottleneck is the main memory, usually a DDR/DDR2 SDRAM. Due to the clock frequency, data bus size, refresh, page open overhead, and other factors, its available bandwidth is limited. A link able to support a much higher bandwidth is useless if the memory is not able to exploit it; this should lead to the development of memories able to guarantee a higher bandwidth, and allow optimizing the efforts in developing high bandwidth links.
- *Power*: according to the consumer market needs, modern SoCs need to reduce power consumption in order to guarantee a higher life to batteries and to reduce the device heating. Low power architecture and design techniques are today widely applied, so the high-power potentially consumed by light emitters is a big enemy for the chip industry.
- *Clocking*: the complexity of modern systems, together with the collaborations between different industries and organizations, is leading to heterogeneous systems where the different intellectual properties (IPs) run at different speeds. This means that the concept of a synchronous system is going to be lost, while the globally asynchronous locally synchronous (GALS) approach will be more and more applied. We will have systems where the different IPs are locally clocked by a specific clock, but at global level they communicate with each other asynchronously. The global clock distribution problem could then disappear, and the application of an optical clock to overcome the usual clock tree distribution problem would be not so significant.

- *Topology*: the choice of the on-chip communication architecture impacts its low-level implementation, also in terms of physical issues. It makes sense to have a ring or a mesh topology where no crossings are required. This would allow using only one wavelength without the need for multiplexing. Also, avoiding crossings would make waveguide implementation easier and would reduce undesired coupling and degrading effects.
- *Technology*: being CMOS compatible is the key to having the optical interconnect solution adopted by industries. The implementation of the optical communication network has to be done keeping the current Si-based CMOS technology in order to reduce the impact on tools and cost. Being able to have standard cells specifically for the driving circuitry of both emitters and detectors will be required as well.
- *Integration capability*: this is another key point. Whatever the material will be (Si, Ge, SiGe, GaAs) there is a need to integrate a huge number of emitters (or modulators) and detectors, as well as a huge number of waveguides according to the selected topology. Current SoCs are composed of a number of IPs of the order of 50–100, so 50–100 emitters (modulators) and detectors would be required on a chip. This must be ensured by the technology.

Of course the materials and the processes required to implement an on-chip optical interconnect must be *CMOS compatible*. This is the key to guarantee industries will adopt such a novel technological solution with no need for completely changing the design and manufacturing processes, which would have a dramatic impact in terms of cost.

To guarantee CMOS compatibility, an accurate choice of the materials used to build an optical interconnect has to be made. The most promising candidates are as follows:

- *III–V compounded semiconductors* (e.g., InP or GaAs), giving the best performance devices, but their integration with silicon-based chips is highly costly at the moment; growing GaAs on Si has, for example, a very high cost. However, the integration of III–V emitters and detectors and SOI waveguides on-chip has been demonstrated.
- *Germanium* and *Si-Ge* alloys, being fairly easy to grow on Si due to their similar lattice structure (they both belong to group IV of Mendeleiev's table); the realization of good detectors, compatible with SOI waveguides, have been demonstrated.
- *Silicon*, giving of course the best advantages in terms of technology since the Si planar technology has been stable and widely used for a long time, but the performance obtained from Si optical devices is currently very low, mainly because of its indirect band gap; moreover, up to now there are no effective Si LASER sources available, not even exploiting the Raman effect, thanks to whom LASER emission from Si has been demonstrated. SOI technology allows the construction of effective optical waveguides exploiting the refraction index difference between Si and SiO_2.

6.5.1 The PICMOS Project

The PICMOS project [5] is a European Commission funded program, completed on March 2007. The main objective was to demonstrate the feasibility of building an optical interconnect layer on top of CMOS.

The photonic interconnect layer exploits waferscale technologies as much as possible; a heterogeneous integration technology based on wafer bonding has been developed. The actual information transport medium, being a passive structure, consists of high index contrast waveguides in silicon; emitters and detectors, being the active devices, have been fabricated in InP-based materials.

For the integration of the photonic interconnect layer with the CMOS chip, two different integration strategies have been investigated (see Fig. 6.9):

- the *wafer-to-wafer bonding* approach, such that the photonic interconnect layer is fabricated in parallel with and independent of the wafer containing the electronic circuits; subsequently, the latter is wafer bonded to the wafer with the photonic wiring circuits and the required electrical connections are provided using wafer scale technologies; finally, the wafer is diced and packaged using standard techniques;
- the *above-IC approach*, consisting in fabricating the photonic interconnect layer directly on-top of the electronic circuits.

The wafer-to-wafer bonding approach (Fig. 6.10) has several advantages; the photonic interconnect layer can be tested before bonding and for its fabrication processing steps (e.g., high temperature steps), which are usually not compatible with CMOS, back-end processing can be employed. Moreover, one could envisage standardized (off-the-shelf) photonic interconnect circuits compatible with multiple electronic circuit designs.

The main advantage of above-IC approach (Fig. 6.11) is the on-line integration ability, including the fact that this approach is similar to current CMOS-fabrication.

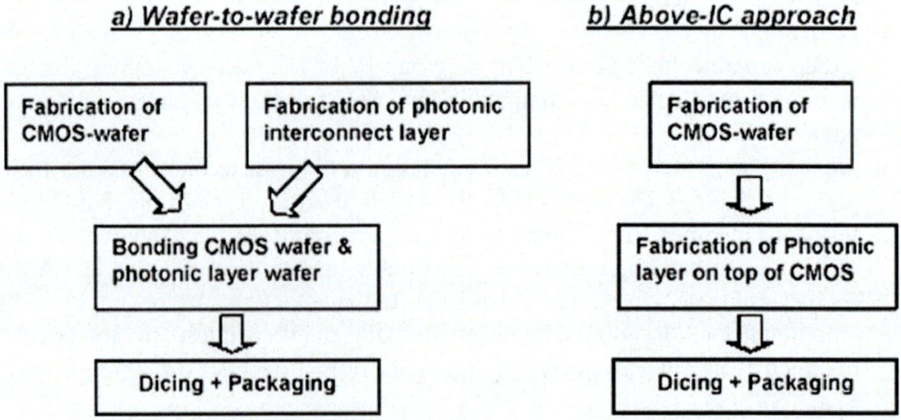

Fig. 6.9 Different approaches for optical interconnect fabrication

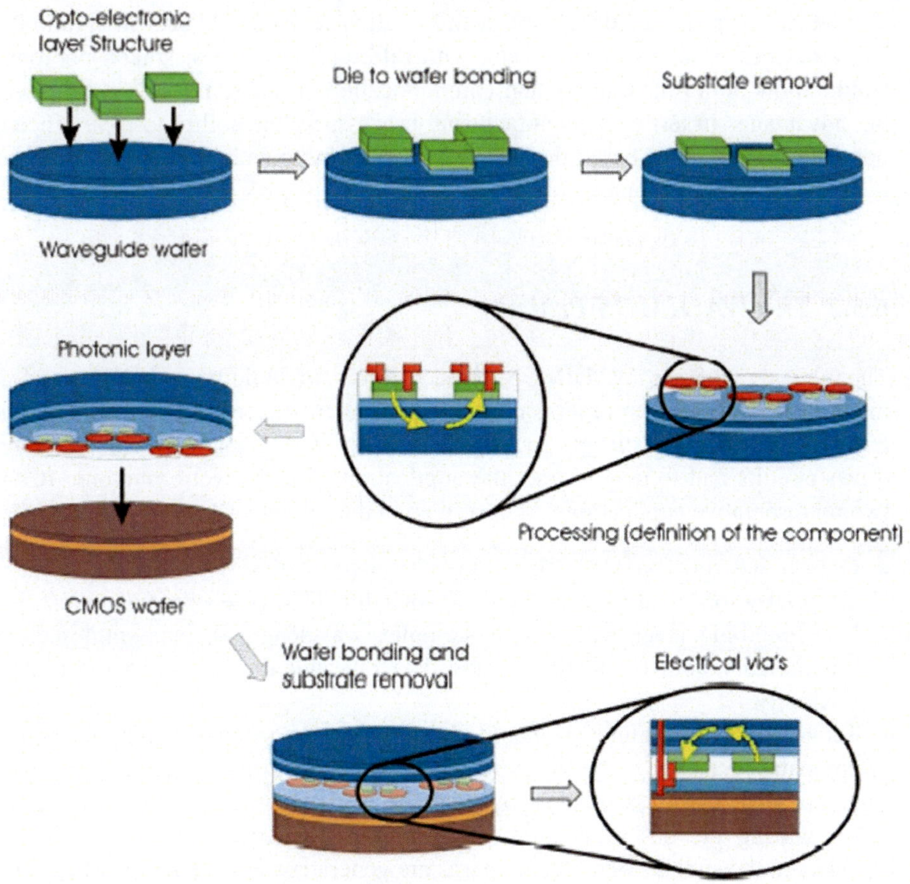

Fig. 6.10 Wafer-to-wafer bonding approach

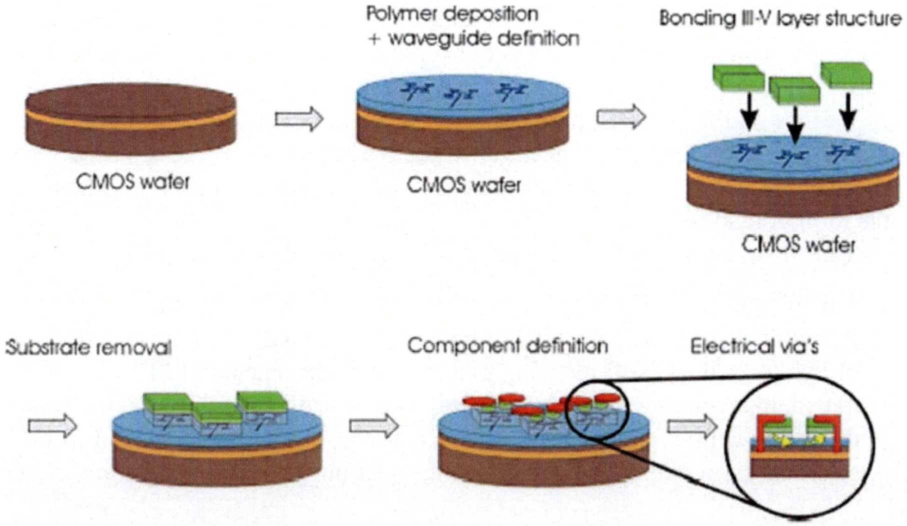

Fig. 6.11 Above-IC approach

Both the approaches described in this section rely on a hybrid integration of III–V devices on a silicon substrate, with different techniques; silicon photonics would allow both functionality and communication on the same silicon-chip, with big advantages in terms of manufacturing process. So the ability to have efficient optoelectronic devices based on silicon is mandatory to embrace silicon photonics in a SoC networking context.

6.5.2 The WADIMOS Project

The objective of the WADIMOS project [6] is to build a complex photonic interconnect layer incorporating multi-channel microsources, microdetectors, and different advanced wavelength routing functions directly integrated with electronic driver circuits and to demonstrate the application of such electro-photonic ICs in two representative applications, as described in the following.

- Optical network on chip: the main objective of the WADIMOS project is the realization of an optical NoC (ONoC) for STMicroelectronics chips. The NoC photonic layer will include complex wavelength division multiplexing functionality both for increasing the data rate and for increasing the routing flexibility.
- Terabit optical datalink: *Mapper Lithography* is a semiconductor equipment company focusing on the development and manufacturing of a new and highly competitive maskless lithography machine using thousands of electron beams for writing the desired patterns. This requires a data-rate of over 100 TB/s between the subfab, where the patterns are generated, and the actual lithography equipment.

The ONoC [7, 8] is a novel on-chip communication system where information is transmitted in the form of light, in opposition to what happens in classical electrical NoC where the information is transmitted in form of voltage levels (and then currents).

It relies on the ISO-OSI protocol stack, and can be seen as being very close to the ST NoC (VSTNoC) architecture, where the physical layer is replaced with a completely new one, exploiting optoelectronics in order to transmit information in the form of light [9].

The ONoC architecture consists of three main sets of building-blocks:

- The initiator optical network interface (IONI) responsible for the conversion of the traffic generated by an initiator into a form suitable to be transmitted in form of light over the ONoC reducing to a minimum the amount of "1"s to be transmitted, thus limiting the time an emitter is required to be turned on and thereby the power consumption, and, for the actual conversion of information from the electrical form into optical form, by means of information serialization, emitter selection, and emitter driving; moreover, to obtain responses (for example, when reading data from memory), a structure replicating the main behavior of the TONI (see later) is also implemented.

- The lambda-router, responsible for the actual propagation of optical information streams from sources to destinations;
- The target optical network interface (TONI) responsible for the conversion of information from optical form into electrical form, by means of photocurrent to voltage conversion, level adjustment, and information deserialization and for the conversion of the traffic generated by the ONoC receiver into a form suitable to be received by the target. Moreover, to send responses, a structure replicating the main behavior of the IONI (see above) is also implemented.

Such building blocks can be assembled together to build proper on-chip communication architectures, as illustrated in Fig. 6.12.

From a physical point of view the electronics implementing the driver of emitters and the control of detectors will be in the CMOS chip, while the actual emitting and detecting devices will be in the photonic chip where the optical interconnect is fabricated. The link between the CMOS electronics and the photonic layer will be a set of vias carrying the information in electrical form from one layer to the other, as shown in Fig. 6.13.

According to that plan, the placement of the ONoC transmitters and receivers within the various IPs in the CMOS chip has to take into account somehow the placement of the photonic layer above the CMOS chip itself, considering that the size of the photonic layer is usually much smaller than the size of the CMOS chip.

This means that for the CMOS chip a special effort is needed in clustering the different IPs, and, within them, clustering the LASER drivers and detectors control logic, in order to place properly the vias responsible for the exchange of information between the CMOS chip and the photonic layer. This is shown in Figs. 6.14 and 6.15.

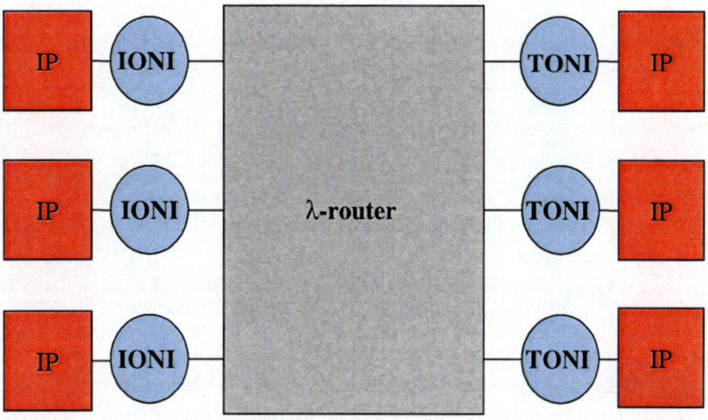

Fig. 6.12 Example of ONoC architecture

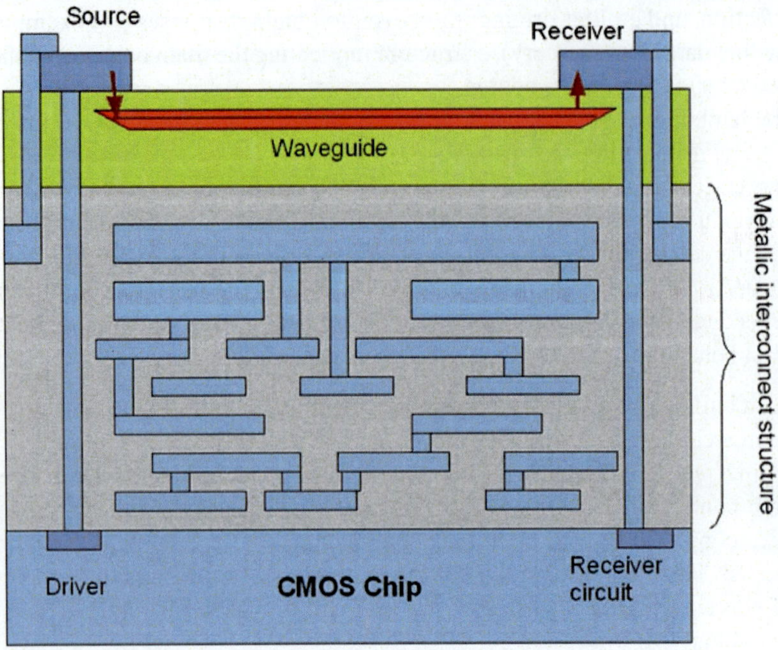

Fig. 6.13 Connection between a CMOS chip and an optical interconnect

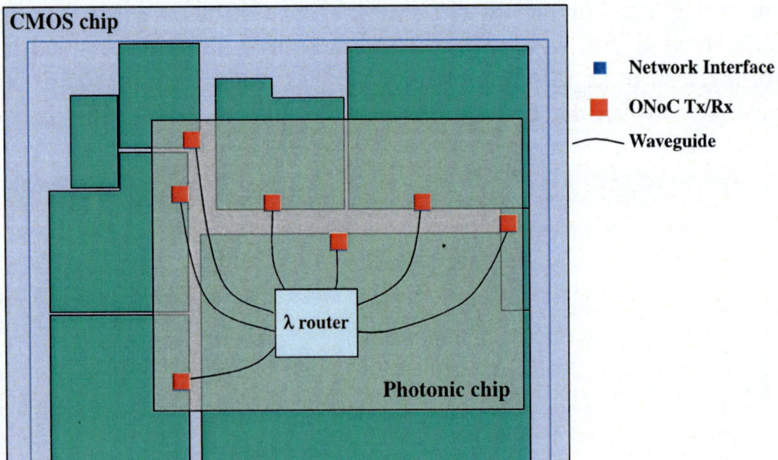

Fig. 6.14 An optical interconnect above a CMOS chip

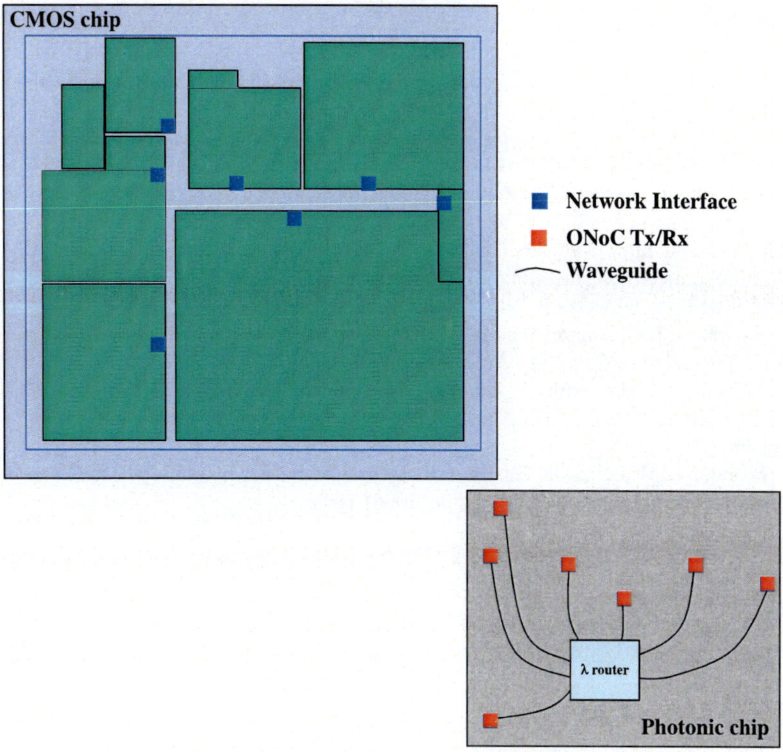

Fig. 6.15 CMOS chip and optical interconnect as two separated chips

6.6 Conclusion

This chapter describes how silicon photonics could help SoC design and implementation from an industrial perspective, based on the vision of STMicroelectronics. The SoC concept is introduced, and current SoC communication systems are described, highlighting their features and limitations.

Moving from such limitations, on-chip optical interconnect is described in terms of its potential advantages according to light properties and technology capabilities, and the European projects PICMOS and WADIMOS are described in order to highlight current work in the field of on-chip optical interconnect.

Overall, this chapter is based on the assumption that, according to current technology capabilities and trends, on-chip optical interconnect will be strongly based on III–V semiconductors, with all the issues coming from the need for ensuring CMOS compatibility. An efficient and reliable silicon photonics would be a big step forward in this field.

References

1. A. Scandurra, STBus communication system: concepts and definitions. (Public specification, STMicroelectronics web site, 2009) 146
2. W.J. Dally, B. Towles, *Principles and Practices of Interconnection Networks* (Morgan Kaufmann, San Francisco, 2003) 149
3. M. Coppola, R. Locatelli, G. Maruccia, L. Pieralisi, A. Scandurra, Spidergon: a novel on-chip communication network, SOC working conference, Tampere, 2004 149
4. ITRS web site, http://www.itrs.net 155
5. PICMOS project web site, http://picmos.intec.ugent.be 162
6. WADIMOS project web site, http://wadimos.intec.ugent.be 164
7. A. Scandurra, I. O'Connor, *Scalable CMOS-Compatible Photonic Routing Topologies for Versatile Networks On-Chip* (NoCArc, Como, 2008) 164
8. I. O'Connor, F. Mieyeville, F. Gaffiot, A. Scandurra, G. Nicolescu, *Reduction Methods for Adapting Optical Network On-Chip Topologies to Specific Routing Applications* (DCIS, Grenoble, 2008) 164
9. A. Scandurra, M. Lenzi, R. Guerra, F.G. Della Corte, M.A. Nigro, *Optical Interconnects for Network On-Chip* (Nanonets, Lausanne, 2006) 164

Index

Chapter 7
High-Speed Photonic Integrated Chip on a Silicon Platform

Ling Liao, Ansheng Liu, Hat Nguyen, Juthika Basak, Mario Paniccia, Yoel Chetrit, and Doron Rubin

Abstract As microprocessor technology advances toward multi-core and many-core architectures, optical interconnect is considered a promising way of meeting the associated demand for giga-scale and tera-scale input/output (I/O). While traditional optical communication systems demonstrate good performance, they are based on discrete components and are not suitable for computing applications, which require solutions with much lower cost and smaller size. Photonic integration, particularly when based on a silicon platform, has emerged as a key approach to realize the required low cost and small form factor optical transceivers. This chapter highlights a recent demonstration of a silicon photonic integrated chip that is capable of transmitting data at an aggregate rate of 200 Gb/s. It is based on wavelength division multiplexing where an array of eight high-speed silicon optical modulators is monolithically integrated with a demultiplexer and a multiplexer. This demonstration represents a key milestone on the way to fabricating terabit per second transceiver chips to meet the demand of future tera-scale I/O.

7.1 Introduction

The computer industry is migrating to a many-core processor architecture to maintain the pace of increasing computational capability while at the same time meet the energy efficiency needs of mobile computing and data center power limits [1]. For these new processor designs, power consumption can be optimized by using multiple types of cores tailored to match the needs of different usage models. Also, cores that are not busy can be turned off temporarily to reduce power consumption during idle times. While quad-core systems are the state of the art today, computers with tens or even hundreds of cores will be the norm over the next decade. One of the challenges with realizing such parallel computing will be the transport of the

L. Liao, A. Liu, H. Nguyen, J. Basak, and M. Paniccia (✉)
Intel Corporation, Santa Clara, CA 95054, USA

Y. Chetrit and D. Rubin (✉)
Numonyx Israel Ltd., Qiryat Gat, 82109, Israel

D.J. Lockwood, L. Pavesi (Eds.): *Silicon Photonics II*.
Topics in Applied Physics **119**, 169–186 (2011)
DOI 10.1007/978-3-642-10506-7_7

massive amounts of data into and out of the chip, into and out of the board, and into and out of the system. Although present implementations of chip-to-chip copper interconnects are being pushed to deliver giga-scale bandwidths with the design of transceivers that use active or passive equalization [2, 3], this is unlikely to be a frequency scalable option. A subset of the future input/output (I/O) interconnects will require an aggregate bandwidth of 100's of Gb/s or even Tb/s for occasions when most of the cores are operating at the same time, so the need to transmit some of these signals optically is a near certainty.

For these interconnections in and around the computer, the optical solution not only needs to meet the performance requirements, it must also be cost effective, compact, and highly integrated. Silicon photonics has emerged as a technology uniquely suited to meet these requirements [4, 5]. It assures low process development and manufacturing costs, as it has ready access to the infrastructure, toolsets, knowledge, and capacity that have been and will be developed for silicon microelectronics. It also benefits from silicon's proven promise of integration, which should lead to further reduction in cost and size. Furthermore, the pace at which the performance of silicon photonic components is advancing demonstrates rapid maturing of this field [6–21]. Integrated Tb/s transmitters and receivers on a silicon platform should not be far on the horizon when using wavelength division multiplexing (WDM) [22]. For instance, an integrated 1 Tb/s transmitter can be realized by using a multiplexer (MUX) to combine 25 laser signals at different wavelengths, each encoded with 40 Gb/s data by a silicon modulator. A 1 Tb/s receiver is possible by using a demultiplexer (DEMUX) connected to an array of 25 photodetectors, each receiving 40 Gb/s data. To date, all the required optical components have been individually demonstrated on the silicon platform. The natural next step is device integration.

This chapter presents a proof of concept demonstration of a silicon photonic integrated chip (PIC) where an array of eight high-speed silicon optical modulators is monolithically integrated with a wavelength DEMUX and a wavelength MUX. The modulators are based on the carrier depletion effect of PN diodes, and the DEMUX/MUX is based on a cascaded asymmetric Mach Zehnder interferometer (MZI) design. This integrated transmitter is capable of sending data at an aggregate rate of 200 Gb/s over a single fiber. Section 7.2 outlines the architecture of this integrated chip and presents design details for both the modulators and the MUX/DEMUX. Section 7.3 describes device fabrication. Section 7.4 first details the testing results of the individual components that make up the integrated chip, and it then discusses the measured DC as well as high-speed performance of the fully integrated device.

7.2 Silicon Photonic Integrated Chip Design

The silicon PIC is based on a silicon-on-insulator (SOI) platform. To target high aggregate data transmission over a single fiber, an eight-channel WDM design is employed. Figure 7.1 shows a top–down schematic of this chip, which includes

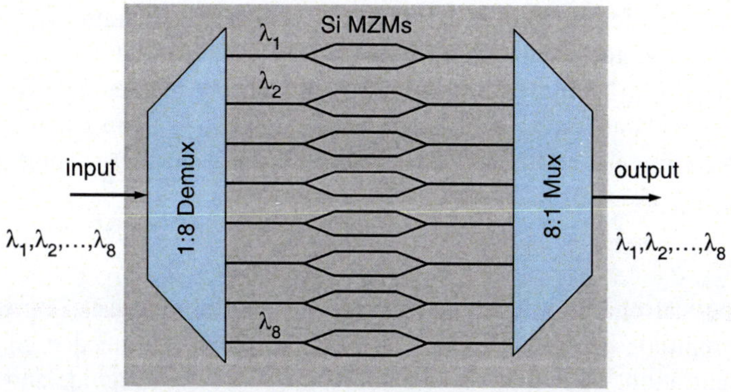

Fig. 7.1 Schematic representation of the silicon photonic integrated chip on SOI substrate. It consists of a 1:8 demultiplexer, eight high-speed silicon MZMs, and an 8:1 multiplexer. A continuous-wave multi-wavelength laser source is outside of the chip. After DEMUX, the individual wavelengths are modulated by their corresponding silicon modulators

a 1:8 DEMUX, 8 high-speed silicon Mach-Zehnder modulators (MZMs), and an 8:1 MUX. As Fig. 7.1 illustrates, a continuous-wave (CW) multi-wavelength laser beam is first split by the DEMUX. Each wavelength then propagates through its corresponding modulator. The CW light in each wavelength channel is amplitude modulated by the MZM so that high-speed signal is encoded onto the optical beam. The modulators are followed by the MUX, where all eight wavelengths are combined into the output waveguide to be coupled off chip. For this transmitter proof of concept demonstration, the laser sources are not integrated onto the silicon chip. While this implementation does not provide the final transmitter solution, it is a first step toward full photonic integration. In the future, the DEMUX in Fig. 7.1 can be replaced by an array of hybrid lasers [21] monolithically integrated on-chip to enable a fully integrated transmitter.

The first key component of the PIC is the high-speed silicon MZM. Its operating speed, together with the number of wavelength channels, fundamentally determines the total data transmission capacity of the chip. To achieve >100 Gb/s, it is critical to design a high-speed modulator with a data transmission capability greater than 10 Gb/s. The silicon modulator design chosen for this demonstration is based on the free carrier plasma dispersion effect [23, 24], where a change in free carrier distribution alters the refractive index of the material and the optical phase of light passing through it. The absorption coefficient change ($\Delta\alpha$) and refractive index change (Δn) in silicon are related to the change in the concentration of electrons (ΔN_{e}) and holes (ΔN_{h}) and are generally described by Drude–Lorentz equations:

$$\Delta\alpha = \frac{e^3\lambda^2}{4\pi^2 c^3 \varepsilon_0 n}\left(\frac{\Delta N_{\mathrm{e}}}{\mu_{\mathrm{e}}(m_{\mathrm{e}}^*)^2} + \frac{\Delta N_{\mathrm{h}}}{\mu_{\mathrm{h}}(m_{\mathrm{h}}^*)^2}\right) \tag{7.1}$$

$$\Delta n = \frac{-e^2\lambda^2}{8\pi^2 c^2 \varepsilon_0 n}\left(\frac{\Delta N_{\mathrm{e}}}{m_{\mathrm{e}}^*} + \frac{\Delta N_{\mathrm{h}}}{m_{\mathrm{h}}^*}\right) \tag{7.2}$$

where e is the electron charge, c is the light velocity in vacuum, ε_0 is the vacuum permittivity, λ is the wavelength in vacuum, n is the refractive index of intrinsic silicon, m_e^* and m_h^* are the electron and hole effective masses, and μ_e and μ_h are the electron and hole mobilities. At the wavelength of 1,550 nm, the refractive index change is empirically related to the free carrier density changes (in units of cm^{-3}) as

$$\Delta n = -(8.8 \times 10^{-22} \Delta N_e + 8.5 \times 10^{-18} \Delta N_h^{0.8}) \tag{7.3}$$

This optical phase shift or modulation can then be converted into the desired optical amplitude modulation using an MZI. To obtain the needed change in carrier concentration, a PN diode is embedded inside a rib waveguide and is driven in reverse-bias to induce carrier depletion [6, 7]. Because the operation of this device relies on majority carrier dynamics that are inherently fast, this modulator is capable of tens of Gb/s bandwidth. Figure 7.2 is a top-down schematic of this modulator, which has a waveguide-based MZI design with PN diode phase shifters placed in both of its two arms. The waveguide has a 0.6 μm rib width, 0.22 μm rib height, and 0.5 μm total height. Both modeling and experiment confirm that these cross-section dimensions yield single mode operation for wavelengths around 1,550 nm. The MZI splitter and combiner are both 1 × 2 multi-mode interference (MMI) couplers.

Figure 7.3 is a cross-section schematic of the phase shifter. It has a horizontal junction where the waveguide slab and part of the rib is doped p-type; the remaining top portion of the rib is doped n-type. In this manner, the p-n junction and built-in depletion region is slightly above the center of the optical mode. This design is aimed at optimizing phase efficiency because when a reverse-bias voltage is applied, the depletion region widens and sweeps across the mode center, allowing for good charge-mode overlap. To electrically contact the n-type region without additional optical loss, a thin silicon layer is grown on top of the rib so that the metal and high concentration doping for Ohmic contact are located close to the waveguide but do not overlap the optical mode. To bias the MZM at the quadrature point, a

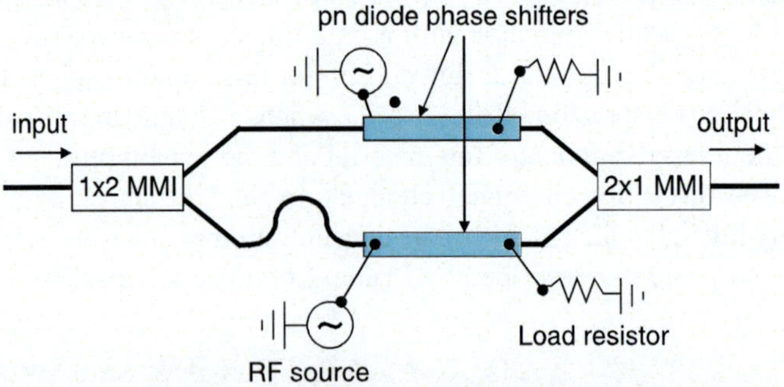

Fig. 7.2 Top–down view of the silicon MZM. It contains a PN diode phase shifter in each arm as the high-speed modulating element and 1 × 2 MMI couplers as the 3-dB splitter and combiner. A thermal heater (not shown) is used to bias the MZI at the quadrature point. On-chip resistors are used to terminate the traveling wave electrodes to reduce RF reflection

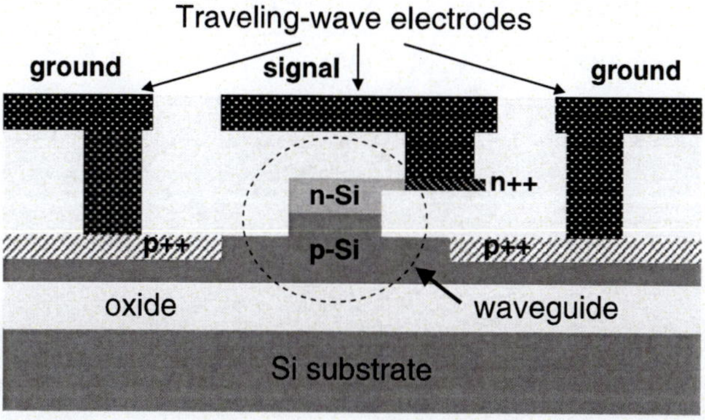

Fig. 7.3 Schematic of the PN diode waveguide phase shifter. A traveling wave electrode based on a coplanar waveguide design allows co-propagation of the electrical and optical waves

thermal heater is included on each arm of the MZI. Note that a forward biased PIN diode in silicon can also be used for the quadrature tuning and should consume less electrical power.

Although the silicon modulator based on the reverse-biased PN diode has high intrinsic speed, the actual demonstrated device speed is usually limited by device parasitics. To reduce the capacitance of the phase shifter, the thin silicon cap layer for metal contact has a single-sided asymmetric design. To further minimize the RC limitation of the frequency response of the modulator [25, 26], a traveling wave electrode based on a coplanar waveguide structure, as shown in Fig. 7.3, is used. The "signal" metal layer is $\sim 6\,\mu m$ wide and is connected to the n-type silicon region through a 1-μm contact. The separation between the signal and ground metal layers is $\sim 3\,\mu m$ and the metal thickness is $\sim 1.5\,\mu m$. The RF traveling wave coplanar waveguide and modulator optical waveguide are carefully designed so that both electrical and optical signals co-propagate along the length of the phase shifter with similar speeds, while, at the same time, the RF attenuation is kept as small as possible. The transmission line loss, impedance, and phase velocity were calculated using the commercial software package HFSS [27]. These parameters strongly depend on the metal trace dimensions as well as the silicon doping concentration and profile. To realize high-speed performance, it is also critical to optimize the termination impedance of the traveling wave electrode to minimize RF reflections that can degrade signal integrity. This termination resistor is monolithically integrated with the modulator and is made of titanium nitride (TiN) thin film.

The MUX/DEMUX is another key component of the integrated chip. In integrated photonics, the MUX/DEMUX is usually based on an arrayed waveguide grating (AWG) [28], a planar Echelle grating [29], or cascaded MZIs [30]. In this work, the cascading MZI design is chosen. Figure 7.4 is a schematic representation of the eight-channel MUX consisting of 3 stages of asymmetric MZIs. Each asymmetric MZI consists of a 1×2 MMI splitter and a 2×2 MMI coupler. Each MZI in the third stage has a length difference between the two arms of ΔL, of $2 \cdot \Delta L$ in the second stage, and of $4 \cdot \Delta L$ in the first stage. To compensate for the MZI phase error that can result from fabrication imperfections, a thermal heater is added to each MZI. Again, the heaters used in the current design can be replaced by a forward biased PIN diode

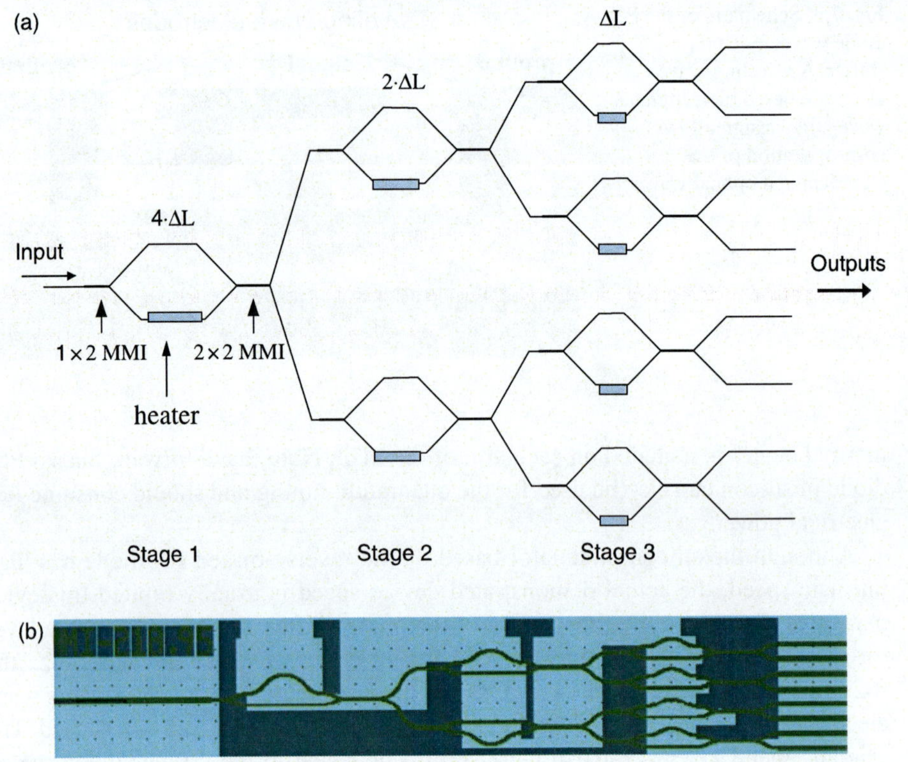

Fig. 7.4 (**a**) Schematic of a 1:8 demultiplexer based on cascaded asymmetric 1×2 MZIs. For each MZI, a 1×2 MMI is used for the input splitter and a 2×2 MMI is used for the output. (**b**) A screen capture of the 1:8 multiplexer device layout. The large *dark* regions are metal traces and pads for the metal heaters

in the future for possibly lower power dissipation. The channel wavelength spacing ($\Delta\lambda$) and ΔL are related according to the following equation

$$\Delta\lambda = \frac{\lambda_0^2}{2 \cdot n_g \cdot (4 \cdot \Delta L)} \tag{7.4}$$

where λ_0 is the central resonance wavelength and n_g is the group refractive index of waveguides, given by

$$n_g = n_{eff} - \lambda \frac{dn_{eff}}{d\lambda} \tag{7.5}$$

where n_{eff} is the effective refractive index.

Because of the small waveguide dimensions used in the silicon PIC, care is taken in modeling n_{eff} as well as n_g of the waveguide using a fully vectorial mode solver based on the local mode expansion method [31]. The MUX is designed with $4 \bullet \Delta L = 97.2\,\mu m$ to target a channel spacing of 3.2 nm (400 GHz) for wavelengths

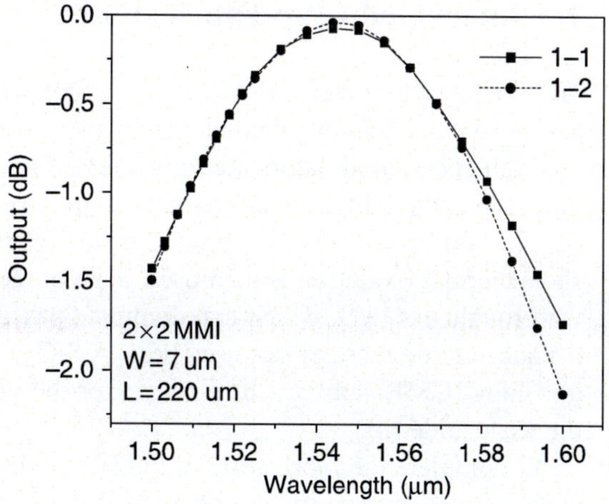

Fig. 7.5 Modeled 2 × 2 MMI coupler loss as a function of wavelength. The MMI width is 7 μm and length is 220 μm. TE polarization was used in the simulation

around 1,550 nm. For both the MZI input (1 × 2) and output (2 × 2) couplers, the MMI width is designed to be 7 μm. The length of the 1 × 2 MMI is 54.4 μm, while the length of the 2 × 2 MMI is 220 μm. The MUX performance is critically dependent on the behavior of the 2 × 2 MMI coupler, so it is carefully modeled to understand its wavelength dependence both in terms of the splitting ratio and device insertion loss. The results are shown in Fig. 7.5. Note that while the splitting ratio is quite insensitive to the wavelength, the insertion loss has strong wavelength dependence. The operating wavelength range of the MUX is therefore primarily limited by the insertion loss of the 2 × 2 MMIs. This simulation result shows that the −0.5 dB bandwidth is ∼39 nm. The MMI coupler fabrication tolerance is also modeled. Figure 7.6 shows the optimal MMI length as a function of MMI width for the wavelength of 1,550 nm. With a ±0.1 μm MMI width variation, the corresponding MMI length change is about ±4%. Such process tolerances can be reasonably met with today's photo-lithography technology.

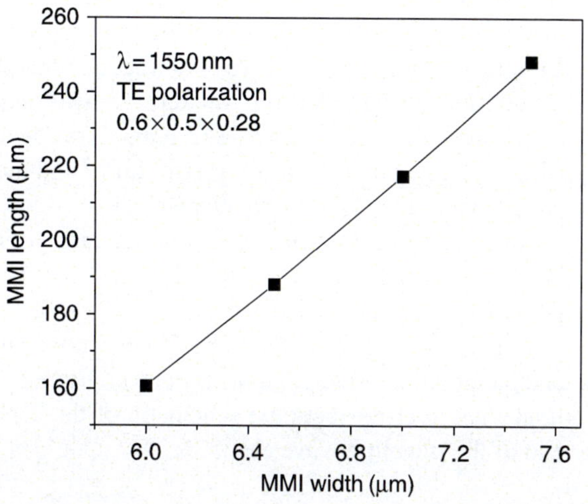

Fig. 7.6 Modeled optimal MMI length as a function of the MMI width for a 2 × 2 MMI coupler at 1,550 nm. TE polarization was used in the simulation

7.3 Integrated Chip Fabrication

To fabricate the integrated silicon PIC, SOI wafers with 1-μm-thick buried oxide and ~0.5 μm thick top silicon layer are used as the starting material. Regions where the high-speed modulators are to be created are first implanted with boron to form the p-type region of the diode. A single silicon etch step then creates all of the MMI couplers and 0.6 μm wide rib waveguides for the DEMUX, MUX, and MZMs. A short thermal oxidation is then used as an oxide smoothing step to reduce rib sidewall roughness [32], which can dominate passive waveguide transmission loss for the sub-micrometer devices used here. A SiO_2 deposition and chemical-mechanical polishing (CMP) subsequently planarizes the wafer, with the CMP stopping above the waveguide rib.

To complete the fabrication of the PN diodes, 2.7-μm-wide trenches are etched in the oxide, asymmetrically positioned above the rib. Silicon is then grown using a non-selective epitaxial process, which fills the trench with single crystalline silicon directly above the rib and poly-Si above the oxide around the rib. To improve the material quality of the poly-Si, the wafers are annealed at 1, 100 °C for 1 h. This same anneal also serves to uniformly distribute the boron dopants that were previously implanted. A silicon CMP then planarized the wafer and ensured that the top silicon layer, also known as the silicon wing, is 0.1 μm thick. Phosphorus is then implanted into the wing and rib to form the n-type region of the diode. The p-doping concentration of the diode is targeted to be $1.5 \times 10^{17}\,cm^{-3}$, and the n-doping concentration varies from $3 \times 10^{18}\,cm^{-3}$ near the top of the wing to $\sim 1.5 \times 10^{17}\,cm^{-3}$ at the PN junction, which is ~0.4 μm above the buried oxide.

To electrically contact the n and p regions, part of the poly-Si wing and silicon slab regions on both sides of the rib are heavily implanted with phosphorus and boron, respectively, to reach surface doping concentrations of $\sim 1 \times 10^{20}\,cm^{-3}$ for Ohmic contact with the metal. Aluminum (Al) is then deposited and patterned to create the traveling wave electrodes and quadrature biasing heaters for the modulators, as well as the thermal heaters for compensating fabrication-induced phase mismatch in the MUX/DEMUX MZIs. A silicon dioxide (SiO_2) layer separates the heaters from the waveguides, so there is no metal induced optical loss. The cross-sectional area of the Al heater is approximately 2.5 μm^2 and the heater lengths for stages 1, 2, and 3 are 797.2, 608.6, and 484.3 μm, respectively. Scanning electron microscope (SEM) pictures of a PN diode phase shifter and the MUX waveguide with Al heater are shown in Fig. 7.7a and b, respectively.

To monolithically integrate the modulator with the termination resistors, which are needed for high-speed operation, a 0.1-μm-thick TiN layer is deposited and patterned to target a resistance value of 15 Ω. To make electrical contact between the traveling wave electrodes and the TiN resistors, another aluminum film is deposited and patterned as an interconnecting layer. Figure 7.8 shows a top–down microscope image and a cross-sectional schematic of the TiN resistor connected to the output end of the traveling wave electrode.

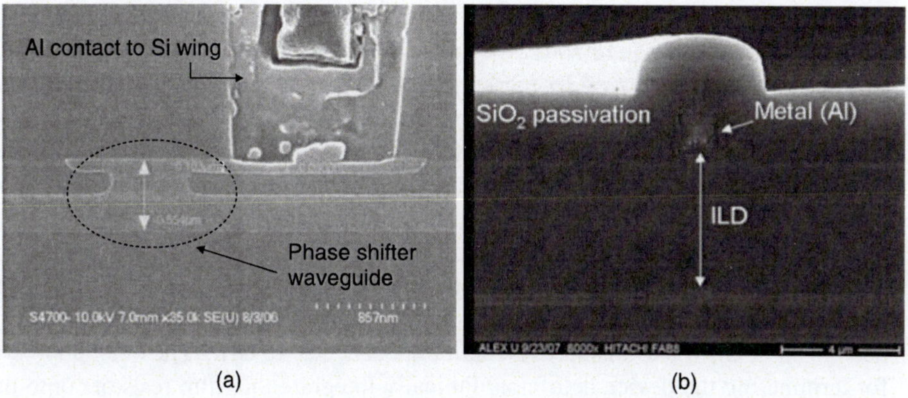

Fig. 7.7 SEM cross-section of (**a**) a PN diode phase shifter, shown with the Al contacting the n-type silicon wing and (**b**) a MUX waveguide with the Al heater for phase error compensation

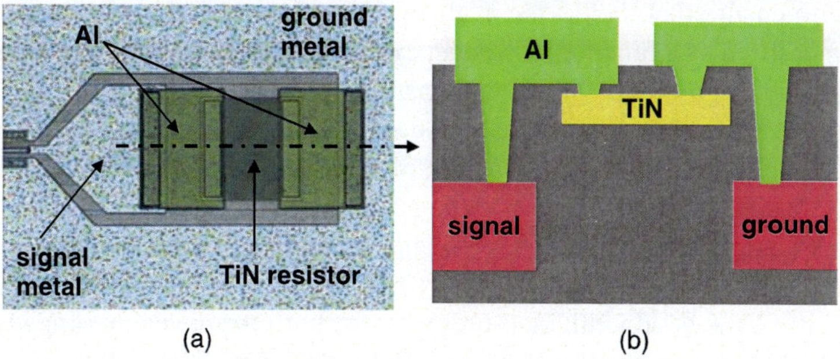

Fig. 7.8 (**a**) Top–down optical microscope image and (**b**) cross sectional schematic of the TiN resistor that is monolithically integrated with the PN diode modulator

7.4 Device Performance

As the performance of the silicon PIC is critically dependent on the performance of its individual components, it is useful to understand each device element first. In this manner, the key results of the high-speed modulators, which have been recently shown to transmit data up to 40 Gbps [7], are presented. This is followed by a discussion of the testing results of a single MUX. Lastly, both DC testing and high-speed data transmission performance of the fully integrated chip are reported.

7.4.1 High-Speed Performance of the Stand-Alone Silicon MZM

One of the device parameters that govern the speed performance of the silicon MZM is the length of the optical phase shifters. Shorter length means higher speed because the device has lower RF loss and less electrical-to-optical signal delay or "walk-off."

To demonstrate ultra high-speed performance, a silicon MZM with a 1-mm phase shifter in each arm was previously designed and fabricated. Its high frequency performance was characterized by measuring both its optical frequency response and data transmission capability. Since the modulator has been designed with traveling wave electrodes, it is imperative to ensure proper termination of the transmission lines to keep the RF signal reflections at a minimum. To enable this, two different approaches have been used for device termination. One approach is to flip-chip bond the silicon MZM chip to a printed circuit board (PCB) and surface mount external resistors on the PCB traces that are connected to the output end of the modulator electrode. The PCB is designed for high-speed performance with low RF loss connectors and PCB traces that have 50 ohm impedance at 40 GHz. The other approach for terminating the device is to monolithically integrate thin film resistors onto the silicon modulator die (see Fig. 7.8).

For the optical frequency roll-off measurement, the signal generator is swept from 100 MHz to 40 GHz. The RF signal is combined with a DC bias using a bias-tee to ensure that the device is always reverse biased. This DC-coupled signal is first measured as a function of frequency using a digital communications analyzer (DCA) electrical module with 63 GHz bandwidth. The RF signal is then connected to either the PCB or the probes that are used to drive the modulator. The modulated optical signal is measured as a function of frequency using a DCA 53-GHz optical module. The frequency response of the MZI modulator is obtained by normalizing the optical output by the input drive voltage for all frequencies. This ensures that the frequency roll-off is independent of any bandwidth limitation of the signal source. The results of the bandwidth measurements for both the device packaged on a PCB as well as that with an on-chip resistor are plotted in Fig. 7.9a. One can see that both devices have a 3-dB roll-off frequency greater than 30 GHz.

The high-speed data transmission performance of the MZI modulator is measured using a pseudo-random bit sequence (PRBS) source with $[2^{31}-1]$ pattern length as the RF input. The output of the PRBS source is sent through a modulator driver, which is adjusted to give an amplified output of 6.2 V_{pp}. Before being

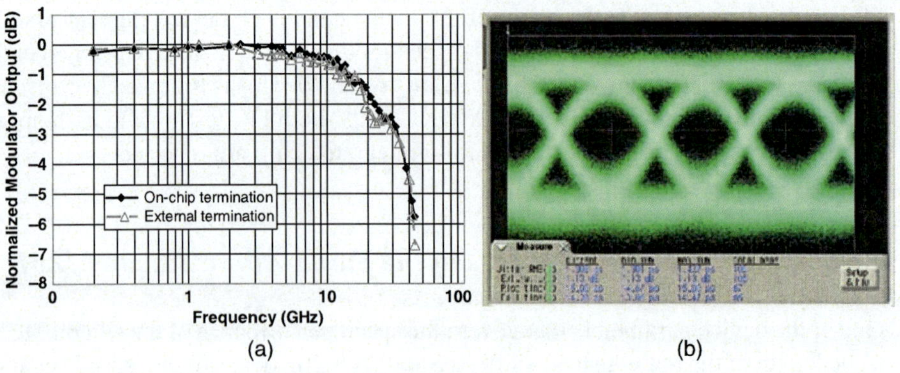

Fig. 7.9 High speed performance of a silicon MZM with 1-mm-long phase shifters: (**a**) normalized optical response as a function of RF frequency and (**b**) optical eye diagram at a bit rate of 40 Gb/s

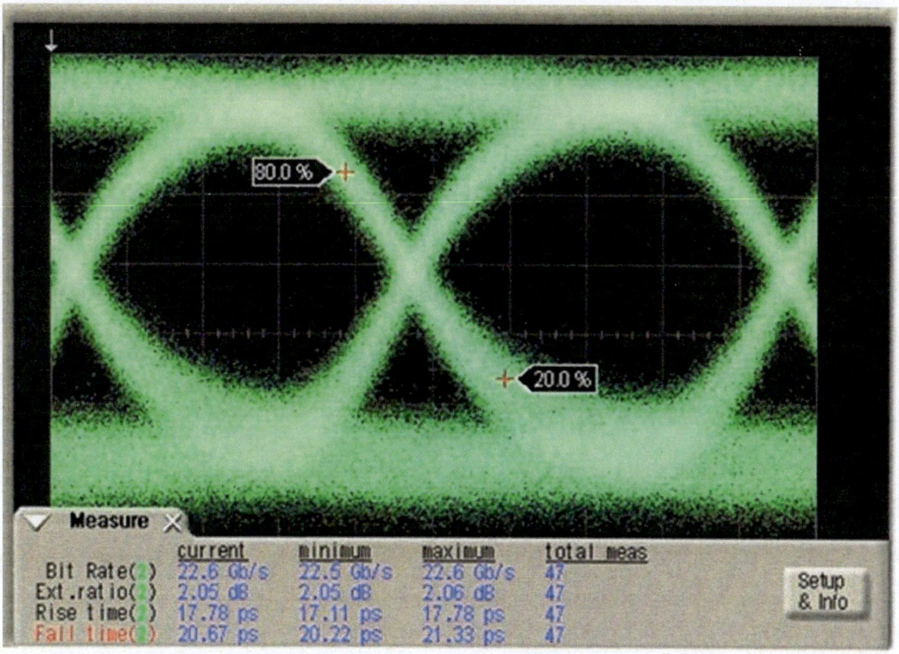

Fig. 7.10 22.5 Gb/s optical eye diagram of a silicon MZM with a 1.5 mm phase shifter in each arm. Based on its fast rise/fall times of 18–20 ps, the device is clearly capable of higher data rates

applied to the modulator, this signal is combined with 3.1 V_{DC} using a bias tee to ensure reverse bias operation of the entire AC voltage swing. Figure 7.9b shows the eye diagram of the modulator optical output at a bit rate of 40 Gb/s. The extinction ratio (ER) is measured to be 1.1 dB with a rise/fall time of ~14 ps. The open eye diagram proves that the modulator is capable of transmitting data at 40 Gb/s, which is consistent with the 3-dB roll-off frequency of >30 GHz.

While this 40 Gb/s performance is unprecedented for a silicon modulator, the device's ~1 dB ER needs improvement. One simple way to achieve higher ER is to increase the phase shifter length, thereby increasing the amount of obtainable phase shift for a given applied voltage. The associated drawback is higher RF loss and lower speed performance. For the silicon PIC, the length of the phase shifters inside the MZMs is chosen to be 1.5 mm as a compromise between speed and ER. Figure 7.10 shows the data transmission capability of this new design. Note that the optical eye diagram is captured at a bit rate of 22.5 Gb/s. It shows ~2 dB ER and a 18–20 ps rise/fall time. Both the open eye diagram and its fast transitions indicate that this device is capable of even higher data transmission rates.

7.4.2 Performance of the Standalone MUX/DEMUX

To characterize the performance of the stand-alone 1:8 MUX/DEMUX, it is placed on a PCB and the Al heaters for phase error compensation are wire-bonded so they

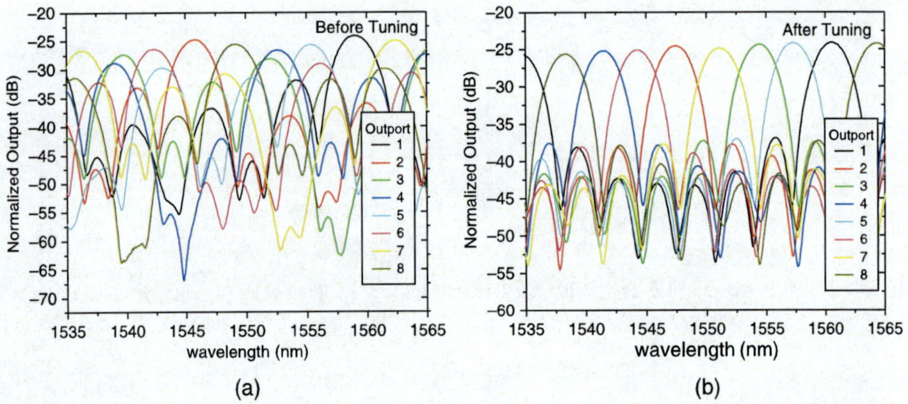

Fig. 7.11 Output spectra of a 1:8 demultiplexer (**a**) before and (**b**) after thermal tuning optimization

can be simultaneously and independently adjusted. Light from a polarized broadband source, a superluminescent light emitting diode (SLED), is coupled into and out of the device using lensed single-mode fibers with a mode field diameter of 3 μm. Because the waveguides and MMIs have optimum performance for transverse electric (TE) polarization, a polarization controller is inserted between the SLED and input lensed fiber to set the desired polarization. An optical spectrum analyzer (OSA) is used to measure the output spectrum.

Figure 7.11a shows the spectra of all outputs of the eight-channel DEMUX before phase mismatch compensation, and Figure 7.11b shows the spectra of the same device after metal heater tuning optimization. The measured on-chip loss is 3.2 dB, of which ~0.6 dB is due to the propagation loss of straight waveguides. The channel spacing is 3.2 nm, 3-dB bandwidth is 3 nm, and adjacent channel isolation is 13 dB. The channel uniformity, defined here as the difference between the best and worst excess losses over all channels, is ~1.5 dB from 1,537 to 1,562 nm. It is found that the loss slowly decreases as the wavelength increases; this is due to the wavelength dependence of the MMI, which was predicted by simulation. In practice, the wavelength dependence is one of the key factors in determining the operating wavelength range. The tuning efficiency of the Al heaters, defined as the electrical power needed for π-phase shift, is approximately 83 mW. Additionally, the thermal crosstalk is investigated by measuring the phase shift caused by adjacent heaters. It is found to be negligible, where >6 W of electrical power is needed to produce a π-phase shift using a heater on an adjacent MZI.

7.4.3 DC Performance of an Integrated DEMUX, MZM, and MUX Chip

An integrated device with a single-input waveguide and a single-output waveguide similar to that depicted in Fig. 7.1 is characterized at DC. It contains a 1:8 DEMUX,

8 symmetric modulator MZIs, and an 8:1 MUX. It differs from the final silicon PIC in that not all of the doping and metal for high-speed modulation have been incorporated. Like the stand-alone DEMUX, it is also wire-bonded to a PCB so that the heaters can be easily adjusted for phase mismatch compensation. To characterize this integrated device, the outputs of distributed feedback (DFB) lasers, with wavelengths 3.2 nm apart and comparable output powers, are multiplexed together off-chip using fiber-based couplers. This combined output is used as the optical source for testing and is coupled into the silicon device using a lensed polarization maintaining fiber. Again, an OSA is used to measure the device output spectrum.

Figure 7.12a shows the spectrum of the integrated device before and after heater adjustment. With phase mismatch compensation, the channel uniformity improved from 13 to 1.6 dB. The on-chip loss is approximately 7.5 dB, of which 6.4 dB is due to the DEMUX and MUX. The remaining 1 dB can be accounted for by the modulator MZIs placed in between the DEMUX and MUX. To quantify the DC ER of the silicon MZMs, the heater for quadrature biasing is tuned to induce a π–radian phase shift between the two arms of each MZI. The introduction of this π–radian phase difference should result in destructive interference and should lead to minimum optical transmission through the modulator. The output spectrum of the integrated device before and after this heater tuning is given in Fig. 7.12b. Note that light passing through the adjusted channel, which corresponds to ~1,533 nm, is at a maximum when there is no phase difference between the two MZI arms and is at a minimum when a π–radian phase is introduced. The resulting ER is in excess of 30 dB.

To verify that the DEMUX is operating correctly such that only one wavelength is propagating through each modulator MZI, the output spectrum the integrated device is monitored while the input DFB lasers are selectively switched on or off. Figure 7.13 shows the device output spectra for when all the lasers are turned on

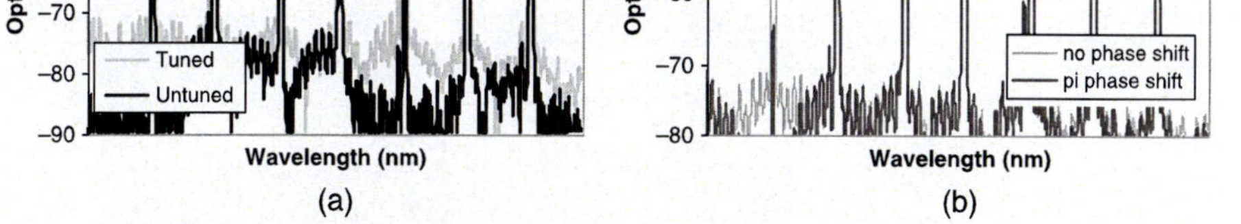

Fig. 7.12 (**a**) Output spectra of the integrated device before and after phase mismatch compensation. Note the significant improvement in device loss and channel uniformity with heater tuning optimization. (**b**) Output spectra of the integrated device when 0 and π–radian phase difference is introduced between the modulator MZI arms for channel 1. A comparison of the spectra indicates that the DC ER of the modulator is >30 dB

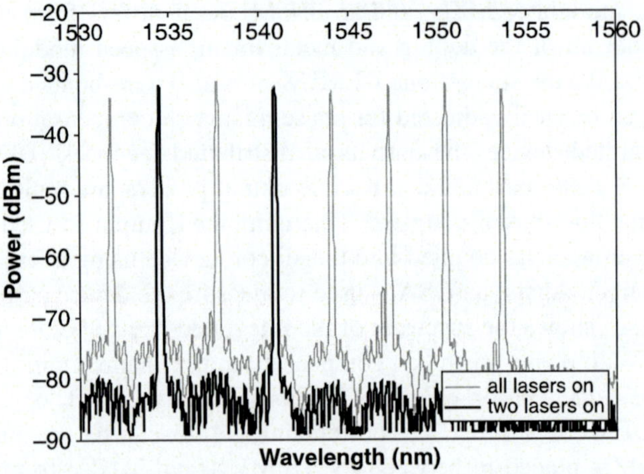

Fig. 7.13 Output spectra of the integrated device for when all input lasers are turned on and when only those corresponding to channels 2 and 4 are turned on. Note that optical cross-talk is better than −40 dB

and when only lasers corresponding to channels 2 and 4 are on. The latter scenario clearly confirms that the input wavelengths are properly demultiplexed because the optical cross-talk, as measured by a comparison of the optical power through "on" channels and "off" channels, is better than −40 dB.

7.4.4 High-Speed Performance of the Silicon PIC

The final silicon PIC, with the DEMUX, the array of eight high-speed silicon MZMs, and the MUX, is wire-bonded to a PCB with low loss RF connectors and routing traces. The PCB, as shown in Fig. 7.14, is also designed for DC bias control of the metal heaters to tune the phase of the MZMs and MUX/DEMUX. For high-speed testing, the differential RF signals from a PRBS generator with $[2^{31}-1]$ pattern length are amplified using a commercially available dual-output driver. The

Fig. 7.14 Photograph of the packaged silicon PIC. It is wire-bonded onto a PCB with low-loss RF connectors as well as DC controls for heater tuning adjustment

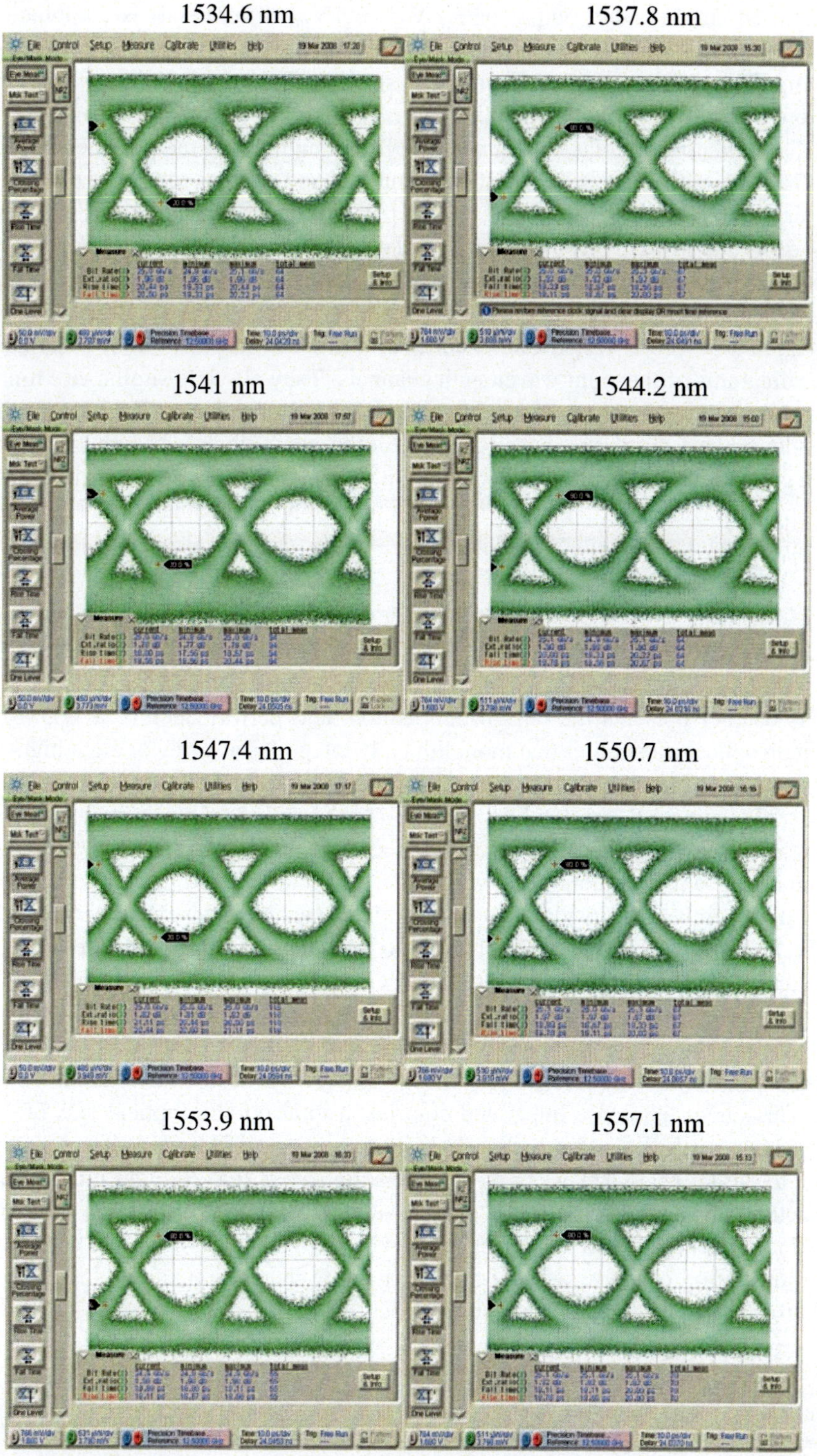

Fig. 7.15 Optical eye diagrams of the eight wavelength channels of the Si PIC. Each is transmitting data at 25 Gb/s with 19 ps rise time, 19 ps fall time, and 2 dB ER

amplified single-ended output of 3.2 V_{pp} (6.4 V_{pp} differential) is combined with 2 V_{DC} using a bias tee to ensure reverse bias operation for the entire AC voltage swing. This DC-coupled drive signal is supplied to each modulator phase shifter through a high-speed coaxial cable that is attached to the PCB RF connector. The MZI modulators are biased at quadrature for all high-speed measurements.

To characterize the silicon PIC, the multiplexed outputs of eight DFB lasers is again used as the optical source. On-chip loss is measured to be 10 dB, of which 6.4 dB is due to the DEMUX and MUX, 2 dB is due to the high-speed phase shifters, and the remaining 1.6 dB is due to the modulator MZIs as well as passive waveguides used for on-chip optical routing. The high-speed data transmission of the integrated device is tested one channel at a time. Figure 7.15 shows the 25 Gb/s eye diagrams of the eight wavelength channels. They all yield similar rise time, fall time, and ER, which are 19 ps, 19 ps, and 2 dB, respectively. These performance results are very similar to those obtained for the stand-alone MZM; this close match is a good indication that device integration did not compromise performance. Clear open eye diagrams at 25 Gb/s suggest that the Si PIC is capable of transmitting data at an aggregate data rate of 200 Gb/s.

7.5 Conclusion

This chapter presents the design, fabrication, and performance of a WDM-based silicon optical transmitter that monolithically integrates an array of eight high-speed silicon modulators with silicon based 8:1 DEMUX/MUX. The inclusion of both a DEMUX and a MUX allows for fiber-to-chip coupling using a single input and a single output fiber. The silicon MZM is based on the free carrier plasma dispersion effect, which is obtained through electric-field-induced carrier depletion of a PN diode embedded inside a rib waveguide. To enable high-speed performance, a traveling wave electrode design is used to allow co-propagation of the optical and electrical waves. Termination resistors are also monolithically integrated with the electrodes to minimize RF reflections that can compromise signal integrity. The MUX/DEMUX is based on cascaded MZIs with integrated metal heaters to allow compensation of MZI phase error that can result from device fabrication.

The silicon photonic integrated chip has a total on-chip optical loss of 10 dB, of which ~3.5 dB is due to the high-speed silicon MZMs and the remaining is due to the DEMUX/MUX. High-speed testing shows that all eight modulators are functional and are each capable of at least 25 Gb/s data transmission. The silicon PIC is therefore capable of an aggregate data rate of 200 Gb/s. This is an important demonstration of the benefits of silicon photonic integration and its feasibility as a platform for future high-speed interconnects.

Acknowledgments The authors thank Fiel Concepcion for sample preparation, D. Li for data collection software, R. Gabay, A. Ugnitz, and G. Nutrica for device fabrication assistance, L. Kulig for material analysis, and T. Mader, S. Q. Shang, G. Sarid, D. D. Lu, H. Braunisch, G. T. Reed, and J. E. Bowers for useful discussions.

References

1. J. Held, J. Bautista, S. Koehl, From a few cores to many: a tera-scale computing research overview (2006), [Online]. Available at: http://download.intel.com/research/platform/terascale/terascale_overview_paper.pdf 169
2. B. Casper, J. Jaussi, F. O'Mahony, M. Mansuri, K. Canagasaby, J. Kennedy, E. Yeung, R. Mooney, A 20 Gb/s Forwarded Clock Transceiver in 90 nm CMOS, Technical Digest of 2006 IEEE International Solid-State Circuits Conference, Session 4/Gigabit Transceivers/4.6, 2006 170
3. G. Balamurugan, J. Kennedy, G. Banerjee, J.E. Jaussi, M. Mansuri, F.O'Mahony, B. Casper, R. Mooney, A Scalable 5–15 Gbps, 14–75 mW low power I/O transceiver in 65 nm CMOS, 2007 Symposium on VLSI Circuits Digest of Technical Papers, 270–271, 2007 170
4. G.T. Reed, A.P. Knights, *Silicon Photonics: An Introduction* (Wiley, Chichester, 2004) 170
5. L. Pavesi, D.J. Lockwood (eds.), *Silicon Photonics* (Springer-Verlag, Berlin, 2004) 170
6. A. Liu, L. Liao, D. Rubin, H. Nguyen, B. Ciftcioglu, Y. Chetrit, N. Izhaky, M. Paniccia, High-speed optical modulation based on carrier depletion in a silicon waveguide. Opt. Express **15**, 660–668 (2007) 170, 172
7. L. Liao, A. Liu, D. Rubin, J. Basak, Y. Chetrit, H. Nguyen, R. Cohen, N. Izhaky, M. Paniccia, 40 Gbit/s silicon optical modulator for high-speed applications. Electron. Lett. **43**(22), 1196–1197 (2007) 170, 172, 177
8. W.M. Green, M.J. Rooks, L. Sekaric, Y. Vlasov, Ultra-compact, low RF power, 10 Gb/s silicon Mach-Zehnder modulator. Opt. Express **15**, 17106–17113 (2007) 170
9. Q. Xu, S. Manipatruni, B. Schmidt, J. Shakya, M. Lipson, 12.5 Gbit/s carrier-injection-based silicon micro-ring silicon modulators. Opt. Express **15**, 430–436 (2007) 170
10. S.J. Koester, G. Dehlinger, J.D. Schaub, J.O. Chu, Q.C. Ouyang, A. Grill, Germanium-on-insulator photodetectors, in Technical Digest of 2005 2nd IEEE International Conference on Group IV Photonics, pp. 171–173 170
11. M. Oehme, J. Werner, E. Kasper, M. Jutzi, M. Berroth, High bandwidth Ge p-i-n photodetector integrated on Si. Appl. Phys. Lett. **89**, 071117–071117-3 (2006) 170
12. O. Boyraz, B. Jalali, Demonstration of a silicon Raman laser. Opt. Express **12**, 5269–5273 (2004) 170
13. H. Rong et al., A continuous-wave Raman silicon laser. Nature **433**, 725–728 (2005) 170
14. A. Liu, H. Rong, M. Paniccia, O. Cohen, D. Hak, Net optical gain in a low loss silicon-on-insulator waveguide by stimulated Raman scattering. Opt. Express **12**, 4261–4268 (2004) 170
15. O. Boyraz, B. Jalali, Demonstration of 11 dB fiber-to-fiber gain in a silicon Raman amplifier. Electron. Express **1**, 429–434 (2004) 170
16. Q. Xu, V.R. Almeida, M. Lipson, Demonstration of high Raman gain in a submicrometer-size silicon-on-insulator waveguide. Opt. Lett. **30**, 35–37 (2005) 170
17. R. Jones et al. Net continuous-wave optical gain in a low loss silicon-on-insulator waveguide by stimulated Raman scattering. Opt. Express **13**, 519–525 (2005) 170
18. R.L. Espinola, J.I. Dadap, R.M. Osgood, Jr., S.J. McNab, Y.A. Vlasov, C-band wavelength conversion in silicon photonic wire waveguides. Opt. Express **13**, 4341–4349 (2005) 170
19. K. Yamada, et al., All-optical efficient wavelength conversion using silicon photonic wire waveguide. IEEE Photon. Technol. Lett. **18**, 1046–1048 (2006) 170
20. H. Rong, Y.H. Kuo, A. Liu, M. Paniccia, O. Cohen, High efficiency wavelength conversion of 10 Gb/s data in silicon waveguides. Opt. Express **14**, 1182–1188 (2006) 170
21. A.W. Fang, H. Park, O. Cohen, R. Jones, M. Paniccia, J.E. Bowers, Electrically pumped hybrid AlGaInAs-silicon evanescent laser. Opt. Express **14**, 9203–9210 (2006) 170, 171
22. G.P. Agrawal, *Fiber-Optic Communication Systems* (Wiley, New York, 1997) 170
23. R.A. Soref, P.J. Lorenzo, All-silicon active and passive guided-wave components for $\lambda = 1.3$ and 1.6 μm. IEEE J. Quantum Electron. **QE-22**, 873–879 (1986) 171
24. R.A. Soref, B.R. Bennett, Electrooptical effects in silicon. IEEE J. Quantum Electron. **QE-23**, 123–129 (1987) 171

25. R.C. Alferness, Waveguide electrooptic modulators. IEEE Trans. Microwave Theory Techn. **30**, 1121–1137 (1982) 173
26. R.G. Walker, High-speed III-V semiconductor intensity modulators. IEEE J. Quantum Electron. **27**, 654–667 (1991) 173
27. For details, see http://www.ansoft.com. 173
28. A. Sugita, A. Kaneko, K. Okamoto, M. Itoh, A. Himeno, Y. Ohmori, Very low insertion loss arrayed-waveguide grating with vertically tapered waveguides. IEEE Photon. Technol. Lett. **12**, 1180–1182 (2000) 173
29. J. Brouckaert, W. Bogaert, P. Dumon, D. Van Thourhout, R. Baets, Planar concave grating demultiplexer fabricated on a nanophotonic silicon-on-insulator platform. IEEE J. Lightwave Technol. **24**, 1269–1275 (2007) 173
30. M. Oguma, T. Kitoh, T. Shibata, Y. Inoue, K. Jinguji, A. Himeno, Y. Hibino, Four-channel flat top low-loss filter for wide passband WDM access network. Electron. Lett. **37**, 514–515 (2001) 173
31. For details, see http://www.photond.com. 174
32. K.K. Lee, D.R. Lim, L.C. Kimerling, Fabrication of ultralow-loss Si/SiO2 waveguides by roughness reduction. Opt. Lett. **26**, 1888–1890 (2001) 176

Index

Chapter 8
CMOS Photonics: A Platform for Optoelectronics Integration

Thierry Pinguet, Steffen Gloeckner, Gianlorenzo Masini, and Attila Mekis

Abstract We describe the intimate relationship between process, devices, and system design by examining the development of Luxtera' s CMOS Photonics technology. We address the challenges of integrating optoeletronic elements, including germanium photodiodes, in a commercial CMOS process without significantly affecting the electronics performance and the manufacturability of the process. The common practices of the electronics design and manufacturing industry are applied to our optoelectronic technology; we discuss our device library and our design and testing infrastructure in support of the engineering of complex optoelectronic circuits and systems. Finally, we use a complete monolithically integrated wavelength division multiplexed 40 Gbps transceiver chip as an example of the application of our complete technology platform and the demonstration of its capabilities for optoelectronic integration.

8.1 Introduction

Silicon photonics as a discipline exploits the optical and electrical material properties of silicon. Various integration schemes that embed the silicon photonic devices into higher level systems have been reported and cover the whole spectrum from hybrid to monolithic implementations [1–3].

Luxtera has taken a complementary metal-oxide-semiconductor (CMOS) first-approach to silicon photonics, incrementally adding process modules to a well-characterized and maintained CMOS-process baseline in order to enable the optical functionalities. This approach is driven by the desire to leverage the advanced fabrication processes run on state-of-the-art toolsets as well as the continuous improvement efforts in all aspects of CMOS-design and manufacturing, but also by the economic reality of the tremendous costs involved in building and maintaining an

T. Pinguet (✉)
Luxtera Inc., Carlsbad, CA 92011, USA
e-mail: thierry@luxtera.com

S. Gloeckner, G. Masini, and A. Mekis (✉)
Luxtera Inc., Carlsbad, CA 92011, USA

D.J. Lockwood, L. Pavesi (Eds.): *Silicon Photonics II*.
Topics in Applied Physics **119**, 187–216 (2011)
DOI 10.1007/978-3-642-10506-7_8

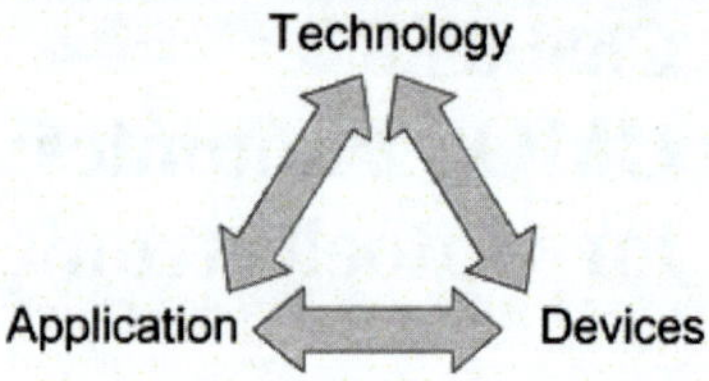

Fig. 8.1 Devices, process, and applications are strongly interrelated

advanced node CMOS-fab, which does not allow building and supporting such a fab purely from an emerging optoelectronic application. Luxtera is a commercial entity and this has strongly impacted the selected integration approach. In our view, applications, technology, and devices cannot be considered in isolation and trade-offs in devices and technology need to be cross-checked against application requirements.

The application defines the required functionalities and to some extent the performance of the devices and we have found in our work here that the coupling between system, devices, and process is extremely tight (illustrated in Fig. 8.1). Just like in electronics, optical device performance is bounded by the capabilities of the CMOS technology and its limitations at a particular process generation (i.e., lithography, etch depth control, etc.). As such, the improvement path for the devices is closely tied to process improvements. Better individual devices can be realized in different material systems, but these so far do not support high levels of system integration. The benefits of integration then will most likely be realized at the system level rather than at the device level. Preferred applications have a predictable roadmap that can be intersected by predictable advances in CMOS technology. Luxtera selected optical interconnects as the target application space due to the required scaling both in performance (speed, power) as well as market-volume (cost) that match scaling characteristics usually associated with CMOS technology.

In Sect. 8.2 we describe how photonic functionality was integrated into a CMOS-process flow. Aside from laser light sources, Luxteras CMOS Photonics technology integrates all components required for optical transceivers. Our approach toward device development was also strongly guided by methodologies used in the CMOS-industry, which we will describe in Sect. 8.3. Section 8.4 gives an overview on design and testing infrastructure, which is a prerequisite for design and design verification of integrated optoelectronic systems. Finally, in Sect. 8.5 we describe an integrated optoelectronic transceiver system, which served as a demonstration vehicle verifying the design flow around which Luxteras CMOS-Photonics technology has been constructed.

8.2 Enabling a CMOS Process for Photonics Integration

The idea of using silicon as a material for photonic integration has been investigated for decades [4] ever since the rise of microelectronics as an industry and the large investment in tools and processes associated with silicon fabrication. Beyond simply using silicon as a source material to build optical-only integrated devices and circuits (like for instance silica-on-silicon photonic lightwave circuits), many different approaches have been proposed to achieve tight integration between photonics and electronics on a silicon substrate [5, 6]. The path followed by Luxtera was to integrate all photonic components in the front-end of a standard CMOS process. In this section we describe the challenges encountered and eventual selections made during the development process and we hope to make evident that there is a large gap between claiming CMOS compatibility, and actually achieving it.

8.2.1 Rationale for Front-End Integration

Luxtera' s philosophy is to extend the general electronics design and fabrication processes to optical components: we want photonics to be just another part of the toolkit that an integrated circuit designer can use to create complex circuits and systems with additional functionality. Thus we want optical components to become basic elements just like transistors, resistors, capacitors, etc. Additionally, we want them to be built in the same silicon, at the same time as the basic electrical devices, so that they are completely compatible with CMOS and can benefit from advancements in silicon processing. Those considerations were the driving factors behind front-end integration.

Figure 8.2 shows a schematic cross-section of the front-end integration approach, showing the monolithic integration of electronics (transistor), passive optical devices (waveguide), active optical devices (modulator), and photodetection (Ge-based waveguide photodetector), all of which are discussed in subsequent sections.

Evidently, the only element of an optical data transmission link that is missing is the light source. While research is ongoing on the monolithic integration of optical

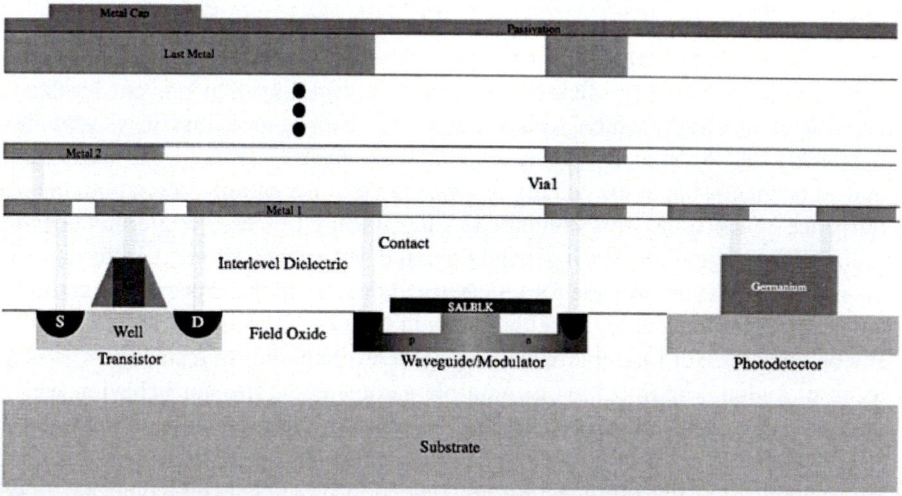

Fig. 8.2 Schematic cross-section of a photonically enabled CMOS process

gain media in silicon [7, 8], the various technologies at the time of writing are far too immature for commercial deployment, so Luxteras products rely on an external laser hybridly co-packaged with the CMOS Photonic chip.

8.2.2 SOI Substrate Design

Silicon-on-insulator (SOI) substrates have traditionally been the wafers of choice for fabricating optical devices, since they provide a high-quality crystalline silicon layer over a silicon dioxide buried layer which acts as a built-in cladding layer for any waveguide-based device. Over the last 10 years, SOI has also become a standard technology for microelectronics, thus promoting the development of SOI substrates. Naturally then, a CMOS SOI process would be an excellent candidate as a starting point for photonic integration.

At the time of the start of development of Luxtera' s process, some SOI process nodes were available and the state of the art was around 130–90 nm (referring to the half pitch of the finest metal lines printable in that process). Freescale (at the time still Motorola) agreed to become Luxteras partner for development, giving us access to their 130 nm SOI process. This process was extensively used for building high-performance PowerPC microprocessors. A 130-nm SOI process was also a good choice because it provides sufficient electronics performance to design 10 Gbps analog circuits used in communication systems (trans-impedance amplifiers, modulator drivers, clock and data recovery circuits, etc.). Unfortunately, while this appeared like a marriage made in heaven, SOI substrate thicknesses designed for electronics tend to have relatively thin top silicon layers, on the order of 100 nm or less in more advanced nodes. This thickness is practically incompatible with designing high-confinement single-mode waveguides: the optical mode at wavelengths of interest (considered to be in the 1,310–1,550 nm range in this chapter) would extend well outside of the silicon, resulting in side-effects: low efficiency of any electro-optic effect where overlap with silicon is necessary, degraded bending radius, and the need to have a very thick buried silicon dioxide layer to prevent leakage to the silicon substrate (which would create high propagation loss in waveguides). Increasing the top silicon thickness was thus needed to enable high performance optical elements, but at the penalty of changing (not necessarily degrading) the performance of the transistors compared to the starting process. While this does not sound like a large issue, it triggers additional development efforts, like for instance the need to re-extract models for all electrical devices in the process. Good models are necessary to design high-yield circuits in CMOS. The selection of the final top silicon thickness for Luxtera (in the range of a few hundreds of nanometers) was the result of a global optimization of multiple parameters: transistor behavior, waveguide loss, grating coupler efficiency (see subsequent section), modulator efficiency, etc., with the goal to build an optical data transmission system. For a different target application the optimization and selection of thicknesses could have been different.

8.2.3 Waveguide Integration

Once the substrate was selected, the next task was to ensure that the most basic of devices, the waveguide, provided sufficient performance to be usable. Much research has been performed on processes to realize low-loss waveguides in silicon [9, 10]. Special processes have been studied, demonstrating a reduction in waveguide loss, for instance by improving the sidewall roughness of etched rib waveguides or other sidewall-related effects by treatment of the etched surface. Unfortunately, few, if any, of these processes have been demonstrated to translate directly to a CMOS process, and additionally, most of the research has failed to deal with other aspects of waveguides in a CMOS process, in particular the effect of the rest of the CMOS process on the waveguide loss and other performance parameters.

In a CMOS process, other effects have an impact on the performance of the optical devices, for instance:

- the presence of other process dielectrics on the silicon layer or near the waveguide
- the oxidation/strip/clean cycles of the exposed silicon surface (used to reduce surface states for electronics performance)
- subsequent steps like implantation, anneals, and other thermal treatments (which could impact for instance the stress in the silicon film)

Extensive characterization of the local environment of the waveguide, in particular the dielectric films in contact with the silicon, is necessary to quantify its impact on waveguide loss.

In Luxtera' s process (Fig. 8.3), the waveguide is a rib waveguide defined by a partial etch of the silicon layer, and is covered with a set of silicon nitride and oxynitride layers that are part of the standard CMOS, to block salicidation locally or form a gate spacer layer. The sidewalls of the waveguide go through the same

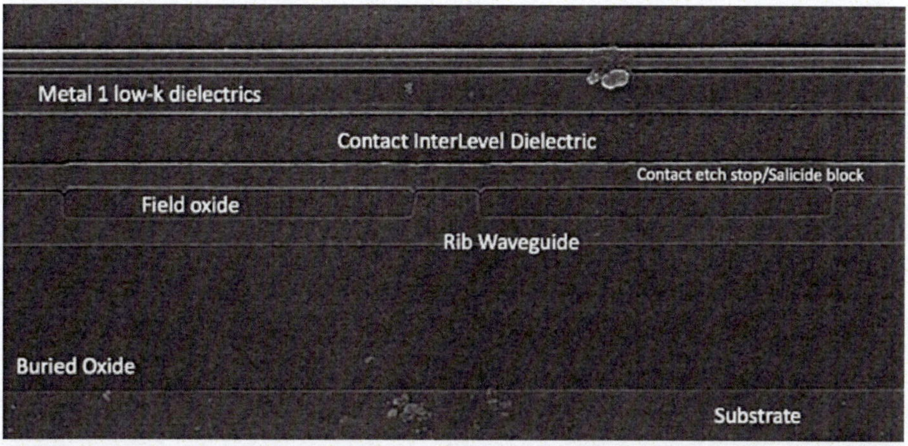

Fig. 8.3 Cross-section scanning-electron microscope picture of rib waveguide and its local environment

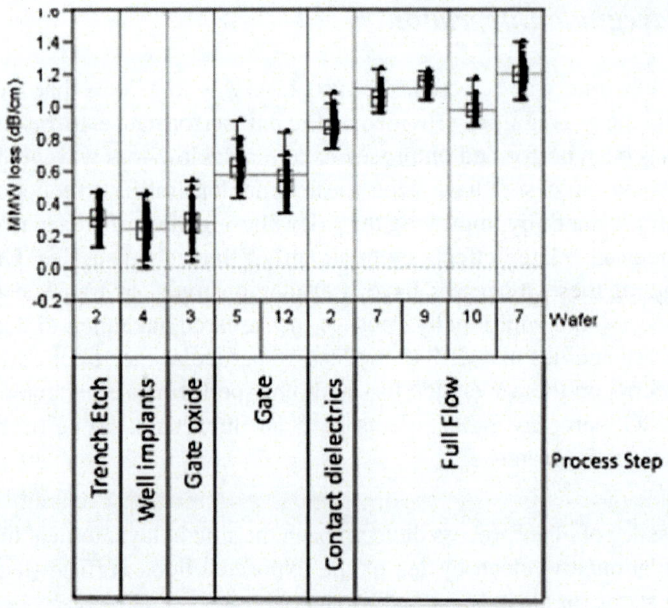

Fig. 8.4 Multimode waveguide loss and its variation through the process flow

oxidation/strip/clean steps as the sidewall of the silicon islands that will eventually form the body of transistors. Beyond the first couple of dielectric layers lie low-k dielectrics used in the back-end. We showed that the first two metal levels and associated dielectrics had an impact on the waveguide loss.

We removed wafers from the process flow at different points of the process, not letting them finish the complete flow, to observe the effects of various process steps on waveguide loss. Figure 8.4 shows the evolution of loss with process steps. There is clear evidence that waveguide loss is strongly affected by many other aggressors besides sidewall formation. Photolithography, due to its impact on the local waveguide geometry, is a key driver of optical loss performance. Figure 8.5 shows a comparison of optical propagation loss for the same waveguide design printed with 248-nm lithography versus 193-nm lithography.

8.2.4 Photolithography

Beyond providing benefits to waveguide loss, as we have just shown, the continual reduction of feature sizes driven by the electronics industry has finally reached a point where it can benefit and even enable novel photonic devices beyond the research laboratory. In the past, very small features (on the scale of a quarter the wavelength of light) could only be printed reliably using an electron-beam tool. But this restricted the designs to small areas and it is not at this point a volume-capable production tool. In Luxteras process, a single photolithography step is used to define all optical devices (except for Ge devices). The lithography is performed

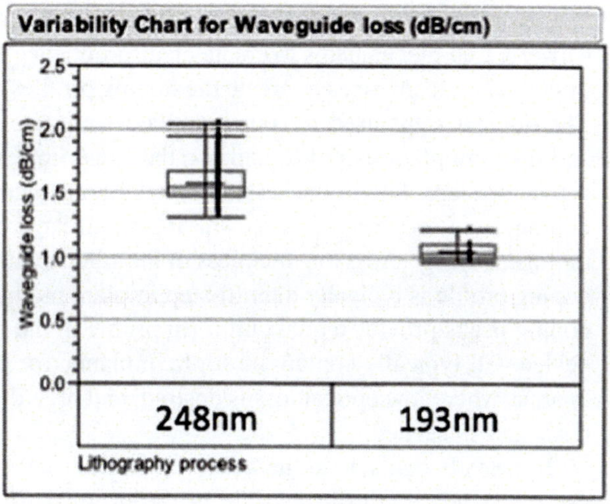

Fig. 8.5 Improvement of waveguide propagation loss with lithography technology

in an advanced 193-nm stepper and is used to print a wide array of different features that are completely different from what is traditionally printed in electronic circuits (often referred to as Manhattan geometries, only oriented along *X* and *Y* axes): straight and curved waveguides, tapers, arrays of holes, arrays of curved gratings, etc. In the 193-nm-based process we can print features as small as 100 nm reliably (see Fig. 8.6), and this also provides a side-benefit: larger feature sizes are much better controlled and have less variability in their dimensions across a field or a wafer, or even from wafer to wafer and from lot to lot.

8.2.5 Active Optical Device Integration

As discussed previously, a single photolithography step is used to define all optical devices in the silicon layer. Ge-based devices require the additional step of growing the Ge film, which is discussed below. Other active optical devices in our

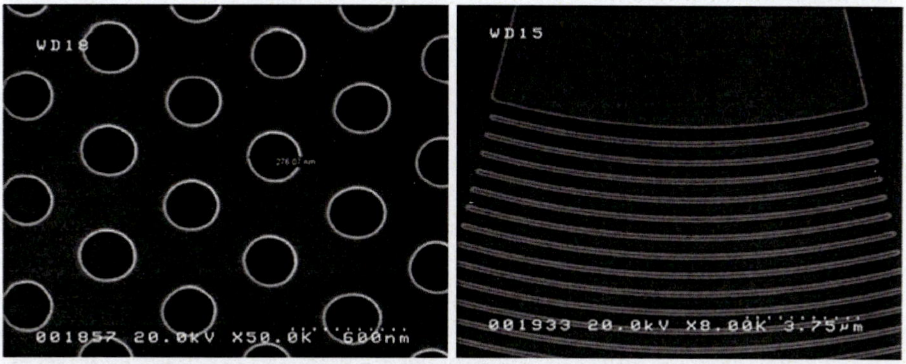

Fig. 8.6 Examples of optical features printed with 193 nm lithography

process (mainly, modulators, either of amplitude or of phase, see Sect. 8.3) rely on carrier-based mechanisms to create electro-optical effects [11, 12]. We use standard implantation techniques to create the doping profiles needed for the various devices. The implant steps used for our modulators are integrated with the standard nwell and pwell implant modules, and use the same implanter tools, as well as the same activation steps, thus not necessitating additional thermal treatment. Various implant dopings are used, most of them sharing a similar profile in an attempt to achieve a uniform doping across the thickness of the silicon with different peak densities. Each doping profile is typically a multi-step implant module, where at least two different doping peaks provide a quasi-uniform profile in the vertical direction. A modulator device will typically contain multiple implants, for instance to achieve low doping in areas where low optical loss is desired, and high doping in areas where low series resistance is needed.

To provide contacts to the active optical devices, including Ge photodiodes, we use the standard contact module of the original CMOS process and the standard source/drain implants and salicidation to provide ohmic contacts to silicon (our Ge module uses different implants and no salicidation to provide contacts to Ge), as shown in Fig. 8.3. Salicidation is blocked over the waveguiding region where the optical mode resides (or else optical propagation loss would be extremely high) by the presence of a dielectric layer. That layer is not present over the contact area. The contact module is designed for a wide range of topography in the original process: contact to the silicon layer, contact to the top of the gate, and contact to the substrate. Thus, it can easily handle contacting our optical devices as well as the Ge layer, as long as it does not differ too much in thickness from the transistor gate poly layer.

8.2.6 Germanium Module

The use of Ge and SiGe alloys has been proposed for photonics applications, as well. SiGe has higher refractive index than Si and its absorption edge extends further in the near infrared allowing efficient light absorption at the wavelengths used in optical communications. Discrete Ge detectors compete with III–V ones in the market of near infrared photodetection by offering a cheaper product though limited in use to the C-band (around 1,550 nm) with the absorption coefficient of Ge being too low in the L-band (1,600 nm) to provide efficient photoresponse. At much lower speed, when the size of the sensitive area is a key factor, Ge detectors are preferred to InGaAs, once again for cost reasons.

Ge-on-Si based photodetectors [13] have been investigated since 1966 [14], and high-speed waveguide SiGe devices were demonstrated in the 1980s at Bell Labs [15]. Those pioneering works evidenced the importance of a buffering technique in the epitaxial growth to reduce and confine the defects (dislocations) far from the active region of the device while enabling a planar, smooth two-dimensional growth of the semiconductor film. The limits of SiGe alloys and superlattices in accessing the III window of optical communications (1,550 nm) even using a waveguide configuration [16] triggered a new effort at the end of the 1990s to identify a suitable epi technique enabling the growth of photodetector-grade Ge films on Si substrates. A few buffering techniques were proposed and investigated in those years. Among them, the most relevant are the gradual buffer approach (sometimes including a chemical-mechanical polishing step about half way from Si to Ge) detailed in [17] and the low temperature buffer originally introduced in [18]. While the former yields films of higher quality (record low dislocation counts have been reported using this technique complemented by cyclic annealing steps [19]), the latter is preferred for integration with CMOS for its lower thermal budget and total thickness/growth time required. Surface illuminated and waveguide photodetectors using variants of the low temperature buffer method have been reported with very high efficiency and speed by several groups [20, 21]. Recently, Luxtera demonstrated waveguide Ge photodetectors integrated in a CMOS process using a low temperature process [22] and we discuss in the next section the challenges of the integration of the Ge module in the CMOS flow.

8.2.6.1 Germanium Module Integration

The integration of Ge to enable photodetectors in a CMOS process requires a careful identification of constraints imposed by the baseline process and the desire to minimize changes and subsequent process development. Such a module consists of a significant number of additional steps, and interactions between these steps and earlier and later stages of the process require a development approach that addresses these challenges from the start.

Three main factors drive the choice of processes and device architecture in the integration process: thermal budget, contamination, and device size. Thermal budget constraints dictate the placement of the Ge epi step after the main doping and the poly annealing step have been performed since Ge cannot withstand their temperature. On the other side, todays low-K backend films are not compatible with the temperatures required during Ge epi. As a consequence, the growth of the Ge film must be placed somewhere before the contact module and after the poly gate formation. Luxteras process inte- grates Ge after salicidation to minimize interactions with the critical transistor spacers (see Fig. 8.7).

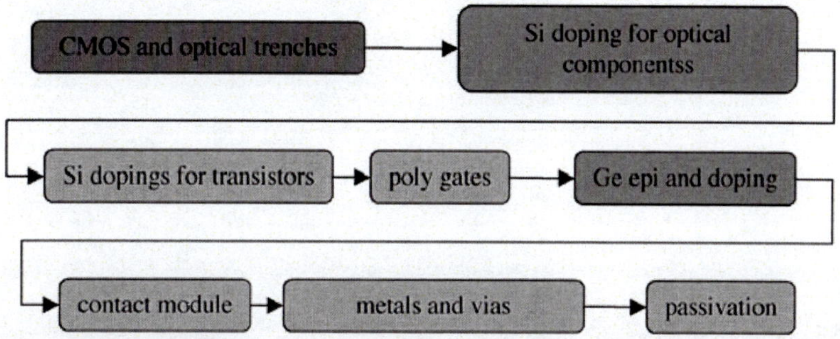

Fig. 8.7 Schematic of location of Ge module within CMOS process flow

Compatibility with CMOS backend processes in essence restricts the device architecture to a waveguide-based photodetector. The headroom limits imposed by the planarization techniques (CMP) in todays technologies force the maximum height of the Ge film used for the photodetector to be less than a few hundred nanometers (the space between the Si surface and the first metal level in a typical CMOS process). This rules out completely the possibility to use surface illuminated devices. The latter, indeed, would require a film thickness of 3–4 μm, at the wavelength of 1,550 nm, to absorb efficiently the light. Fortunately, the waveguide configuration relaxes this constraint due to the light propagation in the wafer plane thus transforming a thickness requirement into length. The film must be thick enough to admit optical modes at the chosen wavelength, which is possible thanks to the very high refractive index of Ge (about 4). The waveguide configuration is appealing also for its potentially higher speed, when compared to the surface illuminated one. By decoupling the absorption from the collection length (the former along the light propagation path, the latter perpendicular to it), the waveguide design allows an almost independent optimization of both. In other words, speed can be optimized without compromising the photoresponse.

The Luxtera process uses the selective growth of Ge in recesses opened in a dielectric stack to form the body of the photodetector (see figure top view SEM). After the Si surface is cleared in the selected areas, the wafer is cleaned using a wet process and introduced in the epi chamber. The selective growth is preceded by an in situ cleaning step aimed at removing the residual oxygen contamination and creates a film 300 nm thick. The wafer is, then, deposited with a dielectric film which protects the semiconductor during the following lithographically defined implant steps. At the end of this process the protective dielectric film is patterned and the wafer proceeds through the standard contact module and the rest of the back-end. The Luxtera process uses the selective growth of Ge in recesses opened in a dielectric stack to form the body of the photodetector (see Fig. 8.8). After the Si surface is cleared in the selected areas, the wafer is cleaned using a wet process and introduced in the epi chamber. The selective growth is preceded by an in situ

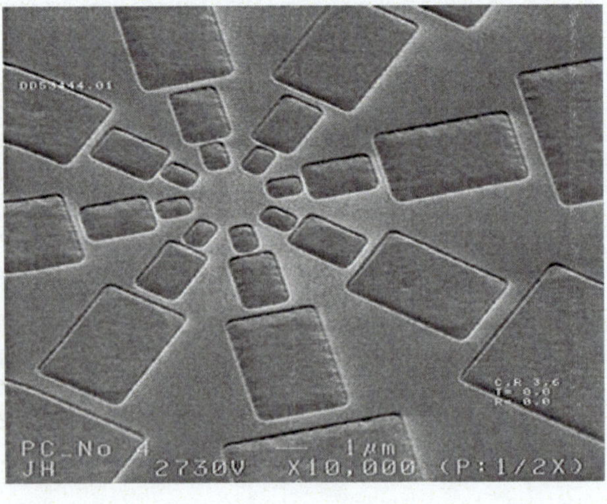

Fig. 8.8 Top view SEM of Ge boxes selectively grown over silicon

cleaning step aimed at removing the residual oxygen contamination and creates a film 300 nm thick. The wafer is then coated with a dielectric film, which protects the semiconductor during the following lithographically defined implant steps. At the end of this process the protective dielectric film is patterned and the wafer proceeds through the standard contact module and the rest of the back-end process.

8.2.7 Process Control and Monitoring

The last critical piece of the process development involved the creation and/or selection of appropriate process control monitors. Typically a CMOS manufacturing line will employ of combination of built-in equipment monitors (pressure gauges, temperature sensors, etc.), metrology (critical dimension measurements, etch depth measurements using SEMs and atomic force microscopes or profilometers), and inline electrical probing (measured when the first metal interconnect layer is available, and then again at the end of the line). There is an extensive infrastructure of tools and measurement equipment that has been developed to support this activity in the semiconductor industry, with two primary goals: control the various process steps and be able to react quickly when the process gets out of control, and screen and scrap bad wafers as early as possible in the line to minimize cost.

During our development efforts, the issue of appropriate process control monitors and wafer acceptance criteria became extremely critical. For instance, the critical dimensions of the optical devices (width of features, etch depth) are primary drivers of the final optical performance, and needed to be tightly controlled. The lithography and etch steps defining the optics occur essentially at the very beginning of the process, so it would be extremely costly to carry out optical probing at the end of the line to determine whether the process yielded in-spec optical devices. Bringing optical probing into the fab quickly proved to be an unfeasible proposition (due to non-standard equipment and lack of trained operators).

Luxtera, in conjunction with our manufacturing partner, ended up developing a set of structures used for inline metrology and inline electrical testing. These structures were carefully designed and measured to generate data that could be directly correlated to final device performance. Examples of such structures included the following:

- arrays of holes and line/spaces of various dimensions measured by critical-dimension (CD)-SEM and wafer-scale AFM,
- inline electrical structures (diffusion resistors, capacitors, diodes) used to validate correct processing of our active optoelectronic devices,
- ring oscillators with optoelectronic devices built in to determine yield of high-speed integrated optoelectronic circuits (for instance a modulator with driver).

It was necessary to run a significant number of wafers and lots (on the order of 30 lots of 25 wafers) to extract the true variability of the process and create statistical models and come to a frozen process with well-understood characteristics and capable process monitors.

8.2.8 Other Integration Elements

Figure 8.9 shows a summary schematic of all the requisite incremental modifications to the original CMOS process to enable photonics. In addition to the critical elements discussed above, the figure shows some other issues to be overcome when integrating photonics in the front-end:

- Ensuring planarity of the silicon and silicon dioxide trench fill after CMP is critical but made difficult by the widely different densities of features on the optical devices versus transistors in our dual-trench approach. Care has to be taken to match densities across designs to ensure compliance with process targets.
- Data processing of the various design layers had to be modified slightly to add compatibility with optical devices and to block nefarious effects of many process steps. For instance, many implant layers are derived from other design layers (for instance, n-type implant mask = NOT p-type implant mask) and could result in the unintentional implantation of optical devices, so blocking layers have to be employed, and boolean operations have to be modified to take this into account.
- Tiling of metal layers (which is usually an automated step that fills empty areas of metal masks with dummy structures to equalize density, again for CMP process window purposes) has to be controlled to keep some empty areas untiled. This is to allow light to travel through the back-end of the chip between the grating couplers etched in the silicon layer and the top surface of the chip where light may couple to/from an optical fiber or a hybridly attached laser.

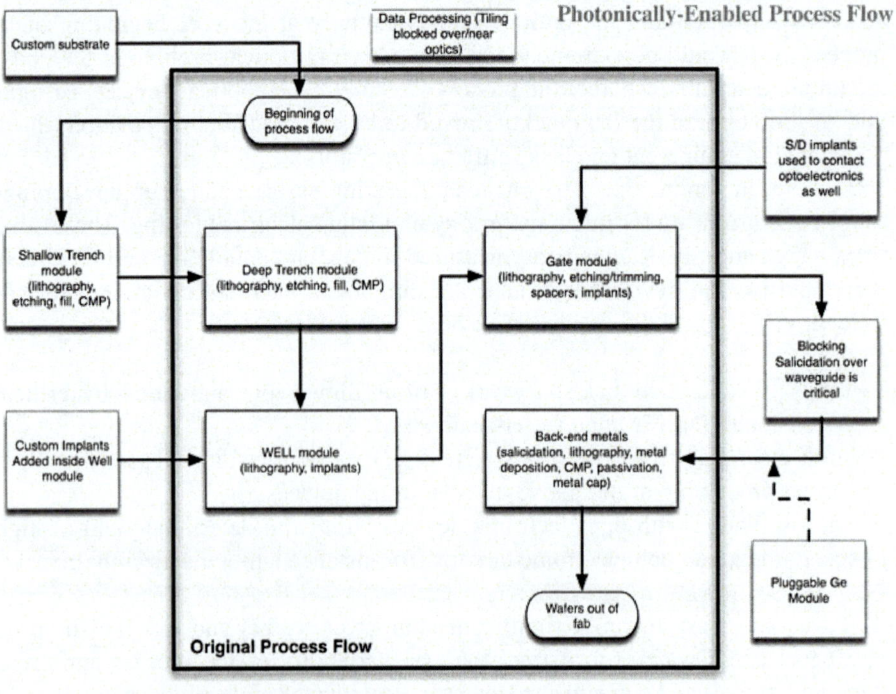

Fig. 8.9 Process flow of photonically enabled CMOS process

In the next section, we discuss the concept of a device library and our development approach.

8.3 Photonic Device Library

8.3.1 Electronics Libraries: The Inspiration

Luxtera' s photonic device library was modeled on electronic circuit design libraries. An electronic library is built on simple elements whose physics is well-understood and whose performance is thoroughly characterized. These elements include transistors, resistors, capacitors, and inductors. The library does not simply contain the layouts for these elements but also models that describe their behavior as a function of inputs to the device, such as voltage or current, as well as their sensitivities to environmental conditions, such as temperature. In addition, the models contain information on how the performance changes with process variability, with best, typical, and worst cases traditionally called process corners.

From the basic elements, simple cells can be built, for example, logic gates for a digital library or amplifiers for an analog library. These lower level elements can then be further referenced in higher level blocks and subsystems, such as driver circuits or digital-to-analog or analog-to-digital converters (DACs and ADCs). Because each of the building blocks has been comprehensively characterized, it is possible to simulate the behavior of the blocks accurately and derive their range of performance due to natural variations in the fabrication process.

Even though complex electronics blocks may contain millions of basic elements, they are always built up in a hierarchical form. This approach allows top-level simulations to be tractables, as they can reference simulation results for smaller blocks in the hierarchy. The hierarchical design based on precise models allows the designer to devise complex systems accurately to given specifications.

8.3.2 Library Hierarchy and Design Flow

Luxtera' s photonic device library is built up in a fashion analogous to its electronic counterpart. The library begins with a collection of optical and opto- electronic elements that serve as a starting point for building complex PICs. These basic photonic elements, such as waveguides, fiber-to-chip couplers, light modulators, and receivers are treated in a similar fashion to transistors and resistors in a library of electronic circuits. The photonic library is fully integrated with the electronic library so that the two types of elements seamlessly combine to construct optoelectronic blocks.

As an example, let us consider a high-speed modulator subsystem, built up hierarchically from elements of different complexities. Figure 8.10 shows a schematic of a simplified hierarchy of the high-speed modulator. The subsystem comprises two main blocks, a Mach-Zehnder interferometer (MZI) modu- lator and a control block

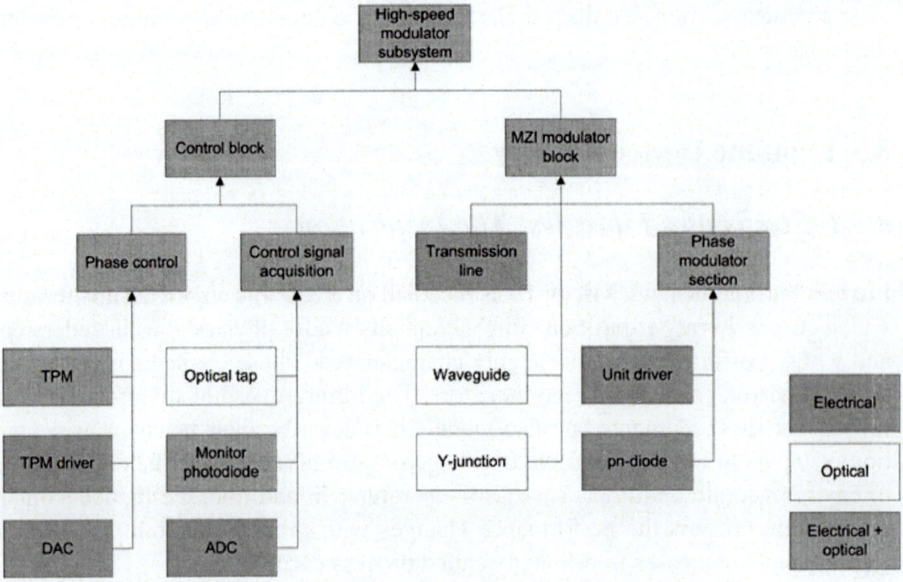

Fig. 8.10 Sample library hierarchy: the high-speed modulator

that controls the relative DC optical phase of the two arms. The MZI modulator block further consists of waveguides, Y-junctions, transmission lines, and several modulator sections, each instance of the latter being made of a pn-diode with a unit driver circuit. The control block con- tains a phase control block and a control signal acquisition block. The former comprises a thermal phase modulator (TPM) phase tuner, driven by a TPM driver controlled by a DAC from the control bus. The latter employs optical taps feeding the light from the waveguides into monitor photodiodes, whose signal in turn is converted to a digital signal by an ADC.

The design of each of the basic components follows a well-defined flow, shown schematically in Fig. 8.11. It begins with a device concept that is naturally constrained by the process. The design selection phase employs a design methodology based on a design-of-experiment (DOE) approach and relies on the testing infrastructure providing an accurate and automated test capability. The outcome of the third phase design verification is a model that describes the performance of the library element. The models provide the performance of each fundamental element as a function of the process variations as well as other inputs to the device, such as wavelength or temperature. In the next subsections we describe the main steps in the design flow and illustrate our approach with an application to the design of a grating coupler.

8.3.3 Device Concepts

The photonic device concepts are enabled by our integration approach, which consists of adding photonics capability into a CMOS process with minimal changes to

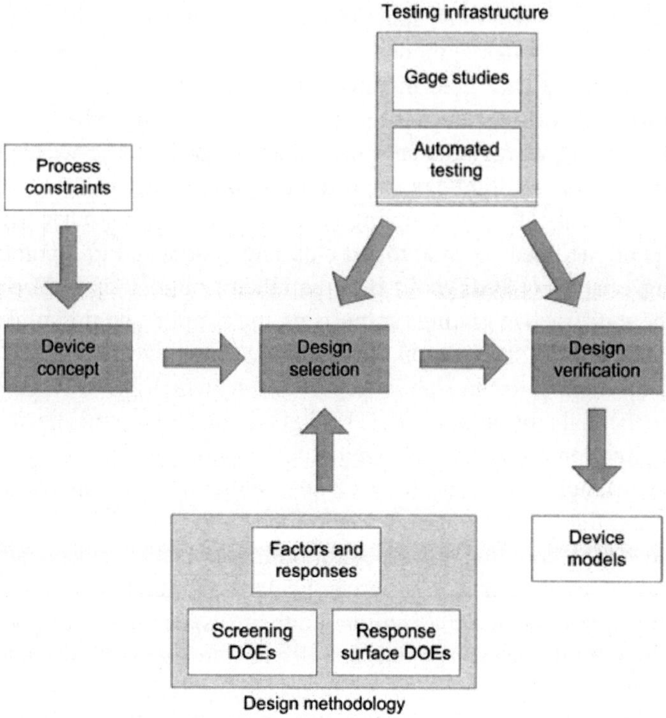

Fig. 8.11 Overview of device design flow

the baseline process. This philosophy allows carrying forward the existing electronic cells with the least amount of modifications. Admittedly, the approach also puts some limits on the component design.

As an example of what constraints the process requirements may place on device concepts, let us consider how to transfer the light signal from a fiber into the CMOS die. The common approach of edge-coupling is not possible in a CMOS process because of the existence of a guard ring around the edge of the chip. This ring comprises metal lines and it has a two fold purpose. It prevents mobile ions from entering the transistors in the chip and bars cracks from propagating to the inside of the chip. Light signals propagating through the guard ring would suffer prohibitively high losses for this approach to be feasible.

One solution is to access the chip surface for light transfer. Just as electrical signals enter the CMOS chip through wirebonds from the top side of the die through bond pads, we can envision "optical bond pads" where the fiber is positioned along the normal to the chip. The light is then "bent" using a photonic library element and directed into waveguides inside the chip. The advantage of such an approach is that the optical interface can now placed virtually anywhere on the two-dimensional surface of the die, providing much more flexibility in the PIC layout compared to the case where optical signal entry is restricted to points along the die edge.

Another issue that one must contend with, whether we use edge or surface coupling, is that the fundamental mode size is orders of magnitude smaller in a silicon-based system on the chip than in standard single-mode fibers. It is again CMOS compatibility that restricts the design space. To be able to utilize the same silicon film for the optical waveguides and the transistor bodies, the waveguide must be built in a thin layer, resulting in a small fundamental mode.

Figure 8.12 illustrates the coupling scheme employed on our CMOS platform. The fiber is aligned near normal to the chip and is attached to a polarization splitting grating coupler (PSGC) built into the silicon optical layer. The curved lines represent bent diffractive gratings comprising the coupler and the solid arrows show the direction of light propagation. The gratings have two functions. They scatter light from the fiber into the slab waveguides into which the waveguide is etched, redirecting the light by about 90°. The light is at the same time focused in the transverse direction inside the slab due to the curvature of the grating. The unetched horn-shaped triangular sections of the coupler allow for a smooth transition between the slab waveguides and the narrow single-mode waveguides used in the PIC.

To achieve high coupling efficiency for both polarization modes in the fiber, two sets of gratings are overlaid in a perpendicular configuration. They redirect light with the appropriate linear polarization into their respective waveguides. The polarization of the light in each waveguide as well as in the fiber is shown in Fig. 8.12 with dashed arrows. As a polarized coherent beam propagating in the fiber can always be written as a linear combination of two orthogonal linearly polarized components, this device can efficiently couple light in any polarization into the chip. By appropriately designing the coupler, it is possible to ensure that light in all polarizations experiences the same coupling loss.

This coupling approach is unlike many other polarization diversity schemes where the two polarization modes in the fiber couple into the two orthogonally

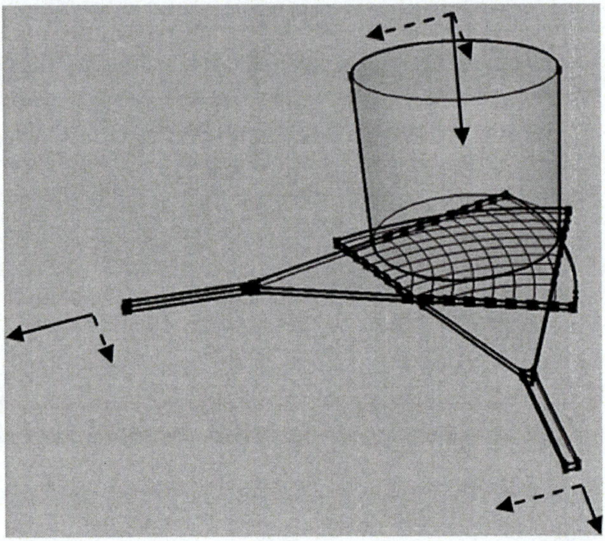

Fig. 8.12 Schematic view of the polarization-splitting grating coupler with a fiber

polarized modes of a single waveguide and then separated out inside the chip to be processed individually. In our approach, the two orthogonal polarizations are separated spatially as soon as light enters the chip. The other difference between this and edge-coupling schemes is that both waveguides carry the light signal in the same fundamental transverse-electric (TE)-polarized mode, not in a TE-mode in one waveguide and a transverse magnetic (TM)-mode in the other waveguide.

8.3.4 Design Selection: The Design-of-Experiment Approach

After formulating the device concept, the next step in the design flow is optimizing the device in the given process. Regardless of whether a library element is relatively simple, like a waveguide, or more complex, like a PSGC, it can be described in terms of a number of parameters, or factors, that determine the response of the device. The factors can be defined by the design, by the process, or by environmental conditions. For instance, some of the design parameters of the PSGC are the number of scattering elements in the grating, their positions in the plane, the width of the input waveguide, the angle of the horn, or the distance of the grating from the input waveguides. Process factors can be the thickness and refractive index of the silicon film as well as that of the backend dielectrics above the coupler. In addition, there are environmental parameters, such as the angle of the fiber attached to the chip or the chip temperature.

Once all the factors are enumerated, the device response parameters are defined, which then fully describe the performance of the particular device. For the PSGC, the response is the transmission efficiency as a function of wavelength and the polarization of the light incident onto the grating. We can further simplify this set of responses by defining a parameterized version of the spectrum. One option is to describe the spectrum in terms of peak wavelength, minimum loss, 1 dB bandwidth, and polarization-dependent loss.

The PSGC then can be considered a black box that takes the factors and turns them into responses, as illustrated in Fig. 8.13. Each factor can vary, be it via design or layout changes, through natural variations in the process, deliberate process changes, or through variations in environmental conditions.

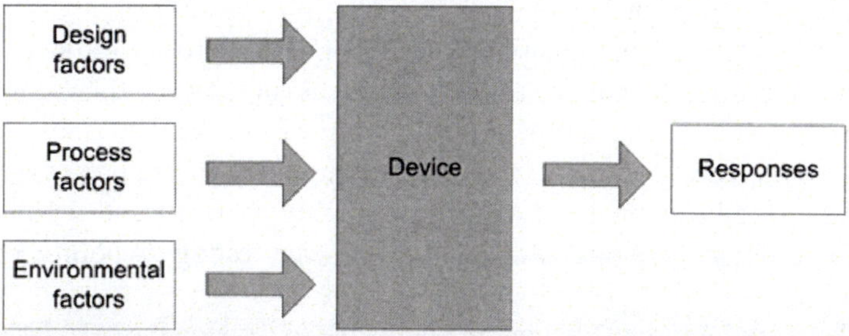

Fig. 8.13 Input factors and response

Armed with a list of parameters, the next step is to quantify the effect of each factor on each response. We can use the design-of-experiment (DOE) approach to map efficiently the effects on the responses as well as interactions between the factors. A systematic approach clears away two main types of stumbling blocks in approaching relatively complex designs like that of the PSGC.

We have avoided the approach that finds an ideal value for the parameters separately because in the presence of interactions, a design determined in this fashion frequently turns out not to be a global optimum. This is because the optimum in some factors is a strong function of the value of other factors, which is due to the interaction between the parameters. In the DOE methodology, the experimental points cover the full N-dimensional parameter space defined by the N parameters. With such an approach, we were able to locate the optimum design point with a single set of experiments.

We used the DOE methodology fully to our advantage when the number of factors is large and all interactions between them need to be mapped out. For the photonic elements, like the PSGC, the number of input parameters is frequently in the dozens when all effects are carefully considered. If each parameter was varied independently, the number of experiment would scale exponentially with the number of factors. The DOE approach can be very economical in that, if appropriately used, the scaling with the number of factors is much weaker. For instance, the parameter space for a design with eight parameters can be fully mapped using a cubic-centered DOE design with only 82 points, whereas a set of experiments where all parameters take three values would contain $3^8 = 6{,}561$ points.

Due to the large number of factors, in the design flow we first run a screening DOE to identify the most important factors. This experiment has two major goals: (1) finding the few input parameters that have the greatest effect on each of the responses and (2) identifying the strongest interactions between the factors. The factors we choose for the experiment include both design and process parameters in order to gain insight into the main drivers of the device performance and their interactions. The screening DOE is also guided by a deep physical understanding of the device as well as the modeling of results in order to reduce the complexity of the experiments.

Once the most important factors have been identified and the approximate optimum design point is located, the device response surface is mapped out. This surface describes the quantitative dependence of each response on its driving factors near the optimum design point.

Even with all the simplifications that the DOE approach affords, we need a reliable testing infrastructure to make it practical (in terms of time, complexity, or setup, and validity of measurement results) to measure the thousands of fabricated devices to obtain necessary information about their design and process sensitivities. What has allowed Luxtera to develop a thoroughly characterized set of devices for the photonic library was our automated wafer scale testing capability, described in Sect. 8.4.

Another very important element in being able to develop the best design for each device is a good understanding of the measurement capabilities of the test systems utilized. Gage studies were run for each type of measurement to understand the capabilities of the test and the errors introduced by variations in measurement. Typically, the gage of a measurement was considered capable if it was about an order of magnitude better than the difference between devices that it was intended to detect.

8.3.5 The Role of Process Variability in Design Selection

Since process factors often turned out to be the strongest ones that determine the variations in device response, these were also exercised in order to obtain a full model for the device: corner lots were run where process parameters were intentionally targeted toward the edges of the normal process distribution. Intrinsic process variations are unavoidable, so to design a robust device, we took into account not only the nominal process points but also its normal variations. Therefore, the primary goal of component library design was not necessarily to achieve a world-class performance for each component but to find a design point with low process sensitivity, while still maintaining sufficiently good performance.

One trade-off that occasionally emerges when selecting a design is the trade-off between producing excellent performance some of the time and designing for low variability, as exemplified in the distributions in Fig. 8.14. A device, whose performance is described using a certain figure of merit (FOM), may have two embodiments. By way of example, one of these designs has low variability, or process sensitivity (dashed line), whereas the other one has high process sensitivity (dotted line). The process may have large enough variations to produce a device with an excellent figure of merit in a small number of cases (a hero device) but at the expense of a very wide distribution over all process corners. To be commercially competitive, it was essential that Luxteras designs were selected with the purpose of having low variability in order to keep the entire population within specifications.

To achieve optimal performance for the device, we can implement process improvements to shift the FOM distribution higher (solid line). In the specific case of the PSGC, these improvements included tightening the depth control of the etch defining the grating features and a smaller linewidth, more accurate lithography

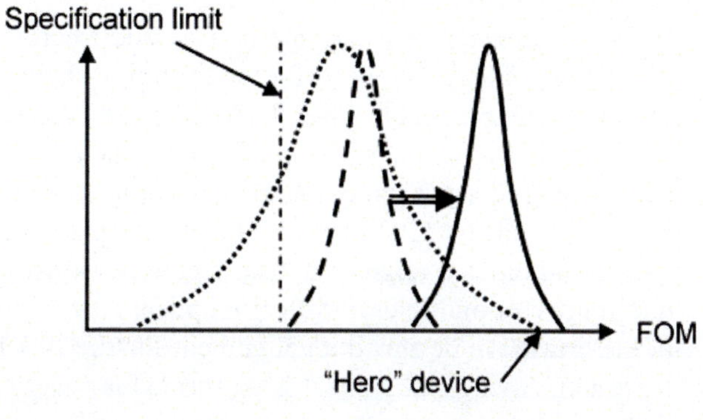

Fig. 8.14 Comparison of the figure of merit (FOM) distributions. *Solid line*: library element optimized for minimal process sensitivity. *Dotted line*: device optimized for hero performance

process. For all process modifications or improvements, one important goal is to retain low process sensitivity for the photonic library elements.

8.3.6 Design Verification and Device Models

With an accurate and automated testing capability in hand, we have characterized each photonic library element similar to the methodology used for creating transistor models for an electronic library. By measuring thousands or tens of thousands of fabricated instances of each device we have verified device performance across many lots and understood their process sensitivity in depth. The outcome of the design verification effort is a faithful model that describes the device response as a function of design, process, and environmental factors.

Once the models have been determined for a library element, it is released for use in a photonic circuit. However, the device development does not necessarily stop here. Automated wafer scale testing enables us to determine individual variance components of the device comprising within wafer, wafer-to-wafer, and lot-to-lot variations. This information can point us toward potential process improvement paths.

8.4 Design and Testing Infrastructure: The Tools of Success

8.4.1 Wafer-Scale Optoelectronic Testing

Beyond the concept of a device library, Luxtera has also extended the philosophy of the microelectronics world to the design and testing infrastructure. As discussed above, the approach taken for the development of the process and the devices would not have been possible without the availability of a similar wafer-scale testing capability to that of the electronics world. The design of grating couplers and the normal coupling of light in and out of the wafers allowed us to use the existing tools and equipment for wafer-scale electrical probing by adding an equivalent fiber-based optical probe to a standard setup.

One of our optoelectronic probestations is shown in Fig. 8.15. We used as the starting platform a prober made by Cascade Microtech, which is capable of loading 8-inch wafers. The optical probe setup is shown in the inset of the figure and consists of a fiber array (typically 16 fibers, with a combination of single-mode and polarization maintaining fibers) mounted at the end of an arm. The arm is attached to a piezoelectric x-y-z stage. The probestation chuck is used to provide large motion across the wafer, while the piezo stage is coupled to an active analog feedback system for automated alignment. The feedback signal can be a combination of sum, differences, or multiples of different optical powers coming from the different fibers in the array. We designed an alignment feature that is placed within each chip on the wafer. A single alignment is performed for every wafer loaded and is sufficient

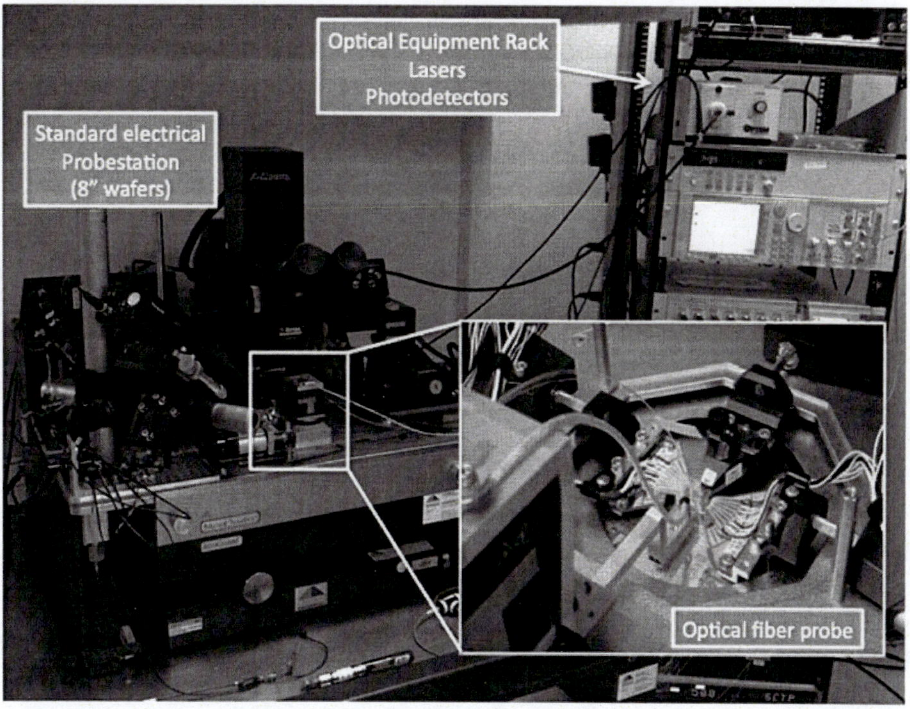

Fig. 8.15 Photograph of wafer-scale optical probestation

to ensure that any subsequent spot on the wafer can be tested. Any combination of wavelength scans and optoelectronic tests (light-current-voltage/LIV, s-parameters, photoresponsivity) can be run automatically across an entire wafer.

This capability is what has enabled Luxtera to approach the development of our device library in the way described in Sect. 8.3. Without such test automation, capturing enough data to obtain statistically relevant information about the response of the performance to the design variations would have been impossible. This capability has enabled Luxtera to approach the development of our device library in the way described in Sect. 8.3 Without such test automation, capturing enough data to obtain statistically relevant information about the response of the performance to design variations would have been impossible.

8.4.2 Design Automation Tools

With the continuous improvement of CMOS process nodes has also come the increase in costs associated with the fabrication of masks and the actual processing in the fab. It is typical for advanced nodes to have costs measured in many millions of dollars for a single tapeout and even at 130 nm the mask costs are far from negligible. So any error in the design can result in the scrapping of an entire maskset and costing enormous amounts. Considering the complexity of VLSI circuits with hundreds of thousands or even millions of gates on a single chip, an error is most likely to happen.

To prevent this, the electronics industry has heavily invested in the development of software tools to aid designers in creating error-free designs capable of first-silicon success. There are three main types of tool categories:

- physical design verification, i.e., does the layout of the chip break any rules making it likely that the yield of the chip will be impacted or even that the chip will be non-functional?
- layout versus schematic, i.e., does the layout actually match the desired circuit architecture?
- simulation, i.e., does the schematic of the circuit provide the expected functionality, performance, power, etc.?

With the capabilities of tight integration of optics and electronics that Luxtera developed came also the dangers of incorrect layouts, so we had to develop additional elements to complement existing electronics tools. The complete toolkit of all these is called the PDK or physical design kit, and Luxtera created a complete optoelectronic PDK capable of performing DRC (design rule checks), LVS (layout versus schematic), and simulations of optoelectronic systems. Our toolkit was based on the initial electronics-only PDK provided by Freescale and was subsequently modified to provide the additions. An example of an optoelectronic schematic (in this case an MZI modulator with electrical driver) and its simulated OPTICAL eye diagram at 10 Gbps is shown in Fig. 8.16.

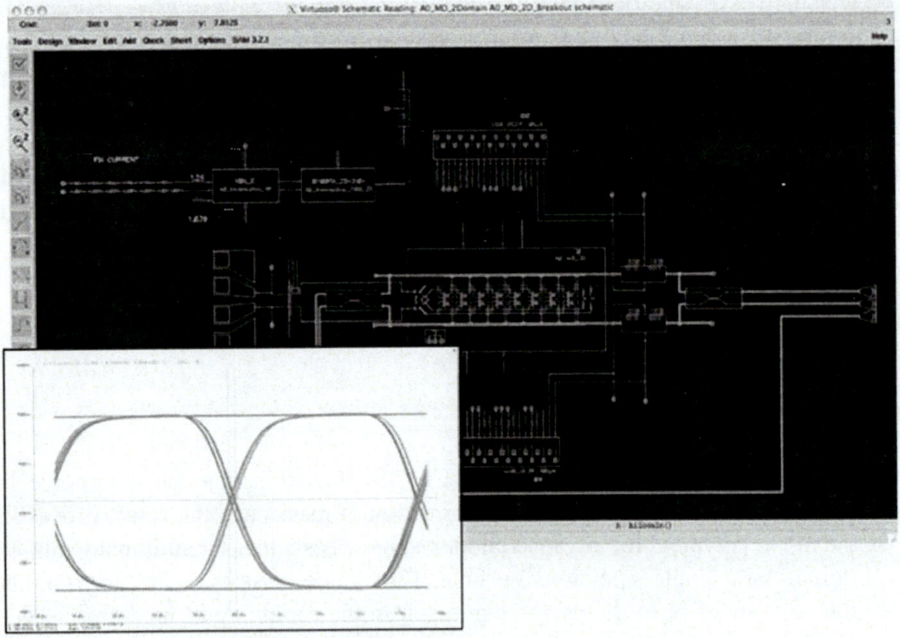

Fig. 8.16 Example of optoelectronic schematic and simulated eye diagram

Without such tools, it would have been inconceivable for Luxtera to successfully design and fabricate the chip that is discussed in Sect. 8.5.

8.5 Example of CMOS Photonic System

To validate the technology platform, we designed, assembled, and tested a technology demonstration vehicle. We decided to build a wavelength-division-multiplexed (WDM) transceiver that integrates:

- Four 10 Gbps transmitters, which are fed by 4 WDM-signals that are spaced at 200 GHz wavelengths.
- A 4-channel WDM-multiplexer (MUX) with 200-GHz spacing.
- A 4-channel WDM-demultiplexer (DEMUX).
- Four 10-Gbps receivers using Ge-PIN photodetectors.

Four external continuous-wave-WDM light sources that are centered at 1,552 nm are used and coupled via grating couplers that serve as optical interface into the chip. The light is routed trough MZI-based modulators. Each modulator block includes optical biasing elements, integrated Ge-monitor photodiodes, and a feedback loop that stabilizes the operation of the transmitter. This is required to guarantee operation throughout the operational temperature range of 0–70°C. After modulation the four signals are multiplexed in the interleaver-based WDM–MUX, which also contains biasing and monitoring elements as well as control electronics. The multiplexed signals are coupled via the optical interface grating coupler into a polarization-maintaining single-mode fiber and transmitted to the WDM-receiver, where an interleaver-based WDM–DEMUX separates the signals and couples into four Ge-PIN photodetectors followed by trans-impedance/limiting amplifier circuitry. We also integrated a BIST-block (build-in self test) that allows testing of the high-speed performance of the transmitters. The SFI interface was chosen as the electrical interface. A complete link transmits and receives 4×10 Gbps bidirectional data streams.

EDA tools are an integral part of the technology platform. This chip was designed using validated electronic and optical libraries, automated design-rule-checking (DRC), and layout-versus-schematic (LVS) verification. Figures 8.17 and 8.18 show a schematic architecture and a top view of the transceiver die highlighting the critical functional blocks. The die has a footprint of 9×8 mm^2.

The die was packaged into a quad-small-form-factor-pluggable (QSFP) module as shown in Figs. 8.19 and 8.20. The optical connections to the die were made via the optical interface, which consists of a linear array of grating coupler spaced at 250 microns to which a fiber array assembly can be bonded. The electrical high-speed signals are routed to the module edge connector via a flex circuit. The flex circuit also accommodates the dense routing of power and control signals to the die and contains necessary decoupling capacitors as well as a connector to a low-speed PCB that contains a micro-controller and voltage regulators.

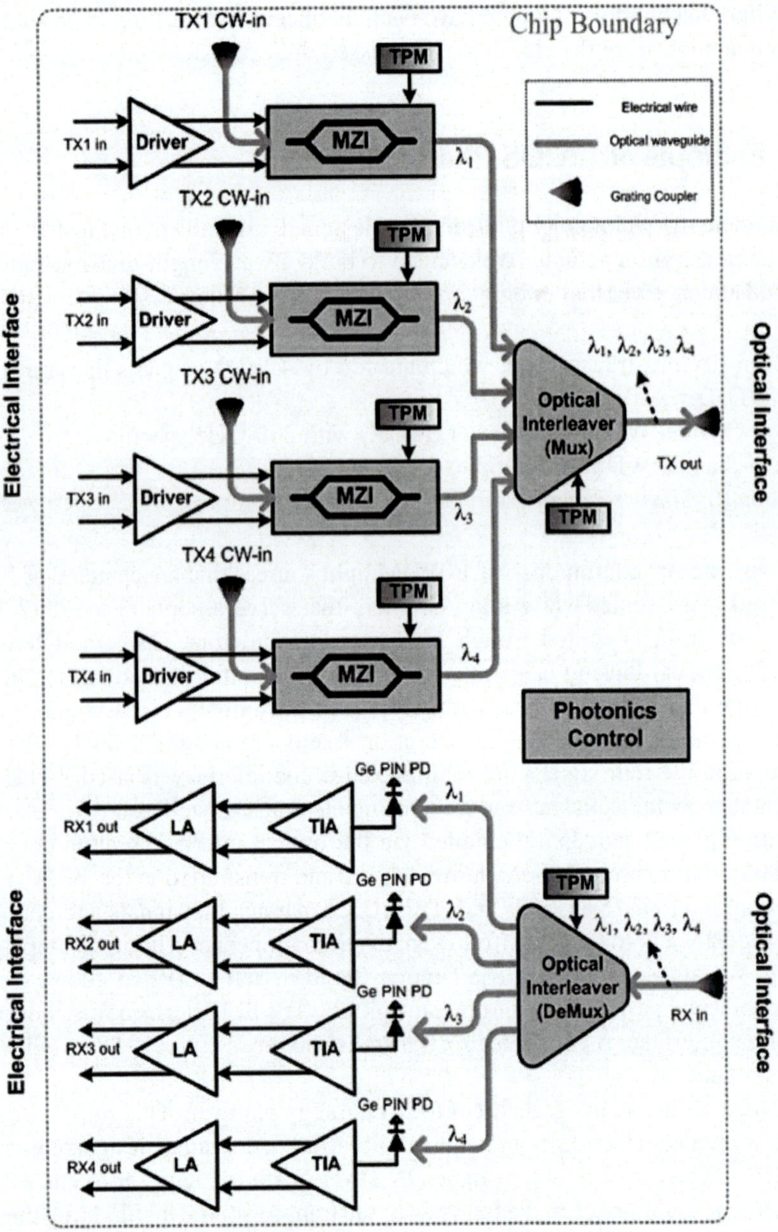

Fig. 8.17 WDM transceiver architecture

We tested transmitter and receiver performance and typical results are shown in Figs. 8.21, 8.22, and 8.23. The receiver subsystem consisting of grating coupler, DEMUX, waveguide routing, and that overall first generation receiver achieved a sensitivity of approximately −10 dBm at BER = 1e-12 with the Ge-PIN receivers achieving a sensitivity of −17 dBm. Link tests were conducted with transmitters

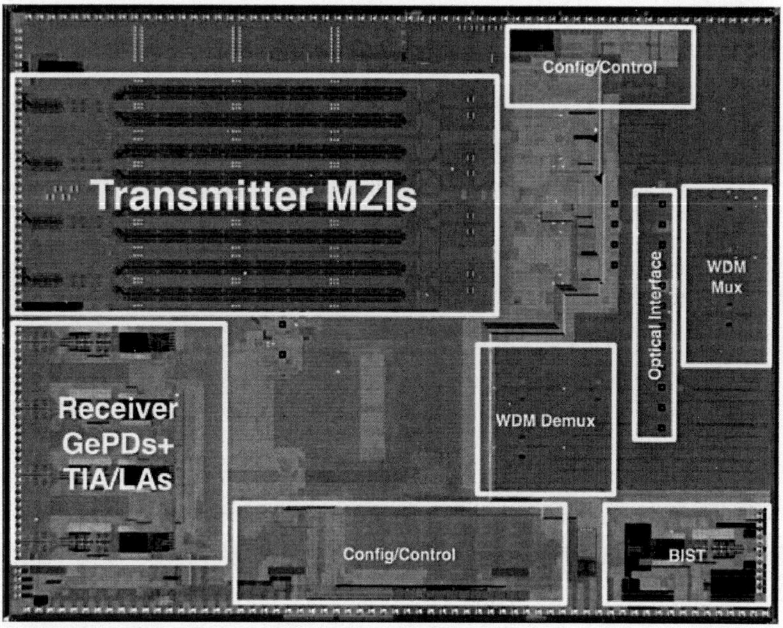

Fig. 8.18 WDM transceiver die shot

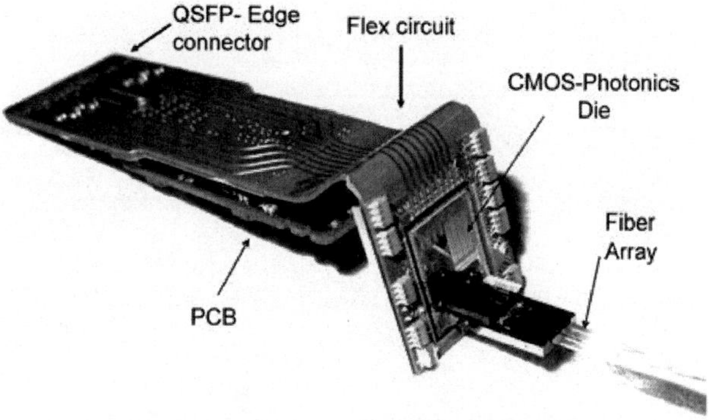

Fig. 8.19 WDM transceiver packaged

configured to emit average powers of −5 dBm per channel with −15 dB crosstalk from adjacent WDM-channels, and error-free bidirectional transmission was obtained. All module-level and subsystem-level test results are in close agreement with system-level simulations conducted during the chip design phase and validate the chosen design flow.

Due to the lack of integrated light sources this demonstrator is of no commercial value. In general the light source is of critical importance for silicon photonics applications and has been delaying the introduction of commercial products. It is not a coincidence that the detector-only optical functionality of silicon has resulted in commercial success in CMOS-image sensors a long time ago.

Historically, new applications and markets in the photonic industry have often emerged whenever new or critically improved light sources became available. For

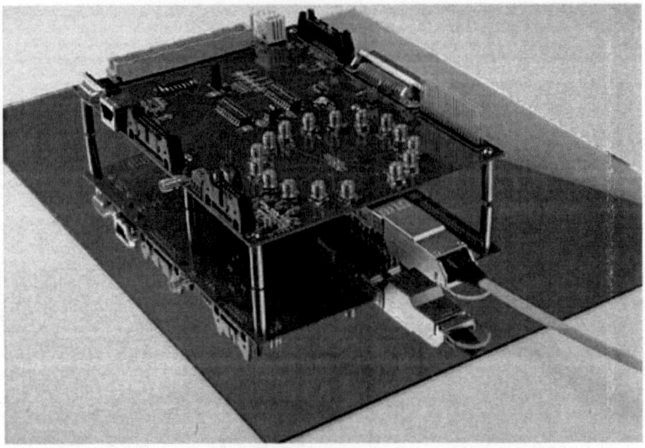

Fig. 8.20 WDM transceiver in test board

CMOS-photonics to become more than a curiosity a convincing solution to the light source problem is required. A cheap, volume-capable laser is required that can be integrated with the CMOS-photonics chip and meets the optical power requirements set by the applications. We believe that this is an engineering problem rather thena science problem and that all critical pieces required to build such a laser are in existence.

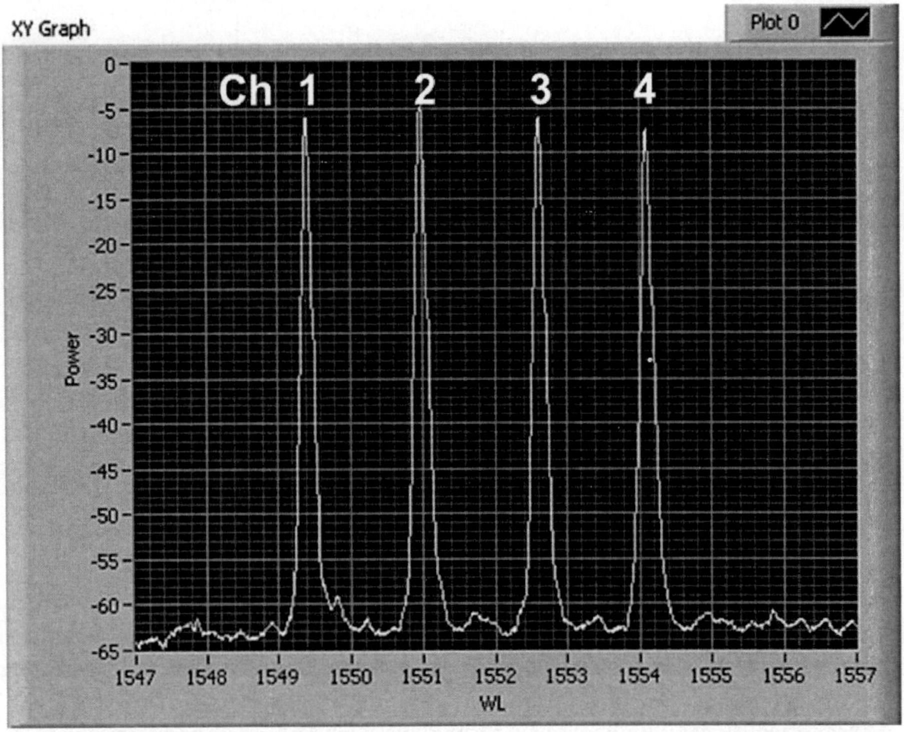

Fig. 8.21 WDM transmitter multiplexed optical output spectrum

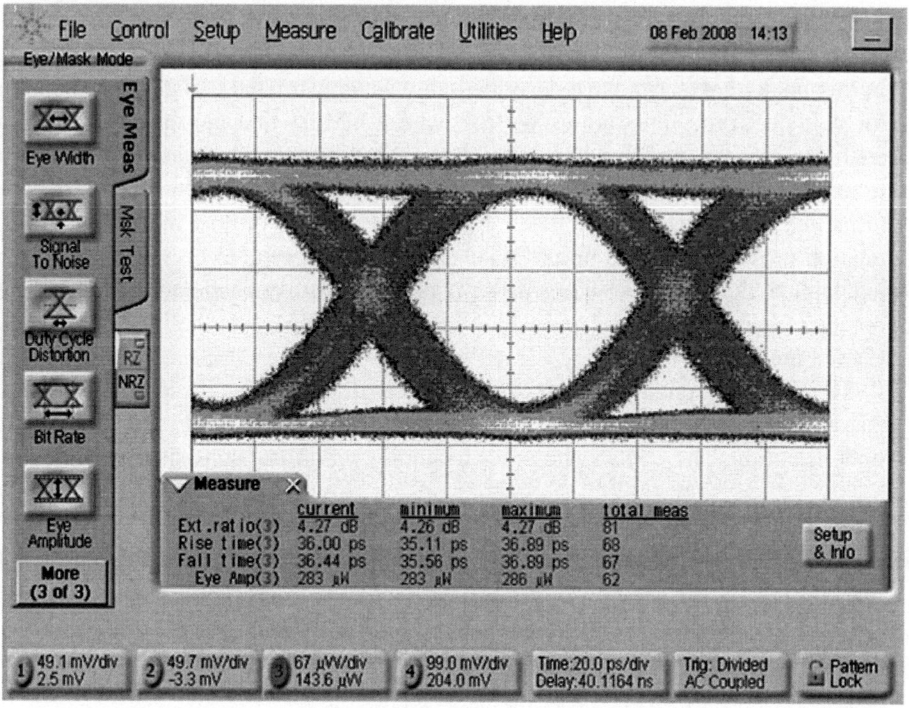

Fig. 8.22 WDM transmitter typical eye diagram for one of the four channels

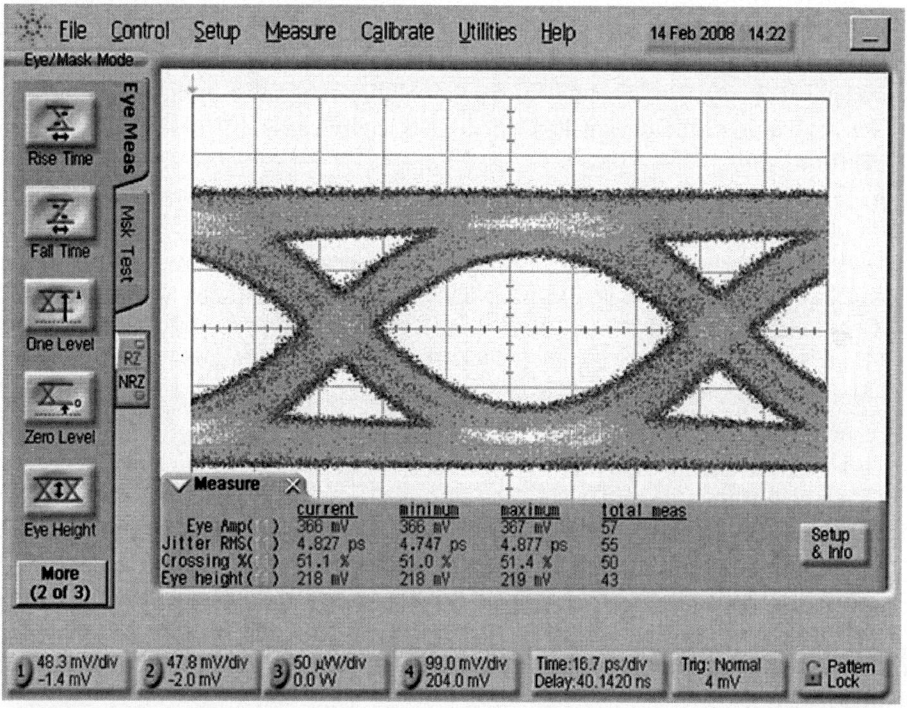

Fig. 8.23 WDM receiver typical performance

8.6 Conclusions

In this book chapter we have described how we constructed a platform technology for optoelectronic integration. We followed a CMOS-first approach by inserting enabling process modules into a baseline CMOS-process flow and by attempting to minimize the number of process modifications and adjustment that are required to maintain the performance of the baseline process. A critical input into the fabrication process is the starting SOI substrate, which needs to be selected in a way to enable optical devices with sufficient optical performance to address the targeted optoelectronic applications and which has a direct impact on the integration of the CMOS modules.

To enable the photonic devices we integrated a trench etch module to form all passive devices, implant modules to form modulators, and a germanium module for light detection. The CMOS photonics technology platform consists of

- photonic and electronic libraries
- design verification and layout verification tools and processes
- system modeling tools
- fabrication of the photonically enabled CMOS-wafers
- wafer-level optical and optoelectronic device and system testing
- module-level photonic packaging and testing.

There is a lot of focus within the scientific and research and development community on the optical devices and their fabrication. We want to emphasize the importance of design and verification tools to create higher level optoelectronic systems that can take advantage of the monolithic integration of the optical devices. We have unified the design flow for optical and electrical CMOS devices in this platform.

We followed a systematic process to develop, characterize, and integrate a photonic library into a chip design flow. All photonic devices were developed with a DOE-based design methodology that exercises design and process factors during development and device verification. The photonic device library with device models is available as input for system designers to model higher-level systems. The library is integrated into a state-of-the-art EDA tool. Wafer-level testing played a crucial rule in the development of the photonic device library as it allowed automated optical testing of the device library elements.

The enabling photonic device for wafer-level testing is the grating coupler, which also forms the optical I/O for all optoelectronic systems. We have demonstrated testing of integrated sub-systems on wafer. From an application perspective CMOS Photonics accomplished monolithic receiver integration by inserting a germanium module into the process flow. Besides enabling high-speed receivers it also provides optoelectronic systems with the critical ability to monitor the light flow and control the operation of higher level subsystems such as modulators and multiplexers.

A demonstration vehicle (40 Gbps WDM transceiver) exercised the complete flow covering die design, design verification, fabrication, wafer test, and module test and validated the Luxtera CMOS photonics technology platform.

Acknowledgments The development of this technology was truly a team effort for which credit goes to the whole Luxtera team as well as our foundry partner Freescale. The technology demonstrator was built under the DARPA-program Electronic and Photonic Integrated Circuit (EPIC) program. We also would like to thank Neil Berglund for critical input and guidance on the path from concept to implementation of the technology.

References

1. B. Jalali, S. Fathpour, Silicon photonics. J. Lightwave Technol. **24**(12), 4600 (2006) 187
2. G. Reed, A.P. Knights, *Silicon Photonics: An Introduction* (Wiley, New York, 2004) 187
3. L. Pavesi, D. Lockwood, *Silicon Photonics* (Springer, New York, 2004) 187
4. R. Soref, B. Bennett, Electrooptical effects in silicon. IEEE J. Quantum Electron. **23**(1), 123 (1987) 188
5. L. Liao, D. Samara-Rubio, M.Morse, A. Liu, D. Hodge, D. Rubin, U. Keil, T. Franck, High speed silicon Mach-Zehnder modulator. Opt. Express **13**(8), 3129 (2005) 189
6. C. Batten, A Joshi, J. Orcutt, A. Khilo, B. Moss, C.W. Holzwarth, M.A. Popovic, H. Li, H.I. Smith, J.L. Hoyt, F.X. Kartner, R.J. Ram, V.Stojanovic, K. Asanovic, Building many-core processor-to-DRAM networks with monolithic CMOS silicon photonics. IEEE Micro **29**(4), 8 (2009) 189
7. R. Jones, H. Rong, A. Liu, A.W. Fang, M. Paniccia, Net continuous wave optical gain in a low loss silicon-on-insulator waveguide by stimulated Raman scattering. Opt. Express **13**(2), 519 (2005) 190
8. A.W. Fang, H. Park, O. Cohen, R. Jones, M. Paniccia, J.E. Bowers, Electrically pumped hybrid AlGaInAs-silicon evanescent laser. Opt. Express **14**(20), 9203 (2006) 190
9. K.K. Lee, D.R. Lim, H. Luan, A. Agarwal, J. Foresi, L. Kimerling, Effect of size and roughness on light transmission in a Si/SiO2 waveguide: experiments and model. Appl. Phys. Lett. **77**, 1617 (2000) 191
10. K.K. Lee, D.R. Lim, L.C. Kimerling, J. Shin, F. Cerrina, Fabrication of ultralow-loss Si/SiO2 waveguides by roughness reduction. Opt. Lett. **23**, 1888 (2001) 191
11. Q. Xu, B. Schmidt, S. Pradhan, M. Lipson, Micrometer-scale silicon electro-optic modulator. Nature **435**, 325 (2005) 194
12. Y. Jiang, W. Jiang, L. Gu, X. Chen, R.T. Chen, 80-micron interaction length silicon photonic crystal waveguide modulator. Appl. Phys. Lett. **87**(22), 221105 (2005) 194
13. G. Masini, L. Colace, G. Assanto, in *Germanium Thin-Films on Silicon for Detection of Near-Infrared Light*, ed. by H.S. Nalwa. Handbook of Thin Film Materials, Vol. 4 (Academic Press, New York, 2002), p. 331 194
14. J.P. Donnely, A.G. Milnes, The photovoltaic response of nGe-nSi heterodiodes. Solid-State Electron. **9**, 174 (1966) 194
15. S. Luryi, A. Kastalsky, J.C. Bean, New infrared detector on a silicon chip. IEEE Trans. Electron. Devices **ED-31**, 1135 (1984) 194
16. L. Naval, B. Jalali, L. Gomelsky, J.M. Liu, Optimization of Si1-xGex/Si waveguide photodetectors operating at lambda=1.3 m. J. Lightwave Technol. **14**, 787 (1996) 195
17. S.B. Samavedam, M.T. Currie, T.A. Langdo, E.A. Fitzgerald, High quality germanium photodiodes integrated on silicon substrates using optimized relaxed graded buffers. Appl. Phys. Lett. **73**, 2125 (1998) 195
18. L. Colace, G. Masini, F. Galluzzi, G. Assanto, G. Capellini, L. Di Gaspare, E. Palange, F. Evangelisti, Metal-semiconductor-metal near-infrared light detector based on epitaxial Ge/Si. Appl. Phys. Lett. **72**, 3175 (1998) 195

19. M.T. Currie, S.B. Samavedam, T.A. Langdo, C.W. Leitz, E.A. Fitzgerald, Controlling threading dislocation densities in Ge on Si using graded SiGe layers and chemical-mechanical polishing. Appl. Phys. Lett. **72**, 1718 (1998) 195
20. S. Fam, L. Colace, G. Masini, G. Assanto, H.-C. Luan, Appl. Phys. Lett. **81**, 586 (2002) 195
21. T. Yin, R. Cohen, M.M. Morse, G. Sarid, Y. Chetrit, D. Rubin, M.J. Paniccia, 31 GHz Ge n-i-p waveguide photodetectors on silicon-on-insulator substrate. Opt. Express **15**(21), 13965 (2007) 195
22. C. Gunn, G. Masini, Closing in on photonics large-scale integration. Photonics Spectra, Dec 2007 195

Index

Chapter 9
Photonics and Electronics Integration

J.-M. Fedeli, B. Ben Bakir, L. Grenouillet, D. Marris-Morini, and L. Vivien

Abstract Different ways to merge photonics devices on an electronic circuit with microelectronics tools on large-size wafers are addressed. This full integration enables improvement of either the functionalities of the electronic circuits or the miniaturization of optical functions. The above-integrated-circuit fabrication is preferred to a combined fabrication process, with two options ruled by thermal constraints. The high-temperature option is based on wafer bonding on an optical silicon-on-insulator module, and the low temperature option relies on the heterogeneous integration of III–V devices. Compatible with these options, several silicon-based building blocks have been developed, such as slightly etched rib waveguides with low propagation loss (0.1 dB/cm), crystalline and amorphous waveguides, efficient fiber couplers, a 42-GHz Ge integrated photodetector, and a 15-GHz Si modulator. Using molecular bonding of InP dice and processing in a microelectronics environment, InP laser sources have been achieved on silicon.

9.1 Introduction

Silicon-based photonics has generated an increasing interest in recent years, mainly for optical telecommunications or for optical interconnects in microelectronic circuits [1]. The development of elementary passive and active components (input/output couplers, modulators, passive functions, and photodetectors) has reached such a performance level that the integration challenge of silicon photonics with microelectronic circuits has been discussed in the literature [2] and active optics cables are now produced. The rationale of silicon photonics is the reduction of the cost of photonic systems through the integration of photonic components and an

J.-M. Fedeli, B. Ben Bakir, and L. Grenouillet (✉)
CEA-Leti, MINATEC, 38054 Grenoble, France

D. Marris-Morini and L. Vivien (✉)
Institut d'Electronique Fondamentale, UMR8622, CNRS – Université Paris-Sud XI, 91405 Orsay Cedex, France

D.J. Lockwood, L. Pavesi (Eds.): *Silicon Photonics II*.
Topics in Applied Physics **119**, 217–249 (2011)
DOI 10.1007/978-3-642-10506-7_9

electronic integrated circuit (EIC) on a common chip, or in the longer term, the enhancement of EIC characteristics with the introduction of optics inside a high-performance circuit. To achieve such a high level of photonic function integration, the light has to be strongly confined in submicron waveguides with a large refractive index contrast ($\Delta n \sim 2$ typically) between the core and the surrounding environment. When one wants to integrate an electronic circuit with some photonic functions in order to build a photonic electronic integrated circuit (PEIC), the question of how to combine the photonic with the electronic parts is raised. The goal of this chapter is to discuss the routes of convergence of photonics and electronics on large wafers and to present the building blocks developed in our laboratories to address this challenge.

9.2 Ways to Integrate Photonics Devices on an Electronic Wafer

Silicon photonics has been developed thanks to the development of high-quality specific silicon-on-insulator (SOI) wafers. The buried oxide thickness was designed to be thick enough to reduce any optical leakage to the substrate. It is comparable to silica-on-silicon technology, which is in production today. Here, the silicon substrate acts only as a convenient and cheap substrate and the difference resides in the waveguiding layer that is made on silicon. In the 1990s, this SOI technology was pioneered by Bookham with waveguide dimensions typically in the micrometer range and is now in production by Kotura for their different products. The rationale for highly integrated photonics is the reduction of the cost and the increase in performance by merging the photonics and the control electronics parts. Different integration technology options are presented in Fig. 9.1. Each one has its own merit and will be discussed. The hybrid solution consists in assembling EICs with silicon

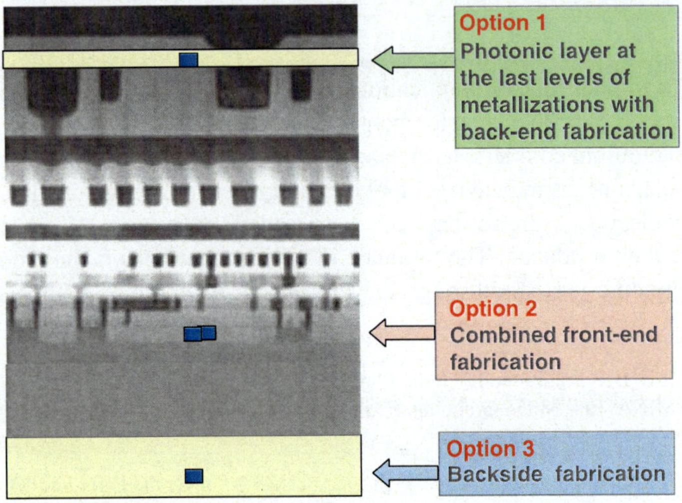

Fig. 9.1 Integration routes

photonic integrated circuits (PICs) by the well developed flip-chip technique. The silicon photonics circuit is considered as a board where the different circuits are attached (laser diodes, drivers, photodetectors, transimpedance amplifiers, etc...). However, full integration offers a reduced assembly and packaging cost reduction and will be only considered in this chapter.

Considering the cross-section of the EIC shown in Fig. 9.1, three options can be considered:

- The first option is often called three-dimensional (3D) or above integrated circuit (AIC) integration, where the photonic layer is realized at the metallization levels with back-end of the line technology (BEOL).
- The second option combines the fabrication of transistors and the photonic devices at the front-end fabrication level (FEOL).
- The third option takes advantage of the flat surfaces at the back-side of EIC wafers, but requires double-side processing.

9.2.1 Above IC Fabrication

With above IC fabrication, the thermal budget is limited to 400 °C in order to avoid any degradation of the EIC, mostly of the interconnections underneath. Two routes were developed, considering the need of introducing devices fabricated with a high-temperature process or using only low temperature fabrication. The high-temperature route takes advantage of the well developed wafer bonding technology. The second one aims to process all the photonic devices with a low temperature and uses the efficient III–V technology.

9.2.1.1 SOI Photonics and Electronic Wafer Bonding

Using the wafer bonding technique, one can introduce a photonic layer at some level in the processing steps of complementary-metal-oxide-semiconductor (CMOS) technology. Since the first metal layers are too densely packed and thin, introduction at the upper metal layers must be considered. The process is described in Fig. 9.2. After fabrication of the last metal layer in an advanced electronic process, a planarized surface can be formed by first coating the surface with a deposited oxide, and then perform chemical mechanical polishing (CMP). On a separate SOI wafer, a photonic circuit is fabricated with silicon waveguides and electro-optical components. After cladding the optical wafer with oxide and planarization with CMP, perfect cleaning of both wafers facilitates their molecular bonding at room temperature. The substrate of the SOI wafer is then removed by mechanical grinding, followed by a chemical etching of Si in order to stop at the buried oxide (BOX) interface. Then, the electro-optical and electrical components are connected by means of silica etching followed by the metal deposition and etch. The alignment between the electrical and the photonic wafers can be performed with present equipment as precise as $\pm 2\,\mu$m. Therefore, the design rules between the metal layers have to take

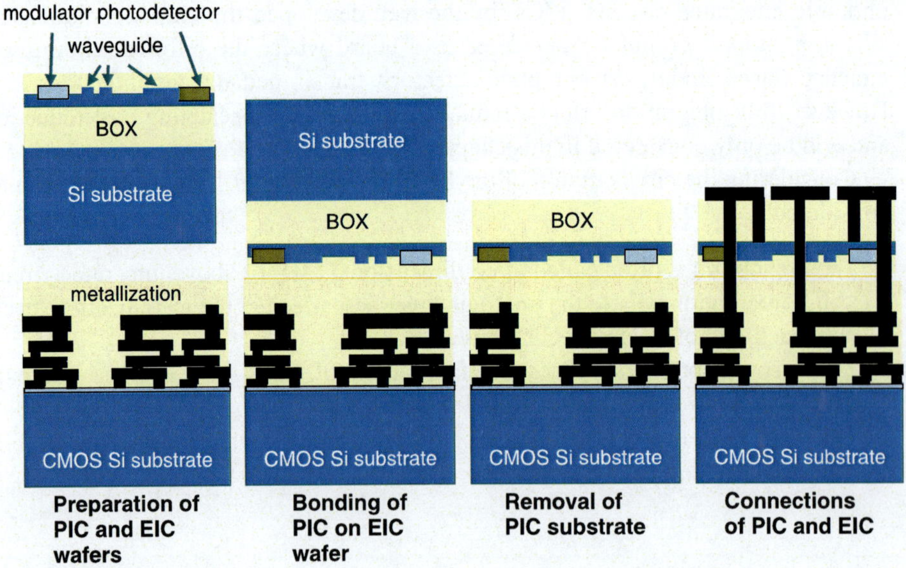

Fig. 9.2 AIC integration with wafer bonding

this alignment margin into account. This technique is often called 3D heterogeneous integration because the electric part is separated from the photonic part without any silicon surface waste at the transistor level. With this approach, any microelectronics technologies can be used for the electrical parts, monocrystalline silicon optical functions such as modulators or Ge photodetectors as well as III–V components can be embedded in the photonic layer and bonded above the integrated circuit.

As a demonstration, on a SOITEC SOI wafer with a 400-nm thick Si layer, we have processed a silicon rib network with cavities filled with germanium. After a SiO_2 cladding deposition, the optical wafer was carefully polished and bonded onto a CMOS wafer coming from STMicroelectronics before substrate removal (Fig. 9.3). A SEM cross-section observation did not reveal any interface defect between the two SiO_2 layers and any degradation of either the metal or the photonic layers (Fig. 9.4).

This wafer bonding technique is a very promising way to integrate a photonic layer into CMOS technology. The wafer bonding technique is mature and the intra connections (3D techniques) between levels of metallization are well addressed by the electronics community. However, the main challenge that needs to be faced is the bonding cost issue compared to the combined fabrication approach.

9.2.1.2 Heterogeneous Integration

As long as the temperature is constrained so that it does not exceed 400 °C, a photonic layer can be defined above the transistors and the dielectric/metallic levels. The obvious way to introduce such a photonic layer is to include it in the same manner as an additional metallic layer in the process flow, i.e., on the top of the stacked layers that have been used for the electrical interconnect. For the passive

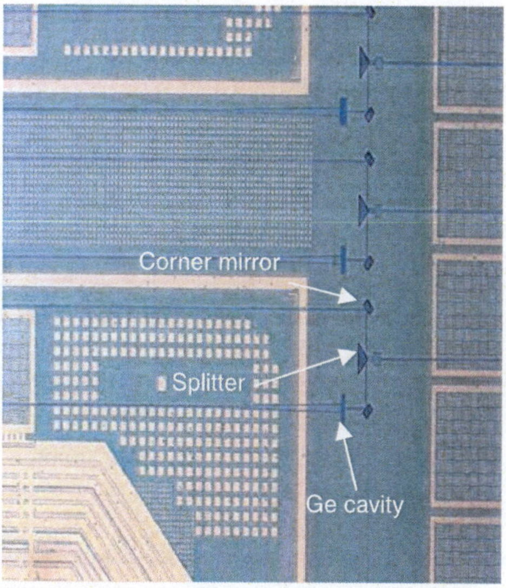

Fig. 9.3 Rib waveguides with splitters, corner mirrors, and rectangular cavities filled with germanium on a CMOS wafer above metallization

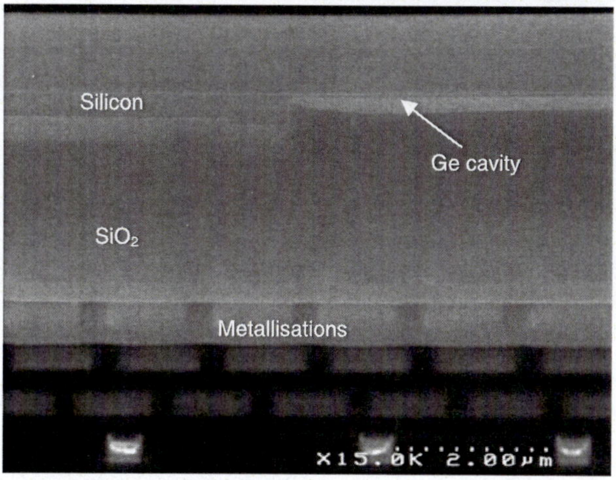

Fig. 9.4 Cross-section of a CMOS wafer with metal levels and a bonded photonic layer with rib waveguides and rectangular cavities filled with germanium

circuitry, one has to rely on a deposited waveguide process, such as SiON, SiN, or hydrogenated amorphous silicon layer, which has a high-contrast index. For the active parts, efficient III–V compounds materials can be bonded and processed to form lasers, modulators and photodetectors coupled to the silicon passive circuitry. Like the introduction of copper in a microelectronics circuit, contamination has to be controlled and some parts of the process have to take place in a dedicated part of the microelectronics clean room. After the amorphous silicon waveguide circuitry is defined, a silica deposition followed by a CMP planarization and a surface functionalization are performed. Active heterostructures grown on InP substrates are bonded on silicon without the need of precise alignment. The InP substrates of these dies are then removed by selective chemical etching and further processing steps are performed that lead to sources and detectors connected to the metallic interconnects of the integrated circuit.

9.2.2 Combined Fabrication

The integration of optical functions that are compatible with microelectronic process technologies presents new and powerful potentialities for integrated circuits. However, a monolithic integration of dissimilar functions still remains a difficult technological challenge. With a combined integration scheme, the new components (waveguides and optoelectronic components) have to be fabricated aside from the transistors. So the process has to take into account both constraints for the electronics and the photonics such as withstanding thermal treatment. Starting with the substrate, photonics components need a separation of larger than 1 μm between the waveguide core (thickness between 200 and 400 nm for a submicron waveguide) and the silicon substrate to avoid light leakage. On the contrary, CMOS technologies are based on either a bulk type substrate or an SOI type with thin BOX and silicon layers (50 nm of Si on a 145-nm BOX, for example). So, as regards the microlectronics technology used, different choices of substrate can be relevant (Fig. 9.5). As the thickness of the BOX is defined by the photonic parts, a modified CMOS technology can be developed using an SOI substrate with around a 1-μm-thick BOX and a thick

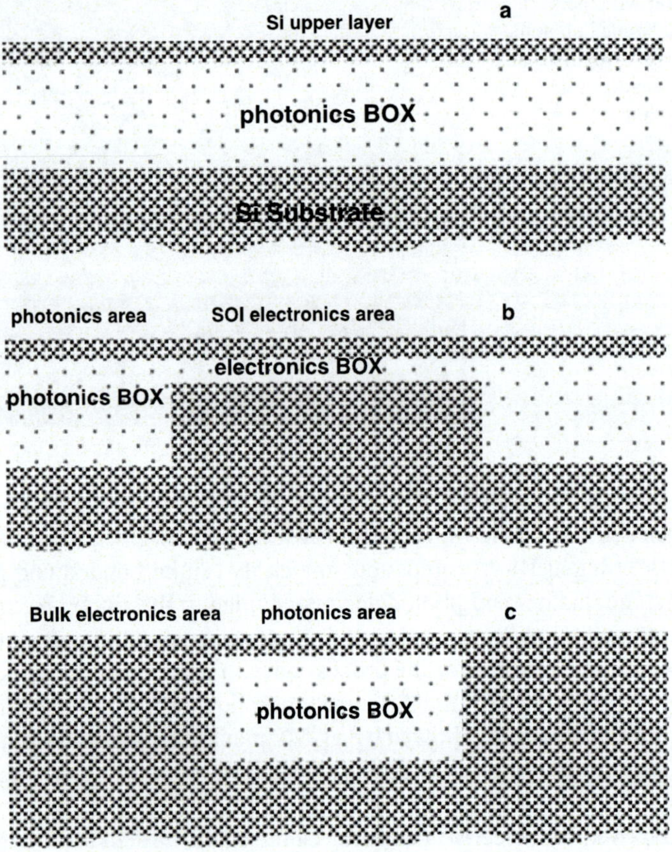

Fig. 9.5 Substrates for combined fabrication

silicon layer (Fig. 9.5a). By engineering the SOI fabrication, a localized thick BOX area for the photonics area can coexist with a thinner BOX for the electronics area (Fig. 9.5b). However the cost could be a stopper. Another option, in order to use bulk electronics technology, consists in defining a photonics area on a bulk substrate where a buried oxide or a cavity (Silicon On Nothing) is formed under a silicon layer (Fig. 9.5c). To combine both process steps, the thermal budget rules the sequence of fabrication. An analysis of the process flow for both technologies reveals that high temperatures ($\approx$1,000 °C) are necessary for the shallow trench isolation and implant activation as well as for the optimization of waveguide losses. Medium temperature ($\approx$700 °C) steps are used for gate oxide, implant anneal and for active photonic layers like a silicon-doped region for modulation and Ge epitaxy for photodetection. BEOL technology is used for metallization on both, with an obvious savings in process steps. So combining steps for the electronic and the photonic parts in order to avoid redundant steps is feasible on an optical SOI substrate, leading to a photonic SOI technology (PSOI), as successfully demonstrated by Luxtera. However, some process steps like higher temperature annealing of material grown at a medium temperature to improve its structural property are forbidden with a consequence of reduced performance. As the microelectronic process is very mature, the introduction of a new photonic part in a large CMOS foundry requires a lot of effort for changing the process. Low and medium scale IC foundries are more suitable to accept such modifications, as they can differentiate their process and address new markets. Moreover, this combined fabrication is fixed for one particular electronic technology and not compatible with other technologies (SiGe, sSOI, GOI, etc.). As an example, a typical 130-nm CMOS technology suitable for 10-GHz components may not be suitable for 40-GHz devices, due to limitations in transistor performance or voltage for amplifiers and drivers.

9.3 Back-Side Fabrication

Fabrication of a photonic layer on the back side of the EIC can be considered and developed thanks to through-silicon vias (TSV). After electronics circuit fabrication on the front side, connections through the substrate (100–200 μm thick typically) allow using the back side of the wafer as another substrate for integration of a different technology (sensors, MEMS, photonics, etc.). Then the solutions for AIC integration described in Section 9.2.1 with either wafer bonding or low temperature processing are applicable. So the main challenge lies in the ability of forming TSV with a high aspect ratio (depth versus width). A low ratio limits the frequency operation in the MHz range, which is incompatible with applications of silicon photonics for communications or optical interconnects with operation in the 10-GHz range. However, for other applications like sensing, the circuit can be connected by flip-chip on a board, leaving the back side accessible for the elements to be detected (biomolecules, gases, ...) or for assembly with microtubes (microfluidics, lab-on-a chip, ...).

9.4 Passive Photonic Circuitry

9.4.1 Guided Structures

According the integration schemes, SOI or amorphous photonic structures should be considered to guide, turn, split, and distribute optical signals.

9.4.1.1 Silicon-on-Insulator Devices

The propagation of the light in a silicon circuit can be done using either rib (partial etching of the silicon film) or strip (full etching of the silicon film down to the buried oxide film) waveguides. The latter allows achieving the highest compactness of the optical mode but propagation losses are higher than those obtained with rib structures. Numerous studies have been done on both kinds of waveguides. Most of the strip waveguides studied in the literature have a height of about 220 nm and a width lower than 500 nm to have single-mode propagation [3–5]. These waveguides allow reaching a very high optical mode confinement and a very low crosstalk between neighboring waveguides separated by 1 μm. Such a strip structure is often used to define optical fiber couplers based on an inverted taper [6, 7] and also for the heterogeneous integration of active III–V material for the provision of an optical source on silicon [8]. However, they suffer from a high-propagation-loss level induced by the lithography and etching processes and the difficulty in reducing these losses to a negligible level.

Rib waveguides are mainly used in the realization of silicon-based optoelectronic devices, such as optical modulators, photodetectors, switches, etc., and also to propagate the light for a long distance on a silicon chip. Indeed, rib SOI waveguides are much less sensitive to scattering losses due to the low interaction between the optical mode and the side-wall roughness [9].Typically, the height and width of such a waveguide are about 400 and 600 nm, respectively, and the etching depth is adjusted to have a single-mode and low loss propagation. Propagation losses as low as 0.1 dB/cm have been obtained using a slightly etched rib waveguide (etching depth: 70 nm) [10]. Such a low loss level has been reached by performing specific process steps to reduce the sidewall roughness, which consists of a 10-nm thermal oxidation at 1100°C, followed by a de-oxidation, and followed again by a second oxidation. Vacuum hydrogen annealing can also be used to reconstruct the silicon edges before thermal oxidation.

From the rib geometry, compact 90° turns and beam splitters can be defined (Fig. 9.6a). To make a direction change of the light, a corner mirror is preferred in comparison to a bend for which the curvature radius has to be large enough to avoid light radiation toward the silicon slab. A corner mirror is formed by fully etching the silicon guide at a 45° angle down to the bottom silicon oxide layer. The theoretical and experimental losses are 0.1 dB and less than 0.5 dB, respectively [11]. The loss can be reduced by improving the ability to vertically etch the silicon waveguide without roughness. To split the light, several possibilities are available: multimode

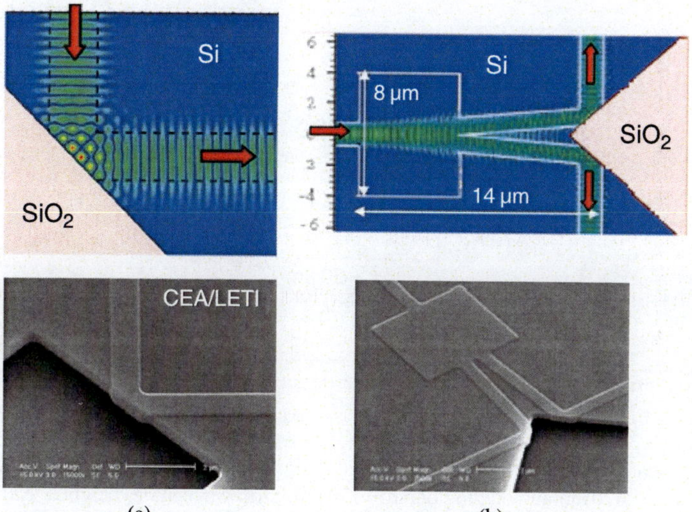

Fig. 9.6 3D-finite-difference time-domain (FDTD) simulations and scanning electron microscope views of (**a**) a corner mirror and (**b**) a compact T-splitter based on rib waveguides

interferometer (MMI), Y-junction, and star couplers, for example. MMI splitters are rather long for rib waveguides and Y-junction splitters are difficult to fabricate because of their small tip. Low loss and compact T-splitters have thus been conceived to be compatible with the minimum size of the deep-UV lithography equipment (Fig. 9.6b). Such a splitter is very compact (8 μm wide and 16 μm long) and the theoretical and experimental excess losses are 0.2 and 0.5 dB, respectively [11]. Such rib structures allow defining passive optical links to distribute light everywhere on a silicon chip [11, 12].

In future photonic circuits, both strip and rib waveguides will be used: strip for fiber coupler and rib for optoelectronic devices, for example. To harmonize such a circuit, optical transitions from strip to rib waveguides and vice versa have to be defined. An example is given in Figs. 9.7 and 9.8.

The optical transition from a rib waveguide (height = 390 nm, width = 420 nm and etching depth = 170 nm) to a strip waveguide (height = 220 nm, width = 460 nm) has been designed, fabricated using deep-UV lithography and characterized at the wavelength of 1.55 μm. The length of the transition is 50 μm, which is long enough to avoid guided mode mismatch. Furthermore, the etching depth of the rib waveguide (170 nm) is chosen to perfectly reach the strip waveguide height. Theoretically, such a transition is lossless (determined by FDTD simulation). Experimentally, losses are lower than 0.1 dB in the wavelength range between 1.45 and 1.6 μm [13].

Passive photonic structures are becoming more and more well-defined and impressive results have been obtained to guide, turn, split, and distribute light on a SOI substrate. This technology can be considered in several integration approaches.

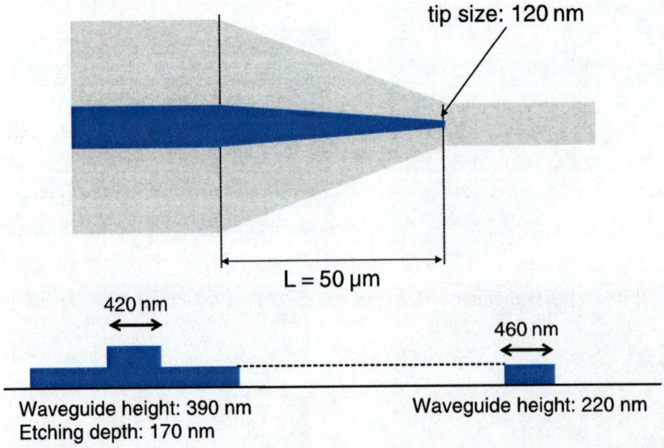

Fig. 9.7 Optical transition from a rib waveguide to a strip waveguide. Rib waveguide: height = 390 nm, width = 420 nm and etching depth = 170 nm. Strip waveguide: height = 220 nm, width = 460 nm. The length of the transition is 50 µm

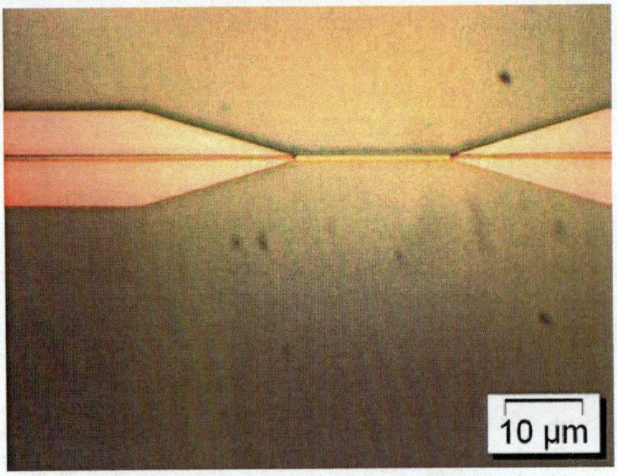

Fig. 9.8 Electron microscope view of a transition from rib to stripe and a transition from stripe to rib waveguides

9.4.2 Amorphous Silicon Waveguide Fabrication

For the low temperature processing integration option, amorphous silicon (a:Si) films were deposited by a capacitively coupled plasma reactor, with a RF excitation frequency (13.56 MHz). The power can be tuned from 30 to 1200 W and the operating pressure can be varied from 0.2 mtorr to a few torr. All films were deposited at temperatures lower than 400 °C to avoid damage to the interconnect layers. Tetraethyl orthosilicate (TEOS) was used as the precursor for oxide deposition and a silane/H_2 mixture for the amorphous silicon. By monitoring the power, the index of the layer was adjusted to be close to the monocrystalline Si one: $n = 3.4913$ at $\lambda = 1310$ nm and $n = 3.4633$ at $\lambda = 1,550$ nm. Sheet optical guided losses were measured during the fabrication process using a prism coupling technique (METRICON 5010) at $\lambda = 1.3$ and 1.55 µm. By optimising the H_2/silane ratio in the deposition chamber, silicon films with losses as low as 0.2 dB/cm at $\lambda = 1.55$ µm

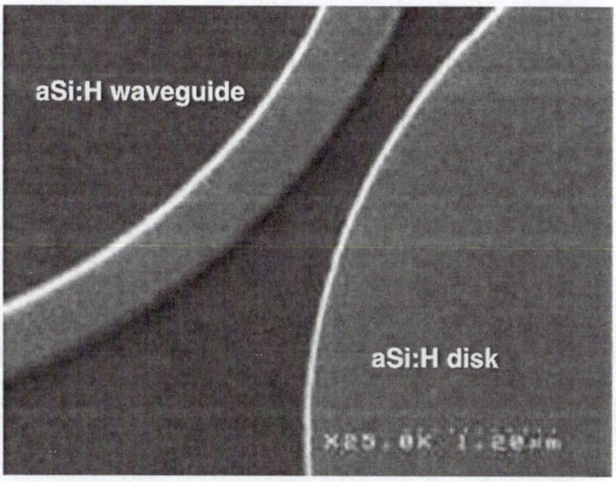

Fig. 9.9 Amorphous Si waveguide with resonating disk

after 350 °C annealing were deposited on silicon wafers covered with 1 μm of TEOS. Deep-UV 193 or 248 nm lithography with or without a hard mask and HBr silicon etching were used to define the waveguide and basic passive functions for optical links (Fig. 9.9). A 1-μm thick SiO_2 oxide layer was deposited to provide an upper cladding. Measurements were performed at a spectral range between 1.25 and 1.65 μm. Results are compared to previous SOI waveguide data [14] and the losses are comparable to that of an SOI waveguide. We notice that for this a:Si waveguide, the losses are respectively equal to 5 and 4 dB/cm for the wavelengths of 1,300 and 1,550 nm. We can consider that these losses are essentially due to the diffraction phenomenon resulting from the side wall roughness of the waveguide. For the waveguide of 800 nm width, the losses become very weak, lower than 1 dB/cm for wavelengths close to 1,300 nm and tend toward the values of a planar waveguide for both types of waveguides (with or without thermal annealing at 350 °C). This shows that the material has a good stability in time. The most important features for a:Si circuitry is the easy possibility to pile up layers and therefore to open up new design concepts or to ease designs such as crossings or coupling. However, the main limitation is due to the limited silicon processing that can be used (lack of ion implantation for doping, silicide, etc.)

9.4.3 Optical Couplers

The coupling of submicron photonic devices and a single-mode fiber (mode diameter = 10 μm) is a real challenge due to their optical mode mismatch. Several strategies are being studied to overcome this issue using either spot size converters [6, 7] or grating couplers [15–19]. The main objective is to adapt the mode sizes between a silicon-based waveguide and a single-mode fiber.

9.4.3.1 Grating Couplers

For several years, grating couplers have been studied for coupling light from and to a submicron waveguide. The main advantages of such a structure are that it is possible to inject light anywhere on an optical circuit and not only at the chip edges, contrary to tapers. Thus, polished facets and anti-reflection coatings are no longer necessary. They can also be used to couple light in every kind of strip or rib waveguide geometries and using any materials, especially silicon-based materials (Si, a-Si, Si_3N_4, etc.) [15–18]. Furthermore, grating couplers can be used to carry out wafer-level testing of photonic devices. This point is crucial for application developments. In the following, theoretical and experimental results for silicon and silicon nitride grating couplers [20] are given at the wavelength of 1.31 μm for TE polarization (Fig. 9.10). The coupling efficiency depends on the bottom and top cladding silica layer thicknesses as well as the silicon-based film thickness. In general, the approach used for the related optimization is based on a differential analysis allowing the determination of the optical field under the grating for the input plane wave [18].

For the two examples presented in Fig. 9.10, the wavelength is 1.31 μm and the duty ratio is 50%. The periods of the silicon (380-nm-thick silicon film) and

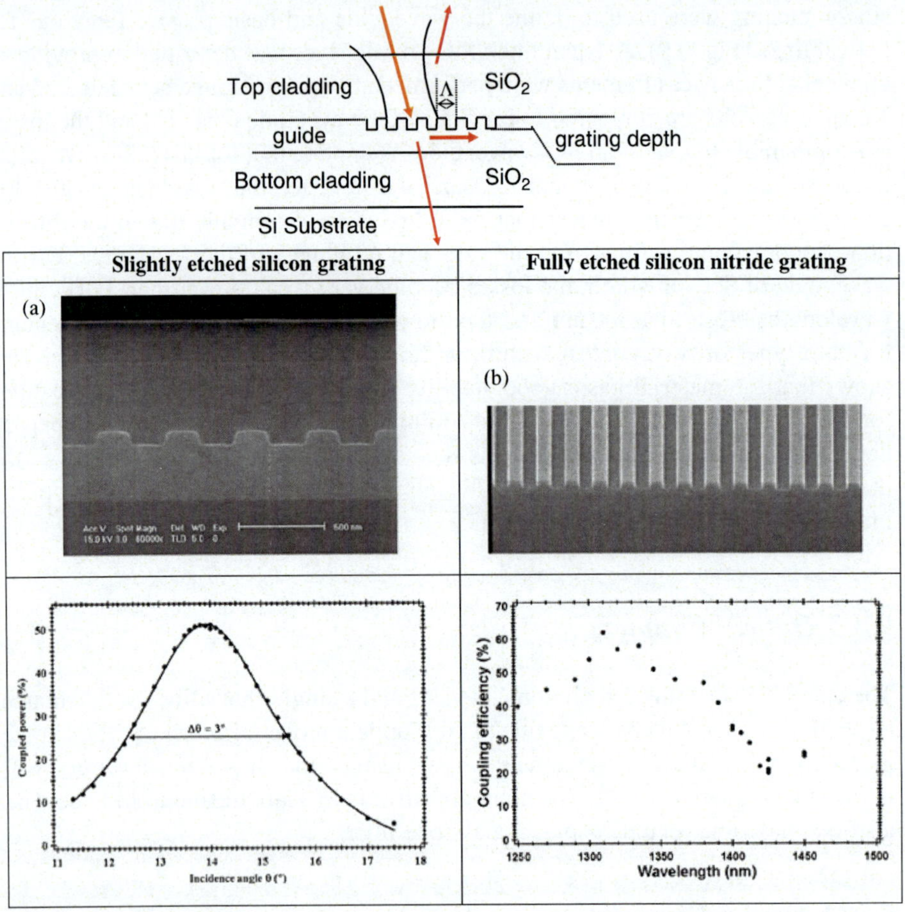

Fig. 9.10 Schematic view of grating coupler: Λ is the period. (**a**) SEM view and coupling efficiency versus incident angle of a slightly etched silicon grating coupler. (**b**) SEM view and coupling efficiency versus wavelength of a fully etched silicon nitride grating coupler

silicon nitride (300 nm thick silicon nitride film) grating couplers are 430 and 1 μm, respectively, and the etching depths are 70 nm (slightly etched) and 300 nm (fully etched), respectively. The resonance angle is chosen to be close to 10° for both couplers.

The coupling efficiencies measured under both silicon and silicon nitride grating couplers are higher than 50 and 60%, respectively, for TE polarization at the wavelength of 1.31 μm. Similar results can also be achieved at a wavelength of 1.55 μm. Furthermore, the angular and wavelength tolerances are higher than 3° and 20 nm, respectively, for both configurations. By optimizing all layer thicknesses, etching depth, grating geometry, etc., an insertion loss lower than 1 dB in the telecom wavelength range can be achieved [2]. The remaining drawbacks of grating couplers are a limited optical bandwidth and their high polarization-dependence.

9.4.4 Efficient and Polarization Insensitive In-Plane Fiber Couplers

Direct in-plane light coupling from a single-mode fiber (mode diameter of 10 μm) into a submicron silicon waveguide is rather difficult because of the weak overlap between the waveguide mode and the fiber mode. Furthermore, another issue is the polarization management. Depending on their cross-section (waveguide height, width, and etching depth), silicon waveguides are generally highly birefringent, and on the other hand, the polarization in fiber-based networks is unpredictable and varies continually with time. The design of an interface building block that enhances light-coupling for all polarization states is therefore a critical issue.

An alternative approach to grating couplers, already extensively discussed in the literature [21, 22], is to use a spot-size converter that transforms gradually the silicon waveguide mode into a wider mode supported by a low-index-contrast waveguide (such as polymers or oxides) and that properly matches a lensed fiber [2, 3], as described in Fig. 9.11. For the common cases where SOI waveguides exhibit a large aspect ratio, silicon couplers are targeted to operate in a single polarization regime (TE, in particular). As it is discussed below, for specific designs, a high-coupling efficiency could be attained for both TE/TM polarizations.

The coupling scheme is presented in Fig. 9.11. The design consists of a tapered nano-wire collector embedded in a low index contrast injector that adiabatically transforms the lensed-fiber mode into a highly confined mode. A silicon strip waveguide of 500 nm width and 220 nm thickness is tapered down to 80 nm by means of deep UV-193 nm lithography and RIE etching techniques. The linear variation of the waveguide width is carried out over a length ranging between 200 and 300 μm, depending on the device design. A 3.5-μm-thick layer of silicon rich oxide (SiOx) is then deposited by PECVD. The amount of silicon nanocrystals in the layer is tailored to obtain a refractive index around 1.6, i.e., close to that of the fiber cores. Then, the thick layer is partially etched (∼1.5 μm) to form a rib-waveguide shaped injector. The unit is encapsulated in silica, and finally planarized by means of CMP.

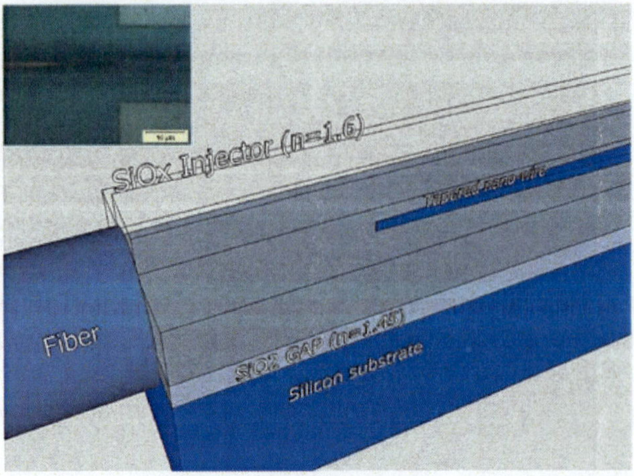

Fig. 9.11 3D representation of the fiber coupler. Insert: Nomarsky image of the silicon nano-tip

This coupling mechanism is based on a phase-matching condition of the fundamental modes of the SOI waveguide and the SiOx injector (for the sake of clarity, only the TE polarization case is presented). Figure 9.12 shows the operating principle of the SOI adiabatic taper structure. The TE effective indices of the supermodes of the entire structure, as well as the effective indices of the local modes, which are the modes of the uncoupled waveguides, are plotted versus silicon waveguide width. When the local mode of the injector excites a supermode and that supermode is transformed adiabatically over the phase matching region, light is coupled to the silicon waveguide. At the tip, the supermode field profile is delocalized from the SOI waveguide core. This delocalization of the field profile ensures a strong overlap with the local mode of the injector. In addition, almost the entire field resides in the

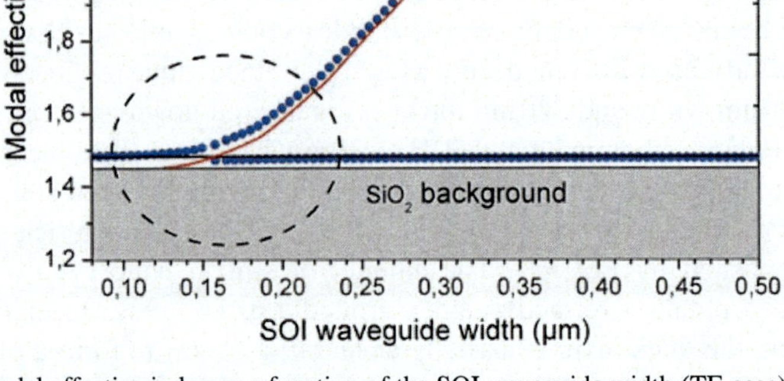

Fig. 9.12 Modal effective index as a function of the SOI waveguide width (TE case). The *dashed circle* is centered on the phase-matching region

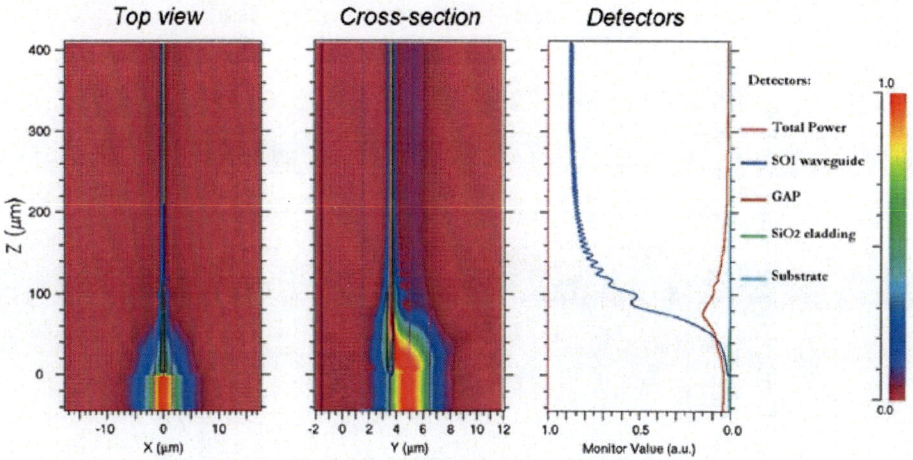

Fig. 9.13 3D beam-propagation-method simulations of the fiber coupler operating at 1.55 μm; electric field patterns and registered power in various detectors (TE case)

SiOx material at the tip, causing the effective index to be close to that of the fiber, which results in negligible scattering losses and back reflections.

Figure 9.13 presents simulation results of a fiber-coupler with a 300 μm taper length, performed at 1.55 μm (TE case). The field patterns clearly demonstrate that the fiber mode easily penetrates into the "injector part" of the device. This gentle transition is ensured thanks to the similar shapes and index contrasts between the lensed-fiber and the injector. The mode is then evanescently coupled into the SOI waveguide along the taper length. It should be pointed out that this design is very robust with respect to variations induced by the fabrication processes. As a relevant example, simulations results indicate insignificant changes of the coupling efficiencies for a taper length variation ranging between 200 and 300 μm. Less than 1 dB coupling losses were calculated over the 1,500–1,600 nm wavelength range.

For the optical characterization of the fabricated devices, a single-mode lensed fiber with a mode field diameter MFD = 3 μm was used as an input reference. The experimental results (Fig. 9.14) demonstrate an excellent agreement with the numerical forecasts. The coupling efficiency remains high in a broad spectral range: the bandwidth at 1 dB is around 100 nm (>300 nm at 3 dB) for both TE and TM polarization states. Future developments will focus on the increase of the cross-section of the injectors in order to achieve an efficient coupling with cleaved single-mode fibers (MFD ~ 10 μm) and on the fabrication of silicon V-grooves at the input for packaging purposes.

9.4.5 *Silicon-Based Optical Modulator*

The optical signal generated by an integrated source has to be encoded to ensure information transmission at high data rates (>10 Gbit/s). This operation can be done in the silicon chip. Indeed, for several years now, impressive progress has

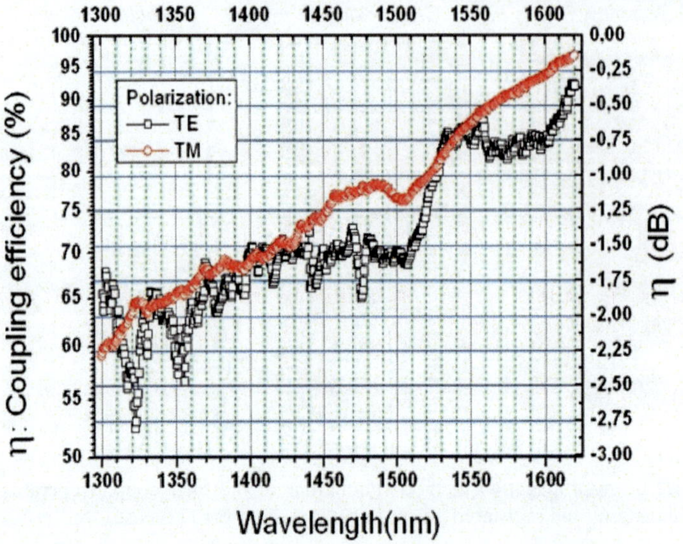

Fig. 9.14 Measured coupling efficiencies for both polarizations

been obtained on silicon-based optical modulators. Several physical effects have been investigated: electro-optical effects in strained silicon [23] and in bulk silicon-germanium [24] or the quantum-confined Stark effect in SiGe/Ge quantum wells [25] have been recently demonstrated. However, most of the fast modulators integrated in a silicon waveguide are based on free carrier concentration variations [26–36]. The first silicon modulator achieving a 1-GHz bandwidth was based on carrier accumulation in a MOS capacitor [26, 27]. Carrier injection has been widely studied, with record data rates of 18 Gbit/s [28, 29]. The most impressive results have been achieved using a carrier depletion phenomenon [30–36]. Indeed, as it is a unipolar effect, it is intrinsically fast (>10 GHz). Furthermore, as a reverse bias is used to sweep carriers out, the electrical current is very low, which leads to low power dissipation. Carrier depletion can be achieved in either a PN or PIN diode using either SiGe/Si modulation-doped quantum wells or all-Si doped layers integrated in a SOI rib microwaveguide.

The principle is the following: at equilibrium (without voltage bias), carriers provided by a doped layer are confined in the microwaveguide. When a reverse bias is applied to the diode, the electrical field sweeps the carriers out of the active region, which is responsible for local refractive index variations. This latter effect induces a phase shift of the guided wave. An interferometer such as a Mach – Zehnder interferometer, a Fabry – Perot microcavity or a microring resonator is used to convert the phase modulation into an intensity modulation.

The main frequency limitation of such a modulator is due to a resistance – capacitor product limitation. Indeed, the reverse biased active diode is a capacitance (C) in series with an access resistance (R) due to the doped silicon regions between the metal and the active region. The device is electrically equivalent to a low pass filter, with a cut-off frequency of $1/(2\pi RC)$: R and C have to be minimized to increase the cut-off frequency. As the modulator capacitance is directly proportional to the active region area, this area has to be as low as possible. Resistances mainly depend on access region geometries and doping levels. However, as the optical loss increases with doping level, a trade-off has to be found between low optical loss and low electrical resistance.

Figure 9.15 presents an example of an all-silicon optical modulator using carrier depletion variation. It is based on reverse bias PIN diode where a thin P^+ doped slit, localized in the intrinsic region, is inserted in the middle of the rib waveguide. With such a structure, a good overlap between the optical mode and the P-doped slit is then obtained to reach good modulation efficiency [30]. Under a reverse-bias, holes located in the slit at equilibrium are sweep out with a rise time lower than 2 ps [36].

The phase shifter structure based on a lateral PIN diode (Fig. 9.15) allows a low capacitance thanks to a low diode area (phase shifter length × waveguide height). As a large part of the active structure is non-intentionally doped, the optical loss is then reduced.

The optical modulator was fabricated on an undoped 200-mm-diameter SOI substrate with a 1-μm-thick BOX layer. The silicon rib waveguide width was 660 nm, the rib height was 400 nm, and the etching depth was 100 nm, leading to a single-mode propagation of the guided mode at a wavelength of 1.55 μm. The doping concentrations in the p- and n-doped regions of the PIN diode were $10^{18}\,cm^{-3}$. A 100-nm-wide p-doped slit, with a doping concentration of about $10^{18}\,cm^{-3}$ was inserted in the intrinsic region. The silicon modulator was based on an asymmetric Mach – Zehnder interferometer, as illustrated in Fig. 9.16. The phase shifter was inserted in both arms over a length of 4 mm, and coplanar waveguide metallic

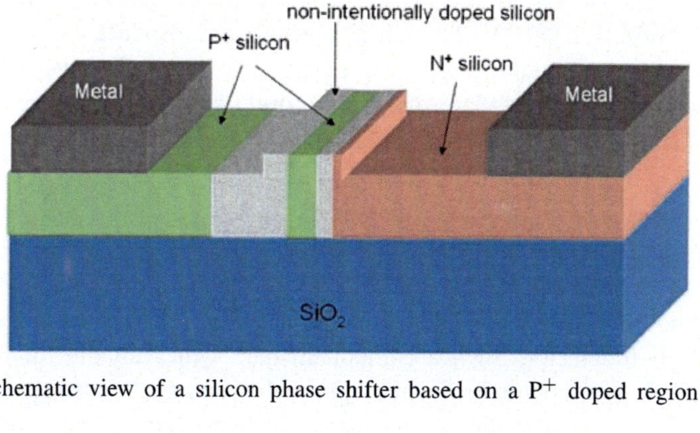

Fig. 9.15 Schematic view of a silicon phase shifter based on a P^+ doped region in a lateral PIN diode

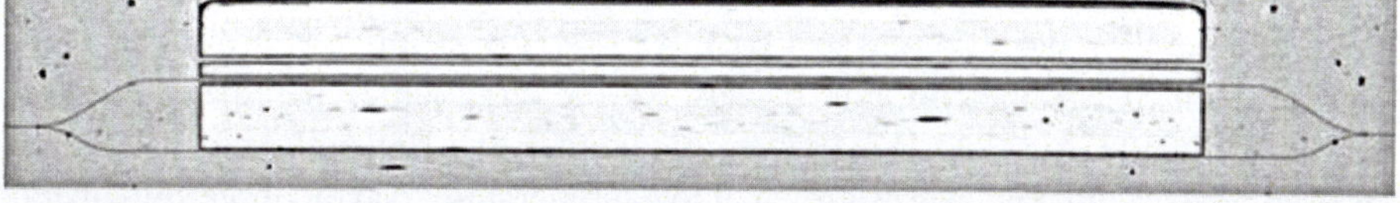

Fig. 9.16 Asymmetric Mach – Zehnder interferometer with the active region localized in both arms

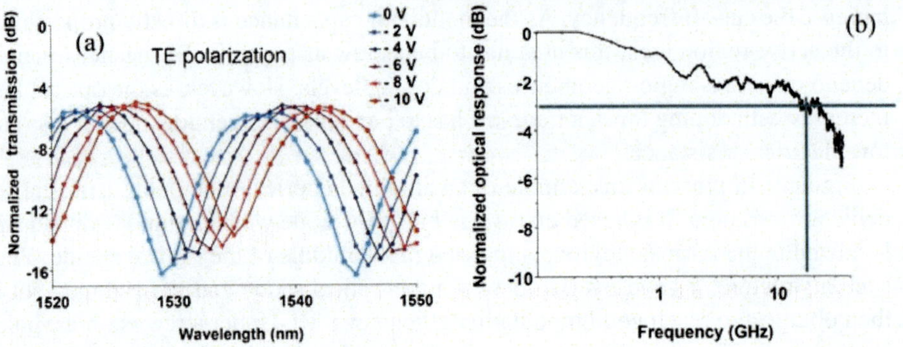

Fig. 9.17 (**a**) Experimental transmission of the modulator as a function of the wavelength, for several reverse bias voltages. (**b**) Normalized optical response of the modulator as a function of the RF signal frequency

electrodes were used to bias one arm [30]. The whole process is fully compatible with microelectronics technology and could be transferred to high-volume microelectronic manufacturing.

Very low values of the reverse current (−2 μA at −10 V) are obtained, which ensures low electrical power dissipation in DC configurations. The transmission of the optical modulator is obtained using a tuneable laser around 1,550 nm for TE polarization. The normalized output spectra of the modulator were measured for a bias from 0 to −10 V (Fig. 9.17a). The effective index variation induced in the phase shifter is due to hole depletion, which creates a red-shift of the spectrum. The insertion loss is as low as 5 dB for a 4-mm-long active region. At a given wavelength, a large transmission variation is obtained that leads to a DC extinction ratio larger than 15 dB from 0 to −10 V. The normalized optical response as a function of the frequency of the modulator is given in Fig. 9.17b. A −3-dB cut-off frequency of 15 GHz has been obtained. The speed limitation of the tested device is not due to carrier transport mechanisms, but rather to the electrical supply.

9.5 Germanium Waveguide Photodetectors

The last photonic element needed to complete the high-speed optical link is the integrated photodetector. Such a component has been available for several years from the III–V semiconductor technology on InP and GaAs wafers. Nevertheless, its integration on large wafers within the mainstream silicon technology requires a hybrid integration approach [36–38]. Within group IV materials, silicon is transparent throughout the required wavelength range (1.25–1.65 μm) and only pure germanium allows light detection up to 1.6 μm [39–41]. Basically, the growth of a thick Ge layer on Si is rather difficult because of the large lattice parameter mismatch between Si and Ge (≈4.17 %). To overcome this, a two-step epitaxial growth of Ge is used. First, a thin Ge layer is grown at low temperature to avoid three-dimensional growth and to concentrate most of the dislocations. This first step allows a growth at a higher temperature of a thick Ge layer with a good structural quality [42, 43]. The absorption coefficient of such a Ge layer is very close to that of bulk Ge ($\sim$10,000 cm^{-1} and 5,000 cm^{-1} at 1.31 and 1.55 μm, respectively). A band-gap shrinkage that leads to absorption up to 1.6 μm is then observed. This red shift of the absorption edge is directly related to a tensile strain induced in the Ge layer. The measured hole mobility is close to 1,300 V cm^{-2} s^{-1} with a residual carrier density smaller than 10^{16} cm^{-3}. Furthermore, the carrier lifetime is estimated to be much higher than the carrier collection time, which avoids the possible recombination of the photogenerated carriers.

Two kinds of germanium on silicon photodetectors have been demonstrated, including MSM (metal-semiconductor-metal) or PIN structures: a surface illuminated photodetector [3–5] and an integrated photodetector in a SOI waveguide [44–49]. Embedding Germanium in a SOI waveguide allows monolithic integration in a photonic circuit. Furthermore, high responsivity and high bandwidth can be simultaneously obtained for waveguide photodetectors.

Two integration schemes for the Ge photodetectors in SOI waveguides can be considered: either evanescent or butt coupling of light from the Si waveguide to the Ge layer. For both integration methods, the direct coupling allows reducing the absorption length down to few microns without any considerations on the Ge layer thickness.

Recently, a 42-GHz waveguide Ge photodetector has been designed, fabricated, and characterized at a wavelength of 1.55 μm (Fig. 9.18) [49].

In such a structure, butt coupling integration has been considered. The rib waveguide width and height are 660 and 380 nm, respectively. The etching depth was about 110 nm. First, a silicon recess is etched down to a 50-nm-thick silicon layer at the end of the rib waveguide. This recess is locally implanted with phosphorus followed

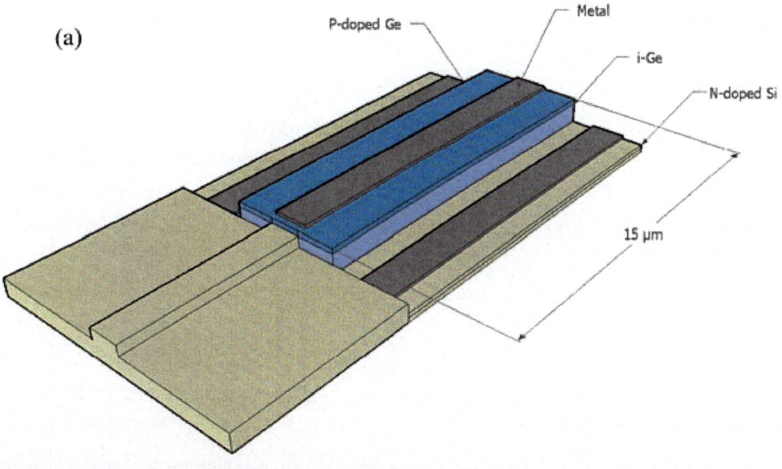

Fig. 9.18 Schematic view of PIN Ge diode integrated in a SOI waveguide

by thermal annealing at 1, 050 °C to form the P^+ ohmic contact. Germanium is then selectively grown by reduced pressure chemical vapour deposition (RP-CVD) in the silicon recess. A 40-nm-thick Ge buffer layer grown at low temperature (400 °C) is followed by a 300-nm-thick Ge film at higher temperature (730 °C) and finally by a 90-nm-thick P-doped Ge layer. An annealing step is performed to reduce the threading dislocation density in the Ge layer (estimated to about 5.10^6 cm^{-2}). Bottom contacts are then defined. A 300-nm-thick silicon dioxide (SiO_2) layer is deposited onto the wafer using plasma-enhanced chemical vapour deposition (PECVD). Openings on both top P^+ and bottom N^+ regions are patterned in the silicon dioxide and a Ti/TiN/AlCu metal stack is deposited to form metallic contacts.

The dark current of the PIN Ge photodiode (15 μm long and 3 μm wide) is as low as 18 nA at a reverse bias of 1 V. That corresponds to a current density of about 60 mA/cm^2. This rather low dark current value is the result of the thermal annealing after the epitaxial growth. However, more efforts should be made to reduce it for low sensing applications.

Under illumination, photocurrent is generated. The responsivity at a wavelength of 1.55 μm is 0.2 A/W without voltage bias and rapidly increases to about 0.9 A/W at 0.3 V reverse bias to finally reach 1 A/W at −4 V. The external quantum efficiency was about 80%. At −0.5 V, the responsivity reaches 1 A/W at a wavelength of 1.52 μm and is still 0.2 A/W at 1.6 μm (Fig. 9.19a).

The −3-dB cut-off point is at a wavelength of around 1.5 μm. The normalized optical responses as a function of frequency, from 0.1 to 50 GHz, are plotted in Fig 9.19b. The −3-dB bandwidths are 12, 28, and 42 GHz at 0, −2, and −4 V biases, respectively.

For several years now, impressive results have been obtained on Ge waveguide photodetectors. The technological process used to fabricate such a photodetector is fully compatible with technologies developed for microelectronic circuits. Future developments will focus on specific requirements for applications such as high-speed optical links between cores of microprocessors or low cost optical communication systems.

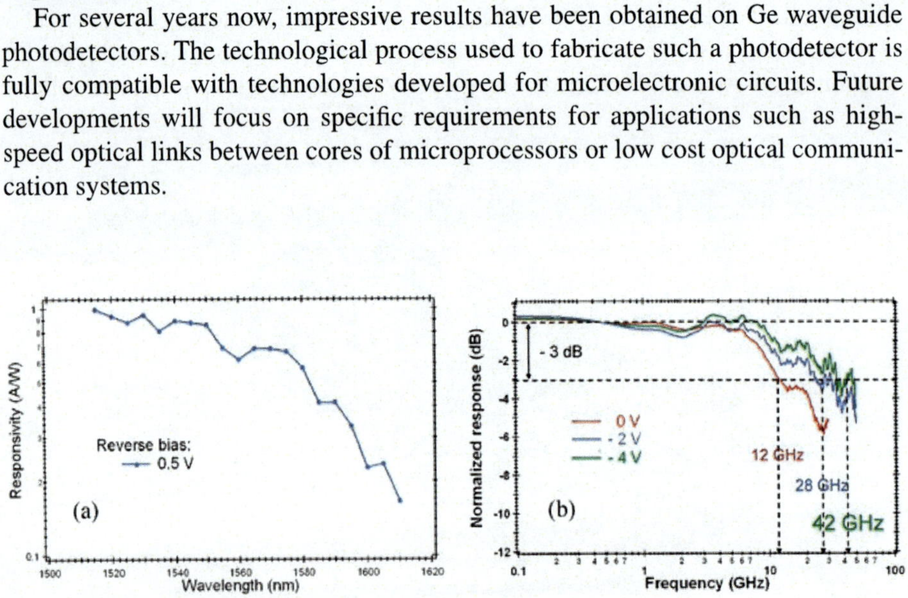

Fig. 9.19 (**a**) Ge photodiode responsivity as a function of wavelength under 0.5 V reverse bias. (**b**) Normalized optical response as a function of the frequency under 0, −2 and −4 V. The photodetector length is 15 μm

9.6 Laser Fabrication on 200 mm Wafer

As efficient modulators can be fabricated, a continuous wave (CW) light source is now needed. A proposed solution is that the light comes from an external InP laser connected with I/O couplers to the passive circuitry of the chip. This requires efficient couplers and expensive packaging. In recently released products [2], the laser source is flip-chipped and the emitted light is injected vertically via a surface grating coupler. The Graal option would be to process a silicon source. Even with the latest developments on Si-based emitters for silicon photonics [50], III–V devices remain today the only solution for lasing. Indeed compared to what happens in silicon – whose (only) disadvantage is to have an indirect bandgap – III–V semiconductors like InP, GaAs, and their associated compounds (GaInAs, GaInNAs, GaInAsP, AlGaInAs) all have direct bandgaps, which means basically that an electron-hole pair created in these kind of active layers has intrinsically around 10^4 times more chance to recombine radiatively into a photon than non-radiatively. Furthermore, InP and GaAs-based materials offer the possibility to address the telecom windows in terms of emission wavelength (1.31 and 1.55 μm). However, the cost of wafers and processing on small diameter InP or GaAs substrates leads to rather expensive components. Therefore, integration of III–V components coupled to passive optical functions on top of an EIC requires new approaches with more integration than the flip-chip approach. To create III–V integration with silicon as intimately as possible, III–V direct epitaxial growth on silicon was studied [51] and lasers on silicon obtained [52, 53]. But due to the high-lattice mismatch between III–Vs and Si, a misoriented Si substrate and/or thick epilayers have to be used, thereby compromising a full compatibility with standard microelectronics. Furthermore, the growth temperature used is not compatible with the above-IC option.

So for the combined fabrication route, light generation is really a difficult issue. With the above-IC route, InP sources can be processed with microelectronics tools after transistor fabrication using bonding of InP dice.

9.6.1 Die to Wafer Bonding of III–V Semiconductor on Silicon

Thus, the issue would be to enable integration of the naked planar III–V-based (active) heterostructures on top of an EIC only where it is required and then to process the III–V in the microelectronics fab. The idea behind this approach is first to reduce the cost of the III–V's introduction on the EIC wafer and second to use the possibilities of microelectronics lines in terms of lithography resolution and alignment to design new generation devices with sub micron design rules. Among these new generation devices, ultra-compact hybrid devices with low power consumption and operating speed $>$10 Gb/s [54] should be mentioned.

Therefore our approach consists in dicing a III–V (usually InP-based) wafer with all the heteroepitaxial layers grown upside down with respect to standard epitaxy. Then the dice are bonded to the required places, and the InP substrate selectively

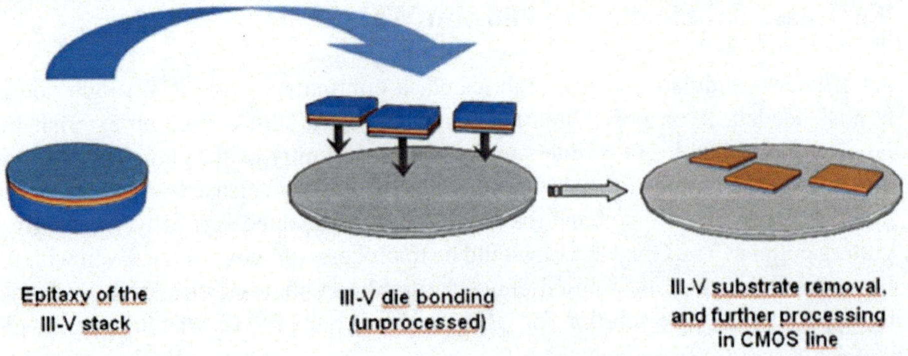

Fig. 9.20 Schematics of the die-to-wafer process

removed in order to leave only the active thin films (~1 μm thick) attached to the EIC wafer, thus enabling processing of InP components on a dedicated 200- or 300-mm fabrication line, as shown in Fig. 9.20. Using this approach where the III–V dice is bonded unprocessed, no precise positioning is required during the bonding step, since the alignment precision between the silicon passive circuits and the III–V active devices is given by the lithography resolution, which is typically ±125 nm for a standard 248-nm deep-UV stepper.

The die-to-wafer bonding technology chosen in our labs is molecular bonding and in particular SiO_2/SiO_2 molecular bonding. This means that SiO_2 is deposited on the III–V dice prior to bonding on a planarized SiO_2 top surface of the EIC. A complete physical model of such a molecular bonding was proposed and presented by Stengl et al. [55] and Gösele et al. [56]. SiO_2/SiO_2 molecular bonding enables good bonding quality without any additional adhesive materials [57], which could inhibit efficient optical coupling. Furthermore, molecular bonding satisfies the requirements better in terms of thermal conductivity and dissipation, transparency at the device working wavelengths and mechanical resistance. SiO_2/SiO_2 molecular bonding was chosen since it is fully compatible with both microelectronics (SiO_2 is a standard material) and the above-IC route in terms of thermal budget (the bonding step occurs at room temperature and the bonding strength can support a thermal budget of 400 °C maximum). However, it requires quasi-perfect interfaces; meaning that surface roughness, defect density, particular contamination, and wafer bowing are critical issues. The required flatness and uniformity can be obtained by use of CMP. The additional role of CMP is to adjust the thickness of the silicon dioxide cladding layer in order to satisfy the optical coupling conditions. The III–V part needs much more "care" since dicing/cleaving of the III–V material can generate a lot of particles, and since III–V epitaxies generally result in emerging defects. The adhesion of the thin SiO_2 layer deposited on the III–V compound has also to be well controlled, as well as its roughness.

Further details on InP-on-Si wafer bonding can be found in [58]. The smallest die size we have bonded is $1 \times 1\,mm^2$. A pick and place apparatus can be used to mount the InP die onto the silicon substrate. As mentioned above, the bonding

Fig. 9.21 A $1.2 \times 1.2\,\text{mm}^2$, 200-μm-thick InP die bonded on an optical layer on a CMOS substrate

Fig. 9.22 Top view of a processed InP heterostructure bonded on silicon passive devices

itself occurs spontaneously at room temperature; however, an annealing at 200 °C for several hours reinforces adhesion. Mechanically thinning the die down to 20 μm was performed after bonding without degrading the quality of the remaining bonded material. The remaining InP substrate and the sacrificial InGaAs layer can then be chemically and selectively back-etched. The additional post-bonding technological steps such as polishing show that the assembled InP dice on the Si substrate can endure many kinds of mechanical maltreatment without debonding. The bond strength between the die and the substrate was measured using Die Shear testing equipment. The shear strength is of 5 MPa ± 1.4 MPa for a 1-mm^2, 360-μm-thick InP dice. Some demonstrations of die bonding are highlighted in Figs. 9.21 and 9.22.

9.6.2 *Design and Fabrication of InP Lasers with Microelectronics Tools*

After the bonding of the III–V heterostructure on 200 or 300 mm diameter EICs, the processing for III–V active devices such as the laser, modulator, and photodetector has to take place inside a microelectronics line. We first discuss in this Section the main building blocks that are currently developed to achieve this goal and then give some examples of InP-based lasers on silicon and investigate in terms of integration which approach will pave the way to a full convergence of photonics and electronics.

9.6.2.1 III–V Process with Microelectronics Tools

The main challenge for the processing of III–V materials with microelectronics tools can be summarized as contamination, which is overcome with careful cleaning procedures. Another issue is the development of III–V dry etching with large-size wafer equipment. Last, but not least, one has also to eliminate the gold-based metals that are usually used to fabricate ohmic contacts on III–V materials, since gold creates a deep level in silicon and is therefore strongly contaminating from a transistor fabrication viewpoint.

To fabricate such lasers on 200 mm wafers, plasma etching of InP and related compounds was investigated on 200-mm processing tools using either HBr, Cl_2/H_2, or CH_4/H_2 based chemistries. The results obtained are shown in Fig. 9.23. HBr etching using inductively coupled plasma (ICP) etching was found to generate no steep profiles, but rather 45° sidewalls (cf. Fig. 9.23a). Furthermore, trenching was observed. These two results are unsuitable for obtaining good quality factors for the lasers and detrimental for electrical injection, respectively. Therefore this kind of chemistry probably needs more optimization. Using Cl_2/H_2 chemistry under ICP plasma mode, steep profiles were achieved, as shown in Fig. 9.23b. However, etching rates of around 2 μm/min were observed, which could induced difficulties in precisely stopping etching on a given epilayer, particularly for thin epistructures. Moreover, etching rates are very sensitive to temperature and the etching quality was found to be very sensitive to the III–V environment (silicon, silicon dioxide), as already reported [59]. This can be a problem when processing an InP dice bonded on a silicon substrate whose SiO_2 versus Si fraction at the surface may vary. Using CH_4/H_2-based chemistry in the RIE plasma mode, a 100 nm/min etching rate was achieved, as well as relatively steep profiles and smooth surfaces (Fig. 9.23c). Furthermore, the etching results seem not to depend on whether the surface exposed to the plasma outside the dice is SiO_2 or Si, making this kind of chemistry suitable for InP/Si laser fabrication.

Ohmic contacts in the world of III–V semiconductors are commonly obtained with Ti/Pt/Au and/or Ni/Ge/Au using a lift-off process. As gold is strictly forbidden in silicon microelectronics, and as lift-off processes are not common in CMOS

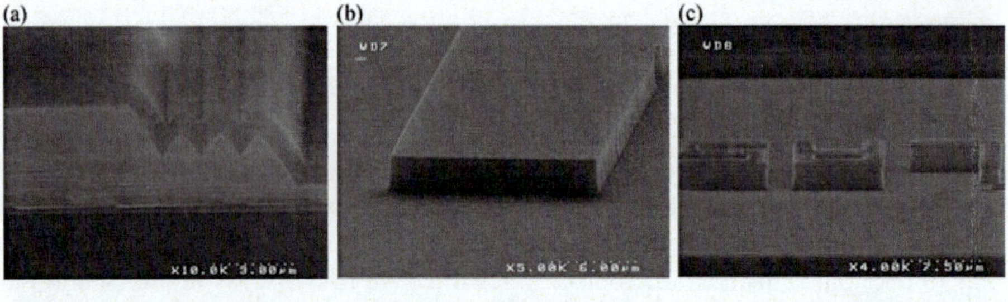

Fig. 9.23 Etching profiles obtained on InP-based dice bonded on silicon using (**a**) ICP HBr plasma, (**b**) ICP Cl_2/H_2 plasma, and (**c**) RIE CH_4/H_2 etching

foundries, an alternative metals deposition and metals etching process must be used for fully CMOS compatible processing. Different metallization schemes, such as Ti/TiN/AlCu or Ni/AlCu, are currently being explored, in terms of process and electrical characterization points of view. The former metallization was used for the microdisk device fabrication detailed below.

9.6.2.2 Fabry – Perot (FP) Lasers on Silicon

Among the different kind of lasers that can be fabricated on silicon, one can distinguish broad area lasers emitting in the milliwatt range and microlasers emitting in the tens of microwatts range dedicated to chip external interconnections or intrachip connections, respectively.

As a demonstration of a laser on silicon, simple broad area FP lasers were fabricated on silicon by means of die-to-wafer bonding, followed by a III–V process technology with facet dry etching. Ti/Pt/Au and Ni/Ge/Au/Ni/Au contacts were deposited by evaporation and lift-off for the p and n contacts, respectively [60]. The 1.55-μm InGaAsP/InP lasers run under continuous wave (CW) operation at room temperature with an output power in excess of 4 mW and a threshold current of 108 mA at 15 °C. Figure 9.24 shows an SEM image of such lasers, whose facets were etched with ICP Cl_2/H_2 plasma, as well as a spectrum obtained at 15 °C and infrared images recorded in stages from below threshold to above threshold. An overall characteristic temperature of 49 K was derived from L-I curves obtained under pulsed operation at different temperatures. We also obtained such FP lasers with facets defined by CH_4/H_2 RIE plasma etching.

Compared to previous results these lasers offer the advantages of having etched facets instead of using dicing and facet polishing, the latter being incompatible with CMOS, and have better thermal characteristics than what can be obtained with benzocylobutene (BCB) as a bonding agent [61].

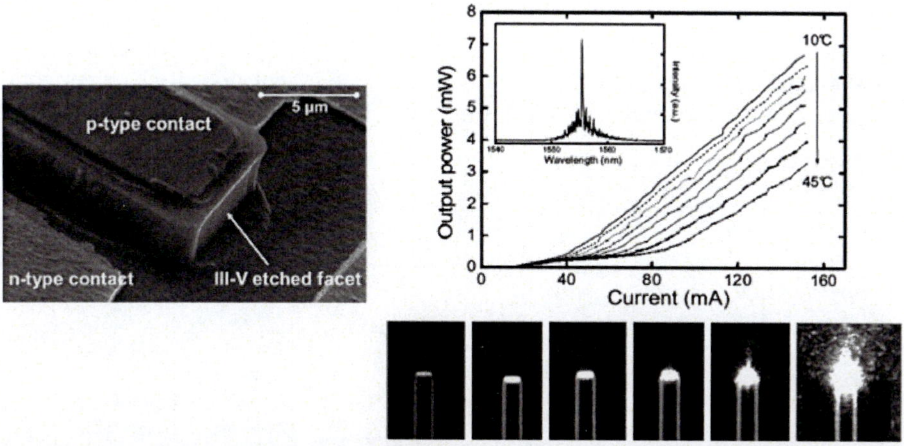

Fig. 9.24 SEM picture of an InGaAsP/InP laser on silicon with ICP Cl_2/H_2 etched facets, L-I curves obtained at different temperatures with a spectrum in the inset obtained at 15 °C, and infrared images of the FP laser obtained at different currents

9.6.2.3 Microdisk Lasers on Silicon

To address intrachip interconnections, microdisk lasers that make use of whispering gallery modes created at their periphery are good candidates, since they are compact (disk radius in the 3–10 μm range) and offer potentially low power consumption. Electrically driven InGaAsP/InP microdisk lasers were successfully obtained [62–64] for the demonstration of a full optical point-to-point link on CMOS [65], within the frame of the European project PICMOS.

Fabrication of an electrically driven microlaser was then studied and consists of microdisks with a vertical P-I-N junction. The process started by a contamination analysis of the 200-mm-diameter wafers after the bonding step and the InP substrate removal, as it was not performed in the same clean room. Then a special decontamination of the rear face of the wafers was performed, in order to avoid any contamination of the chucks of the clean-room equipment. A SiO_2 hard mask of 100 nm was deposited by PECVD. Microdisks were defined with 248-nm deep-UV lithography. Special attention has to be given to the optical focus, due to the presence of the dice on a limited area of the wafer. The hard mask is then etched with InP as the stopping layer. The partial etching was performed with 200-mm ICP equipment using HBr RIE, as it was the only one available at that time. A second lithography step followed by an InP etching defined the slab necessary for the bottom contact. Then 1.5 μm of SiO_2, which is a low index and electrical isolating material, was deposited in place of the BCB used for planarization in the PICMOS demonstration. CMP was then performed to get a planar surface with a 400-nm separation from the upper surface of the InP disk. Even if gold-based contacts have well known properties on InP, CMOS processes are not compatible with such a metal (except for back-end metallization) because of contamination risks. Ti/TiN/AlCu contacts were an alternative solution since they allow a low resistive contact. TLM measurements were performed on a trial InP wafer with a 500-nm-thick $5.10^{18}\,cm^{-3}$ N^+ Si doped layer and showed that the contacts were of ohmic types. Top and bottom electrodes were formed via openings in the SiO_2 to the bottom and upper InP N doped surface and by patterning the electrodes after the Ti/TiN/AlCu deposition. Figure 9.25 shows the final device. Light emission produced by continuous wave (CW) electrical injection at room temperature was observed (see Figs. 9.26 and 9.27), but optical characterization proved that the structure was not lasing, even in pulsed mode, due to the slanted

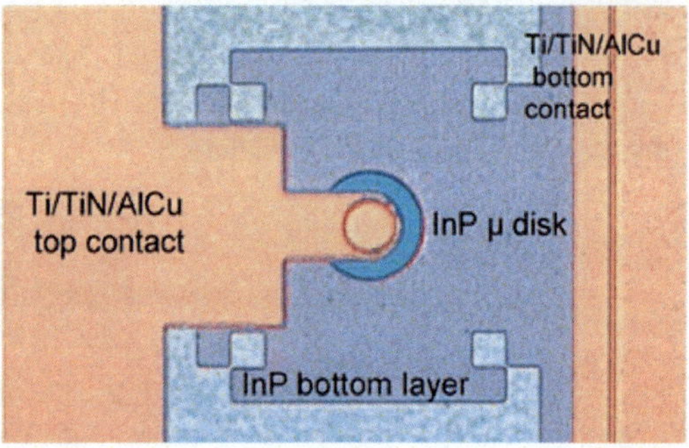

Fig. 9.25 Optical top view of a micro disk

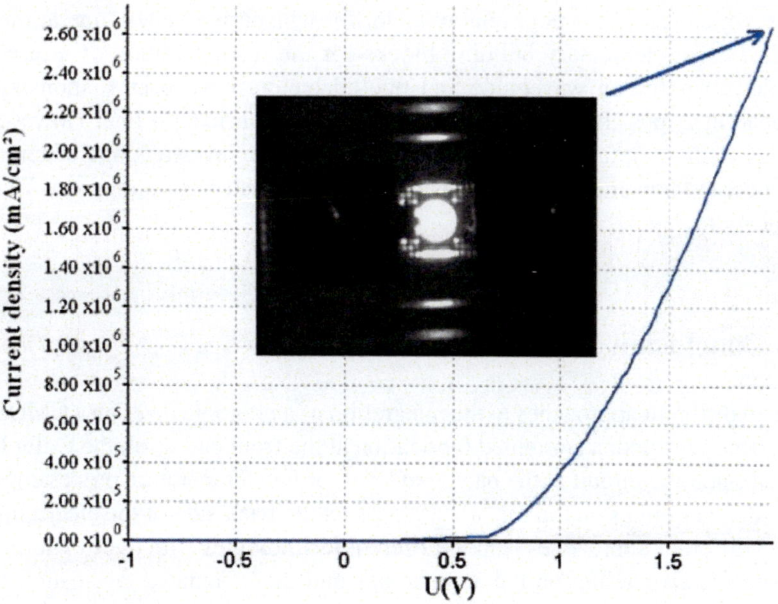

Fig. 9.26 PIN characteristic for Ti/TiN/AlCu contacts

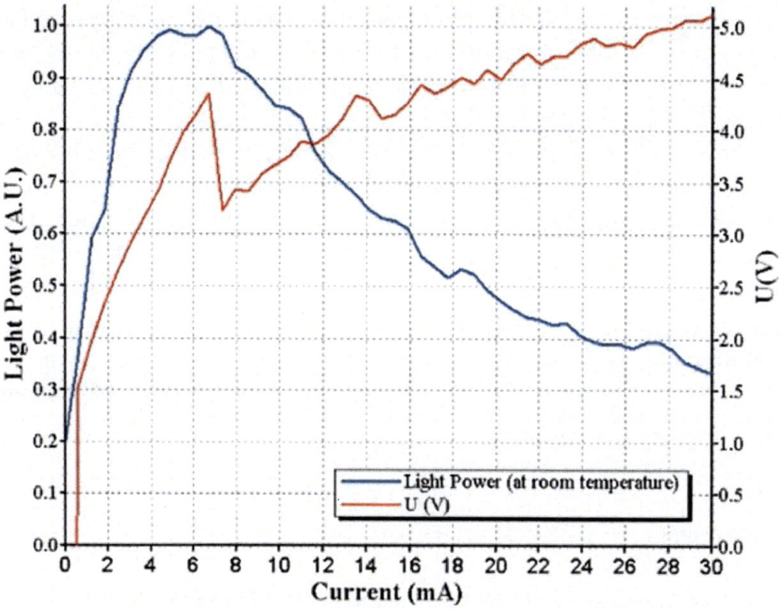

Fig. 9.27 Light power emission at room temperature

edges which result in a poor quality factor for whispering gallery operation. Nevertheless, basic elementary building blocks for the demonstration of a laser source coupled to a silicon waveguide and photodetectors have been demonstrated and fabrication is possible on a 200-mm-diameter Si fabrication line. However, more studies such as optimization of the etching process, investigations of temperature dependency, power range, etc., have to be performed before they can be used in applications

9.7 Conclusion

Several different approaches to the integration of a photonic layer on a CMOS circuit have been reported: a combined fabrication at the front end level, the wafer bonding of SOI photonic circuit at the back-end level, or low temperature processing of photonic devices either on top metallization or on the back side of the electronic wafer. These different approaches lead to different technologies with their own merits and drawbacks. Depending on the applications and the associated fabrication volumes the system designers would be able to choose the best way to make their desired system with the necessary building blocks. With the goal of fabrication with microelectronics, we developed submicron silicon photonics functions using either SiGe technology or the InP heterogeneous bonded one. Due to the cost and intellectual property issues, it is rather difficult to get electronic wafers for research. Thus only one demonstration of integration could be achieved, which was the bonding of a SOI wafer with a silicon rib waveguide and cavities filled with germanium above the metallization of a dummy CMOS wafer. As long as contamination was well controlled with appropriate cleaning process steps, the fabrication of InP devices on top of a silicon wafer with low temperature processes was established with microelectronics tools. This opens the way to the integration of efficient laser sources as well as photodetectors and modulators.

Acknowledgments This paper summarizes the work of numerous other colleagues who could not be all cited and the authors want to acknowledge their contributions. This work is supported by the European community projects FP6-2002-IST-1-002131-PICMOS and FP7-2008-IST-2-HELIOS and by the French RMNT programs "CAURICO."

References

1. G.T. Reed, The optical age of silicon. Nature **427**, 595–596 (2004) 217
2. T. Pinguet, B. Analui, E. Balmater, D. Guckenberger, M. Harrison, R. Koumans, D. Kucharski, Y. Liang, G. Masini, A. Mekis, S. Mirsaidi, A. Narasimha, M. Peterson, D. Rines, V. Sadagopan, S. Sahni, T.J. Sleboda, D. Song, Y. Wang, B. Welch, J. Witzens, J. Yao, S. Abdalla, S. Gloeckner, P. De Dobbelaere, Monolithically integrated high-speed CMOS photonic transceivers, *Proceedings of the 5th International Conference on Group IV Photonics*, Sorrento, 17–19, Sept 2008 217, 229, 237

3. Y.A. Vlasov, S.J. McNab, Losses in single-mode silicon-on-insulator strip waveguides and bends. Opt. Express **12**, 1622–1631 (2004) 224, 229, 235
4. W. Bogaerts, D. Taillaert, B. Luyssaert, P. Dumon, J. Van Campenhout, P. Bienstman, D. Van Thourhout, R. Baets, V. Wiaux, S. Beckx, Basic structures for photonic integrated circuits in Silicon-on-insulator. Opt. Express **12**, 1583–1591 (2004) 224, 235
5. S. McNab, N. Moll, Y. Vlasov, Ultra-low loss photonic integrated circuit with membrane-type photonic crystal waveguides. Opt. Express **11**, 2927–2939 (2003) 224, 235
6. V.R. Almeida, R. Panepucci, M. Lipson, Nanotaper for compact mode conversion. Opt. Lett. **28**(15), 1302–1304 (2003) 224, 227
7. G. Roelkens, P. Dumon, W. Bogaerts, D. Van Thourhout, R. Baets, Efficient silicon-on-insulator fiber coupler fabricated using 248 nm deep UV lithography. Photon. Technol. Lett. **17**(12), 2613–2615 (2005) 224, 227
8. P. Rojo Romeo, J. Van Campenhout, P. Regreny, A. Kazmierczak, C. Seassal, X. Letartre, G. Hollinger, D. Van Thourhout, R. Baets, J.-M. Fedeli, L. Di Cioccio, Heterogeneous integration of electrically driven microdisk based laser sources for optical interconnects and photonic ICs. Opt. Express **14**(9), 3864–3871 (2006) 224
9. F. Grillot, L. Vivien, S. Laval, E. Cassan, Propagation loss in single-mode ultrasmall square silicon-on-insulator optical waveguides. J. Lightwave Technol. **24**, 891 (2006) 224
10. S. Lardenois, D. Pascal, L. Vivien, E. Cassan, S. Laval, R. Orobtchouck, M. Heitzmann, N. Bouzaida, L. Mollard, Low-loss submicrometer silicon-on-insulator rib waveguides and corner mirrors. Opt. Lett. **28**, 1150–1152 (2003) 224
11. D. Marris, L. Vivien, D. Pascal, M. Rouvière, E. Cassan, A. Lupu, S. Laval, J.-M. Fedeli, L. El Melhaoui, Ultralow loss successive divisions using silicon-on-insulator microwaveguides. Appl. Phys. Lett. **87**, 211102 (2005) 224, 225
12. L. Vivien, S. Lardenois, D. Pascal, S. Laval, E. Cassan, J.-L. Cercus, A. Koster, J.-M. Fedeli, M. Heitzmann, Experimental demonstration of a low-loss optical H-tree distribution using SOI microwaveguides. Appl. Phys. Lett. **85**, 701–703 (2004) 225
13. Results from European Helios project: http://www.helios-project.eu/ 225
14. R. Orobtchouk, N. Schnell, T. Benyattou, J.-M. Fedeli, Compact building block for optical link on SOI technology, *Proceeding of 12th European Conference on Integrated Optics ECIO*, Grenoble, 221–224, 2005 227
15. D. Taillaert, H. Chong, P. Borel, L. Frandsen, R.M. De La Rue, R. Baets, A compact two-dimensional grating coupler used as a polarization splitter. IEEE Photon. Technol. Lett. **15**(9), 1249–1251 (2003) 227, 228
16. L. Vivien, D. Pascal, S. Lardenois, E. Cassan, F. Grillot, S. Laval, J.M. Fedeli, L. El Melhaoui, Light injection in SOI microwaveguides using high-efficiency grating couplers. J. Lightwave Technol. **24**, 3810–3813 (2006) 227, 228
17. G. Masanovic, G. Reed, W. Headley, B. Timotijevic, V. Passaro, R. Atta, G. Ensell, A. Evans, A high efficiency input/output coupler for small silicon photonic devices. Opt. Express **13**(19), 7374–7379 (2005) 227, 228
18. N. Landru, D. Pascal, A. Koster, Modelling of two-dimensional grating couplers on silicon-on-insulator waveguides using beam propagation method, Opt. Commun. **196**, 139–147 (2001) 227, 228
19. R. Orobtchouk, A. Layadi, H. Gualous, D. Pascal, A. Koster, S. Laval, High-efficiency light coupling in a submicrometric silicon-on-insulator waveguide, Appl. Opt. **39**, 5773–5777 (2000) 227
20. G. Maire, L. Vivien, G. Sattler, A. Kazmierczak, B. Sanchez, K.B. Gylfason, A. Griol, D. Marris-Morini, E. Cassan, D. Giannone, H. Sohlstrom, D. Hill, High efficiency silicon nitride surface grating couplers, Opt. Express **16**(1), 328–333 (2008) 228
21. V.R. Almeida, R. Panepucci, M. Lipson, Nanotaper for compact mode conversion, Opt. Lett. **28**(15) 1302–1304 (2003) 229
22. G. Roelkens, P. Dumon, W. Bogaerts, D. Van Thourhout, R. Baets, Efficient silicon-on-insulator fiber coupler fabricated using 248-nm deep UV lithography. IEEE Photon. Technol. Lett. **17**(12), 2613–2615 (2005) 229

23. R.S. Jacobsen, K.N. Andersen, P.I. Borel, J. Fage-Pedersen, L.H. Frandsen, O. Hansen, M. Kristensen, A.V. Lavrinenko, G. Moulin, H. Ou, C. Peucheret, B. Zsigri, A. Bjarklev, Strained silicon as a new electro-optical material. Nature **441**, 199–202 (2006) 232
24. J. Liu, M. Beals, A. Pomerene, S. Bernardis, R. Sun, J. Cheng, L.C. Kimerling, J. Michel, Waveguide-integrated, ultralow-energy GeSi electro-absorption modulators, Nat. Photonics **2**(7), 433–437 (2008) 232
25. Y. Kuo, Y. Kyu Lee, Y. Ge, S. Ren, J.E. Roth, T.I. Kamins, D.A.B. Miller, J.S. Harris, Strong quantum-confined stark effect in germanium quantum well structures on silicon. Nature **437**, 1334–1336 (2005) 232
26. A. Liu, R. Jones, L. Liao, D. Samara-Rubio, D. Rubin, O. Cohen, R. Nicolaescu, M. Paniccia, A high speed silicon optical modulator based on a metal-oxide-semiconductor capacitor. Nature **427**, 615–618 (2004) 232
27. L. Liao, D. Samara-Rubio, M. Morse, A. Liu, D. Hodge, D. Rubin, U. Keil, T. Franck, High speed silicon Mach-Zehnder modulator, Opt. Express **13**(8), 3129–3135 (2005) 232
28. S. Manipatruni, X. Qianfan, B. Schmidt, J. Shakya, M. Lipson, High speed carrier injection 18 Gb/s silicon micro-ring electro-optic modulator, *Proceedings of Lasers and Electro-Optics Society IEEE*, 537–538, 2007) 232
29. W. Green, W.M. Green, M.J. Rooks, L. Sekaric, Y.A. Vlasov, Ultra compact, low RF power, 10 Gb/s silicon Mach-Zehnder modulator. Opt. Express **15**(25), 17106–17113 (2007) 232
30. D. Marris-Morini, L. Vivien, J.-M. Fedeli, E. Cassan, P. Lyan, S. Laval, Low loss and high speed silicon optical modulator based on a lateral carrier depletion structure. Opt. Express, **16**(1), 334–339 (2008) 232, 233, 234
31. D. Marris-Morini, G. Rasigade, L. Vivien, E. Cassan, S. Laval, P. Rivallin, P. Lyan, J.-M. Fedeli, Recent progresses in fast silicon modulators, *Proceedings of the 5th conference on Group IV Photonics*, Sorrento, 17–19 Sept 2008 232
32. T. Pinguet, V. Sadagopan, A. Mekis, B. Analui, D. Kucharski, S. Gloeckner, A 1550 nm, 10 Gbps optical modulator with integrated driver in 130 nm CMOS, *Proceedings of the 4th conference on Group IV Photonics*, Tokyo, 19–21 September 2007 232
33. L. Liao, A. Liu, D. Rubin, J. Basak, Y. Chetrit, H. Nguyen, R. Cohen, N. Izhaky, M. Paniccia, 40 Gbit/s silicon optical modulator for high speed applications. Electron. Lett, **43**(22), 1196–1197 (2007) 232
34. D. Marris-Morini, X. Le Roux, L. Vivien, E. Cassan, D. Pascal, M. Halbwax, S. Maine, S. Laval, J.-M. Fedeli, J.-F. Damlencourt, Optical modulation by carrier depletion in a silicon PIN diode. Opt. Express **14**, 10838–10843 (2006) 232
35. D. Marris, E. Cassan, L. Vivien, Response time analysis of SiGe/Si modulation-doped multiple-quantum-well structures for optical modulation. J. Appl. Phys. **96**, 6109–6112 (2004) 232
36. D. Marris-Morini, X. Le Roux, L. Vivien, E. Cassan, D. Pascal, M. Halbwax, S. Maine, S. Laval, J.-M. Fedeli, J.-F. Damlencourt, High speed all-silicon modulation-doped optical modulator. J. Luminescence, **121**, 387–390 (2006) 232, 233, 234
37. G. Roelkens, D. Van Thourhout, R. Baets, R. Nötzel, M. Smit, Laser emission and photodetection in an InP/InGaAsP layer integrated on and coupled to a SOI waveguide circuit. Opt. Express **14**(18), 8154–8159, September (2006) 234
38. J. Bourckaert, G. Roelkens, D. Van Thourhout, R. Baets, Compact InAlAs_InGaAs MSM photodetectors integrated on SOI waveguides. IEEE PTL **19**(19), 1484–1486 (2007) 234
39. M. Jutzi, M. Berroth, G. Wöhl, M. Oehme, E. Kasper, Ge-on-Si vertical incidence photodiodes with 39-GHz bandwidth. IEEE Photon. Technol. Lett. **17**, 1510–1512 (2005) 234
40. M. Rouvière, L. Vivien, X. Le Roux, J. Mangeney, P. Crozat, C. Hoarau, E. Cassan, D. Pascal, S. Laval, J.-M. Fedeli, J.-F. Damlencourt, J.-M. Hartmann, S. Kolev, Ultra-high speed germanium-on-silicon-on-insulator photodetectors for 1.31 μm and 1.55 μm operation. Appl. Phys. Lett. **87**, 231109–231111 (2005) 234
41. O.I. Dosunmu, D.D. Cannon, M.K. Emsley, B. Ghyselen, J. Liu, L.C. Kimerling, M. Selim Ünlü, Resonant cavity enhanced Ge Photodetectors for 1550nm operation on reflecting Si substrates. IEEE J. Sel. Top. Quantum Electron. **10**, 694–701 (2004) 234

42. M. Halbwax, M. Rouviere, Y. Zheng, D. Debarre, L.H. Nguyen, J.L. Cercus, C. Clerc, V. Yam, S. Laval, E. Cassan, D. Bouchier, UHV-CVD growth and annealing of thin fully relaxed Ge films on (001)Si. Opt. Mater. **27**, 822–826 (2005) 235
43. J.-M. Hartmann, A. Abbadie, A.M. Papon, P. Holliger, G. Rolland, T. Billon, J.-M. Fedeli, M. Rouvière, L. Vivien, S. Laval, Reduced pressure–chemical vapor deposition of Ge thick layers on Si(001) for 1.3–1.55-μm photodetection. J. Appl. Phys. **95**, 5905–5907 (2004) 235
44. L. Vivien, M. Rouvière, J.-M. Fedeli, D. Marris-Morini, J.-F. Damlencourt, J. Mangeney, P. Crozat, L. El Melhaoui, E. Cassan, X. Le Roux, D. Pascal, S. Laval, High speed and high responsivity germanium photodetector integrated in a silicon-on-insulator microwaveguide. Opt. Express **15**(15), 9843–9848 (2007) 235
45. D. Ahn, C. Yin Hong, J. Liu, W. Giziewicz, M. Beals, L.C. Kimerling, J. Michel, J. Chen, F.X. Kärtner, High performance waveguide integrated Ge photodetectors. Opt. Express **15**(7), 3916–3921 (2007) 235
46. T. Yin, R. Cohen, M. Morse, G. Sarid, Y. Chetrit, D. Rubin, M.J. Paniccia, 31GHz Ge n-i-p waveguide photodetectors on silicon-on-insulator substrate. Opt. Express **15**(21), 13965–13971 (2007) 235
47. G. Masini, G. Capellini, J. Witzens, C. Gunn, *Proceedings of the 4th International Conference on Group IV Photonics*, WB2, 28, Tokyo, 19–21 September 2007 235
48. J. Wang, W.Y. Loh, K.T. Chua, H. Zang, Y.Z. Xiong, T.H. Loh, M.B. Yu, S.J. Lee, G.-Q. Lo, D.-L. Kwong, Evanescent coupled Ge p-i-n photodetectors on Si-waveguide with SEG-Ge and comparative study of lateral and vertical p-i-n configurations. IEEE Electron. Dev. Lett. **29**, 445–448 (2008) 235
49. L. Vivien, J. Osmond, J.-M. Fedeli, D. Marris-Morini, P. Crozat, J.-F. Damlencourt, E. Cassan, Y. Lecunff, S. Laval, 42 GHz p.i.n Germanium photodetector integrated in a silicon-on-insulator waveguide. Opt. Express **17**, 6252–6257 (2009) 235
50. N. Daldosso, D. Navarro-Urrios, M. Melchiorri, C. Garcıa, P. Pellegrino, B. Garrido, C. Sada, G. Battaglin, F. Gourbilleau, R. Rizk, L. Pavesi, Er coupled Si nanocluster waveguide. IEEE J. Sel. Top. Quantum Electron. **12**(6), 1607–1617 (2006) 237
51. M.N. Charasse, J.P. Hirtz, III–V on silicon achieving the perfect match. Phys. World, 28–31, January (1990). 237
52. G. Balakrishnan, S.H. Huang, A. Khoshakhlagh, P. Hill, A. Amtout, S. Krishna, G.P. Donati, L.R. Dawson, D.L. Huffaker, Room-temperature optically-pumped InGaSb quantum well lasers monolithically grown on Si(100) Substrate. Electron. Lett. **41**(9), 350–352 (2005) 237
53. Z. Mi, P. Bhattacharya, J. Yang, K.P. Pipe, Room-temperature self-organised $In_{0.5}Ga_{0.5}$ As quantum dot laser on silicon. Electron. Lett. **41**(13), 742–744 (2005) 237
54. A.W. Fang, H. Park, O. Cohen, R. Jones, M.J. Paniccia, J.E. Bowers, Electrically pumped hybrid AlGaInAs-silicon evanescent laser. Opt. Express **14**(20), 9203–9210 (2006) 237
55. R. Stengl, T. Tan, U. Gösele, A model for the silicon wafer bonding process. Jpn J. Appl. Phys. **28**, 1735 (1989) 238
56. U. Gösele, Y. Bluhm, G. Kastner, P. Kopperschmidt, G. Krauter, R. Scholz, A. Schumacher, St.Senz, Q.-Y. Tong, L.-J. Huang, Y.-L. Chao, T.H. Lee, Fundamental issues in wafer bonding, J. Vac. Sci; Technol. **A17**(4), 1145–1152 (1999) 238
57. M. Kostrzewa, P. Regreny, M.-P. Besland, J.-L. Leclercq, G. Genet, P. Rojo-Romeo, E. Jalaguier, P. Perreau, H. Moriceau, O. Marty, G. Hollinger, High Quality Epitaxial Growth on New InP/Si Substrate, *Conference on Indium Phosphide and Related Materials*, 325–328, 12–16 May 2003 238
58. M. Kostrzewa, L.Di Cioccio, J.-M. Fedeli, M. Zussy, P. Regreny, J.-C. Roussin, N. Kernevez, Die-to-wafer molecular bonding for optical interconnects and packaging. Sens Actuators A: Phys. **125**(2), 411–414, (2006) 238
59. S. Guilet, S. Bouchoule, C. Jany, P. Chabert, Optimization of a Cl2—H2 inductively coupled plasma etching process adapted to non thermalized InP wafers for the realization of deep ridge heterostructures, J. Vac. Sci. Technol. B **24**(5), 2381–2387 (2006) 240

60. T. Dupont, L. Grenouillet, P. Philippe, M. Perrin, P. Gilet, B. Ben Bakir, J.-M. Fedeli, P. Grosse, L. Di Cioccio, A. Chelnokov, Continuous wave InGaAsP/InP Fabry-Perot lasers on silicon, *Proceedings of the 20th International Conference on Indium Phosphide and Related Materials*, IPRM 2008, 1–2, 25–29 May 2008. 241
61. J. Van Campenhout, P. Rojo-Romeo, P. Regreny, C. Seassal, D. Van Thourhout, L. Di Cioccio, J.-M. Fedeli, R. Baets, Electrically-injected InGaAsP microdisk lasers heterogeneously integrated on and coupled to a silicon-on-insulator waveguide circuit. Opt. Express **15**(11), 6744–6749 (2006) 241
62. P. Rojo-Romeo, J. Van Campenhout, P. Regreny, A. Kazmierczak, C. Seassal, X. Letartre, G. Hollinger, D. Van Thourhout, R. Baets, J.-M. Fedeli, L. Di Cioccio, Heterogenous integration of electrically driven microdisk based laser sources for optical interconnects and photonic ICs. Opt. Express **14**(9), 3864–3871 (2006) 242
63. J. Van Campenhout, P. Rojo-Romeo, P. Regreny, C. Seassal, D. Van Thourhout, L. Di Cioccio, J.-M. Fedeli, R. Baets, Electrically-injected InGaAsP microdisk lasers heterogeneously integrated on and coupled to a silicon-on-insulator waveguide circuit. Opt. Express, **15**(11) 6744–6749 (2006) 242
64. P.R.A. Binetti, J. Van Campenhout, X.J.M. Leijtens, M. Nikoufard, T. de Vries, Y.S. Oei, L. Di Cioccio, J.-M. Fedeli, C. Lagahe, R. Orobtchouk, X. Letartre, P. Regreny, P. Rojo-Romeo, C. Seassal, P.J. van Veldhoven, R. Notzel, D. Van Thourhout, R. Baets, M.K. Smit, An optical interconnect layer on silicon, *Proceedings of the 13th European Conference on Integrated Optics*, 2007 242
65. D. Van Thourhout, J. Van Campenhout, P. Rojo-Romeo, P. Regreny, C. Seassal, P. Binetti, X.J.M. Leijtens, R. Nötzel, M.K. Smit, L. Di Cioccio, C. Lagahe, J.-M. Fedeli, R. Baets, PICMOS – a photonic interconnect layer on CMOS, *Proceedings of the 33rd European Conference on Optical Communication* (ECOC '07), 16–20 Sept (2007). 242

Index

Index

D.J. Lockwood, L. Pavesi (Eds.): *Silicon Photonics II.*
Topics in Applied Physics **119**, 251–253 (2011)
DOI 10.1007/978-3-642-10506-7

P

Q

R

S

T

U

V

W